The Fossil-Fuelled Climate Crisis

"A major innovation for the subfields of environmental sociology and ecological social theory. Building on the Weberian theoretical framework of social closure, coupled with a social practices approach, Murphy presents the climate crisis in a new, and dare I say even hopeful, light."

—Michael S. Carolan, *Ph.D., Associate Dean for Research & Graduate Affairs, College of Liberal Arts, Professor, Department of Sociology, Colorado State University, USA*

"The threat of the climate crisis is global, creeping and urgent like the current COVID-19 virus. This is the long-awaited first book on the climate crisis to use Murphy's social closure framework. He contributes a brilliant and candid sociological analysis of structures, impacts, and solutions of climate change. Murphy stresses the importance of visibility and concreteness to raise our awareness in order to efficiently mitigate the problem."

—Koichi Hasegawa, *Professor-emeritus of Tohoku University, Japan. Past President of International Sociological Association's Research Committee on Environment and Society*

"*The Fossil-Fuelled Climate Crisis* helps us to understand the severity, causes, consequences, and potential solutions to the greatest challenge facing humanity and the other creatures of the world. Professor Murphy, a deep-thinking scholar with a long track-record of environmental research, guides us through the complex issues and daunting conundrums we face and points the way toward advances in research and action. Senior researchers and students alike will find this an engaging and insightful book."

—Richard York, *Professor of Sociology and Environmental Studies, University of Oregon, USA*

"*The Fossil-Fuelled Climate Crisis* deals with the most pressing issue of our age. Drawing from a rich body of social sciences, Raymond Murphy demonstrates not only why climate change is hard to tackle but also explores how it can be tackled. This book is a must read for those who not only would like to understand the climate change issue but also what is needed in order to break current emission trends."

—Rolf Lidskog, *Professor in Sociology, Director for Environmental Sociology Section, Örebro University, Sweden. Vice-president of Research Committee Environment & Society, International Sociological Association*

"This is an innovative contribution that uses the concept of social closure to push us to reconsider our thinking about climate change and its possible solutions. Murphy builds a critical social science analysis on climate science foundations in ways that will help readers connect the sociological and climatological dimensions of the environmental crisis. Through a close analysis of the range of climate solutions currently on offer, Murphy warns against a 'Faith 2.0' in the human mastery of nature whereby technological solutions will come to the rescue and let us 'ride out' our environmental crises. A real strength of the book is Murphy's ability to use historical examples (such as asbestos) to illuminate the social choices that are necessary to meet the imperative for transformative climate action."

—Mark C. J. Stoddart, *Professor, Department of Sociology, Memorial University of Newfoundland, Canada*

"In the wake of the recent global economic, political, and biological crisis, Raymond Murphy's theoretically rich, incisive analysis of climate change and consequent social closure provokes reflection about what needs to be done to avert a more profound, lasting catastrophe for humanity and the other life with which we share the planet."

—Robert J. Antonio, *Professor of Sociology, University of Kansas USA*

"Murphy brilliantly shows how the causes of anthropogenic climate change are deeply embedded in our social practices, and how unequally the resulting burdens and benefits are distributed. But even more importantly, he shows how a number of other environmental problems were mitigated, and how the lessons thus learned can help us deal with the problem of climate change."

—Tuomas Ylä-Anttila, *Associate Professor of Political Science, University of Helsinki, Finland*

"Murphy is as relentlessly focused on making his point using clear language and familiar examples as he is on identifying the root cause of the climate crisis. This book is not a polemic against fossil fuels but a considered analysis of how social relations drive dangerous environmental outcomes and where opportunities lie for genuine social and political reform. It should be widely read."

—Distinguished Professor Stewart Lockie, *Director, The Cairns Institute, James Cook University, Australia*

Raymond Murphy

The Fossil-Fuelled Climate Crisis

Foresight or Discounting Danger?

palgrave
macmillan

Raymond Murphy
School of Sociology and Anthropology
University of Ottawa
Ottawa, ON, Canada

ISBN 978-3-030-53324-3 ISBN 978-3-030-53325-0 (eBook)
https://doi.org/10.1007/978-3-030-53325-0

This Palgrave Macmillan imprint is published by the registered company Springer Nature Switzerland AG
The registered company address is: Gewerbestrasse 11, 6330 Cham, Switzerland

To my grandchildren in the hope that their generation may continue to benefit from the indispensable services of nature that my generation enjoys.

Preface

In late 2019 and early 2020, as I was putting the finishing touches on this book, flights from Wuhan, China to other countries carried two invisible enemies of humanity. First, epidemiological science had long predicted that, with humans invading and manipulating nature, a local outbreak anywhere of an infectious virus could be rapidly carried worldwide by some of the four billion passengers flying annually and become a pandemic everywhere. The prediction came true with the novel COVID-19 virus, which was lethal not only for human health but also for the economy and social life where a failure of foresight led to discounting warnings of danger and defences were not implemented promptly. This outcome was not, however, inevitable. A few countries heeded warning signs of danger, foresight prevailed, they acted quickly and decisively, thereby largely containing the virus and mitigating the economic impact.

The second unseen enemy was carbon dioxide from all these flights. The combustion of jet fuel lifting and propelling heavy planes produces a chemical reaction with oxygen in air resulting in huge amounts of carbon dioxide imperceptible to the senses. Researchers have calculated the carbon dioxide emissions for the 17,300 km. round trip between Wuhan and Rome attributable to each economy-class passenger to be 2.8 tonnes, or 5600 pounds, which is almost thirty times the weight of a 200-pound passenger. Multiply those emissions by the billions of passengers flying each year and the result is a massive quantity of carbon dioxide emissions, which remains in the atmosphere for a century causing a greenhouse effect and global warming. And aviation is just one of many causes of fossil-fuelled climate change, albeit an important one without a foreseeable technological solution. Scientific warnings of dangerous climate change have been even more numerous than warnings of pandemics, but they too are being discounted, emissions continue to far exceed carbon withdrawal by forests, and the problem is worsening.

There are similarities and differences between these two invisible enemies. Both are global threats to everyone, with the poor being the most vulnerable. They can be prevented at an early stage, but if not, the consequences and measures to mitigate them become highly disruptive of normal social practices. If the pandemic and fossil-fuelled climate change are not contained, the harm threatens to be enormous not only to health and the environment respectively but also to the economy and standard of living. Both are multifaceted threats. In the two cases, future technological innovations are relied upon: vaccinations and anti-viral drugs for pandemics and carbon capture and storage for fossil-fuelled climate change. If fossil-fuelled aviation continues to carry many billions of passengers annually and if relations with nature are not improved, other viruses could emerge into pandemics and the climate crisis will worsen further. Societies enthusiastically accept the applications of science but are reluctant to act on its troubling conclusions about danger and to implement solutions that require cost and social change.

There are nevertheless major differences. First, although both result in slow-onset harm, the time lags between cause and effect are very different: a couple of weeks for the most serious consequences of a virus like COVID-19 on the human body, but a century for the worse effects of fossil-fuelled climate change. Earth is a big planet with an enormous atmosphere and oceans, so it takes time to pollute them by extracting carbon safely stored by nature underground and transferring it to the sky and oceans. If degraded, imagine the time it would take to restore the environment to its present beneficial habitat for humanity. Second, pandemics often peter out on their own, for example the 1918 Spanish flu disappeared after it killed 38 million people and the survivors had become immune. Fossil-fuel combustion emitting greenhouse gases, on the contrary, makes global warming progressively worse by melting the permafrost releasing methane, a potent greenhouse gas, and melting the ice cover of the Arctic Ocean reducing its reflective capacity. It threatens to unleash runaway feedback loops of nature's dynamics. Third, although the 2020 pandemic was really bad for human health, the economy, and social life, human-made global climate change will likely be much worse if scientific warnings of danger continue to be discounted and available remedies not implemented.

Both of these dangers resulting from our relations with the dynamics of the natural world could incite blaming the other's shoddy response and ignoring our own failings. This is true on both the individual and societal levels. It is necessary to investigate and learn lessons from disastrous failures of foresight, but pointing fingers to dodge one's own responsibility is counterproductive. Pandemics and global warming could instead foster a more cosmopolitan outlook that we are all in this together. As American President John F. Kennedy stated: 'Our most basic common link is that we all inhabit this planet. We all breathe the same air. We all cherish our children's future. And we are all mortal'. The atmosphere and oceans are the most fundamental commons. Fossil-fuelled practices of present and past generations are causing an intensifying greenhouse effect. This is inadvertently creating an environmental debt to be paid belatedly by our children and grandchildren, and risks closing off resources and opportunities to those future generations. The present golden age of fossil-fuelled practices threatens to result in pollution-caused downward intergenerational mobility. Cherishing our children's and grandchildren's future gives hope that action will be taken to meet the challenge of the climate crisis. It is this spirit of cooperation that needs to be cultivated. To paraphrase Bank of England Governor Mark Carney's wise conclusion: The more we practise foresight, the less we will regret in hindsight.

This book seeks to construct a sociological analysis of fossil-fuelled climate change on the solid foundation of natural science and assess solutions that have been proposed. Its principal objective is not to convince climate change deniers that it exists and results from human activities, although the book may contribute to that. Such efforts have been done extensively in many other excellent books. Rather it aims to move the analysis forword by advancing a social science understanding of how fossil-fuelled social practices threaten to undermine nature's services to humanity and to unleash dangerous dynamics of nature, as well as explain what can be done about it.

Ottawa, Canada Raymond Murphy

Acknowledgements

This book is my attempt to understand the human-made climate crisis and assess possible solutions. Its arguments have been forged in the fires of presentations and debates at conferences of the Environment and Society Research Committee of the International Sociological Association, the Environmental Sociology research cluster of the Canadian Sociological Association, the Environmental Sociology section of the American Sociological Association, and the Environment & Society research network of the European Sociological Association. I am indebted to the influence and wisdom of participants at these conferences, reading their insightful work, and having the distinct pleasure of associating with them over the years. I want to thank Riley Dunlap for integrating me into American and international environmental sociology networks when I was transitioning from writing about social closure and for his stimulating research on climate change. I also would like to pay tribute to environmental sociologists whose work influenced this book but have now passed away: Frederick Buttel, William Freudenburg, and Ulrich Beck. Another deceased sociologist who has had a lasting effect on my thinking and thereby on this manuscript was the British academic

Frank Parkin, to whom I am indebted for his creative conceptualization of social closure and his vivid writing style. Probing further back, my teachers for my undergraduate degree in physics instilled in me a strong appreciation of natural science, including its logic, the need for assertions to be empirically grounded, and science's self-admitted fallibility. Readers will surely notice this throughout the book.

I want to thank the previously anonymous readers chosen by Palgrave Macmillan to evaluate the book project and four chapters I initially submitted. The insightful comments and encouragement by Robert J. Antonio, Professor of Sociology at the University of Kansas in the United States, and Tuomas Ylä-Anttila, Associate Professor of Political Science at the University of Helsinki, Finland were enormously helpful at this crucial stage of writing.

It takes many years for an author to write a book like this. The most important person I want to acknowledge is my wife, Ruth Marfurt. I am thankful for her encouragement, her timely advice about writing when I needed it, her support in so many ways, and her love. My thanks at Palgrave Macmillan go to Rachael Ballard at the initial stage for her encouragement, to Joanna O'Neill for overseeing production, to Preetha Kuttiappan and her team for the production, and to all of them for their exemplary professionalism.

Contents

List of Figures

1

Introduction

The 2018 Nobel laureate for economics William Nordhaus (2013: 19) illustrates how deceptive are the consequences of fossil-fuelled social practices that cause climate change. If he drives 100 miles in his car, 'I consume 5 gallons of gasoline. This will produce about 100 pounds of CO_2, which will come out of the tailpipe and go into the atmosphere. I cannot see or hear it or smell it, and I generally do not even think about it. If I am like most people, I will probably assume that my trip will have no effect on the world's climate, and so I will ignore the consequences'.[1] If the trip by Nordhaus were unique, the consequences would be negligible. However, it is typical of much more numerous sets of social practices. An easy way to remember the consequences of fossil-fuelled social practices is as follows. If you drive the average 3500-pound midsize car a distance in miles equal to your weight in pounds, you are emitting your weight of carbon dioxide into the atmosphere on that one trip: a 200-pound person driving 200 miles emits 200 pounds of carbon dioxide. Hence, think about the weight of CO_2 emitted by millions of people commuting to and from work in cars, usually one person per car, in urban sprawled metropolitan areas. This has to be understood in its global context. The average car will be driven about 10,000 miles yearly,

R. Murphy, *The Fossil-Fuelled Climate Crisis*,
https://doi.org/10.1007/978-3-030-53325-0_1

so it will emit about 10,000 pounds of CO_2 annually. There are about one billion cars in the world. Therefore, by simple arithmetic the global fleet of cars is emitting about 10 trillion pounds of CO_2 every year.

In addition, imagine the weight of CO_2 being pumped into the atmosphere by the combustion of jet fuel lifting and powering a heavy plane, multiplied by the distance travelled per plane every year and the number of planes. Moreover, there are ships of all sorts and fossil-fuelled industries, including those that combust fossil fuels to extract, transport, and refine fossil fuels. Cement is the essential constituent used worldwide in concrete buildings, roads, bridges, dams, pipes, etc. Its manufacture by combusting fossil fuels emits a weight of carbon dioxide into the atmosphere almost equal to the weight of cement in these heavy structures. All this carbon pollution of the atmosphere continues year after year. The scale of carbon being taken from safe storage in the ground and emitted into the atmosphere is enormous. Carbon dioxide, which the US Supreme Court in 2007 ruled a pollutant in terms of the Clean Air Act, remains in the atmosphere for on average one hundred years and piles up, then descends to acidify the oceans.[2] Our planet and its atmosphere used as a carbon waste dump are huge so it takes time to pollute them, but that implies it will take much time for them to be restored to their beneficial state, if that proves possible.

Fossil fuels have been and remain the energy source that has powered modern society since industrialization. Even ecological saints with the best intentions can not avoid using fossil fuels, either directly by gassing up their vehicles or indirectly by flying in a plane, using air conditioners and social media. Every individual, every company, and every country is a carbon polluter and has contributed to global warming. But this fact should neither legitimate carbon pollution in terms of a we-are-all-sinners ideology nor lead to fatalism. The significant variables are the amount of emissions and whether action is being taken to decrease them or increase them. This varies enormously. There are colossal polluters—huge even relative to their industry (Freudenburg 2006)—and groups that promote fossil fuels and therefore emissions. Nevertheless, there are also small polluters, and groups that try to reduce emissions. A small number of gigantic corporate carbon polluters have major consequences for global warming and climate change, but so does a massive number of much smaller individual polluters, and the latter's pollution loading varies greatly according to their social class and specific practices. Hence, the fossil-fuelled practices of both huge oligopolistic companies and those of ordinary people and institutions need to be examined.

Many readers might be tempted to assume that Nordhaus is mistaken because five American gallons of gasoline only weigh about 31 pounds. Where do the remaining 69 pounds of CO_2 come from? But Nordhaus is right. His example demonstrates how insidious fossil-fuelled social practices are, and how important it is for the population, decision-makers, and social scientists to ground their understanding of global warming on natural science's comprehension of biophysical dynamics. This is one of the reasons why Chapter 2 gives a brief outline of the natural science understanding of fossil-fuelled climate change,[3] including an explanation of the Nordhaus example and the cement illustration. Pielke (2010: 46–50) claims there is an 'iron law of climate change' that describes the population's refusal to support a price on carbon pollution anywhere near what is needed to prevent fossil-fuel combustion causing greenhouse-gas emissions and global warming. This book argues that the refusal is at least in part contingent on lack of a practical understanding of how ordinary fossil-fuelled practices are contributing to the problem. Scientists communicate in the language of gigatonnes, but non-scientists think in pounds or kilograms. Hence, an everyday grasp of the fossil-fuelled climate crisis is being lost by lack of translation of the abstract theoretical knowledge of science into commonplace units. Safran Foer (2019a: 13; 2019b: R14) states that 'most of us would find it difficult to explain how our individual and collective behavior is boosting hurricane winds by almost thirty miles per hour or contributing to a polar vortex that [sometimes] makes Chicago colder than Antarctica'. This book strives to promote as much as possible a practical comprehension of the fossil-fuelled climate crisis.

Is It a Crisis?

The term 'global warming' was suitable when Wallace Broecker of Columbia University introduced it in 1975, and 'climate change' was fitting when the National Academy of Sciences announced it in 1979.

But subsequent massive emissions have made the carbon accumulation in the atmosphere and the greenhouse effect much worse (IPCC 2018; US Global Change Research Programme 2018; UNEP 2018). Because of those emissions, the global surface temperature has increased 1 degree Celsius since the pre-industrial period and already hurricanes, deluges, droughts, wildfires, etc., are more intense. Most nations agreed in the Paris Accord of 2015 to limit the increase to 2 degrees. The 2018 emissions gap report documented, however, that few countries are on track to achieving the goal, that most of the highest emitting countries are not meeting their near-term targets much less the more ambitious ones to start later, and that the present global trajectory is leading to an increase of 3 degrees (UNEP 2018). This would result in foreseeable dire consequences and likely some unforeseeable ones. There is strong backlash against the Paris Accord: the USA withdrew; a carbon tax in France resulted in riots in 2018; the Canadian government is promoting the extraction of high-emissions bitumen by purchasing a pipeline, and conservative parties oppose a price on carbon pollution; Australia is combusting and exporting coal; etc. In 2011, Victor (2011) published a book entitled *Global Warming Gridlock*. Nine years later in 2020, the gridlock persists. The Global Carbon Project (2018) documented that emissions, far from decreasing, have increased 2% in 2018 to a new record high.

Government policies to control greenhouse-gas emissions are often not implemented, or are overturned by the next government in the name of stimulating economic growth. This lurching to-and-fro contrasts with the unidirectional accumulation of carbon in the atmosphere and the growing hazard of global warming. In the USA, the Republican Administrations of George W. Bush and Donald Trump rolled back the climate change mitigation initiatives of the Democratic Administrations of Clinton/Gore and Obama/Biden. The 1997 Kyoto Protocol was signed by 192 countries, but did not require any restriction on emissions by big carbon polluters like China, India, Brazil, and South Africa. Therefore, the highest carbon polluter at the time the USA refused to ratify it, and the big per capita carbon polluter Canada reneged on its ratification when a conservative government took power.

One significant reason why anthropogenic global warming is so challenging consists of the speed of greenhouse-gases accumulating in the atmosphere compared to the slowness of the transition to clean, renewable energy. Implementing new energy sources requires, in addition to innovative technologies, new infrastructures and social learning. Rhodes (2018: O1) documented that 'across the past 400 years, as the world has transitioned from wood to coal, to oil, to natural gas and nuclear power, the average transition time from zero to 50-percent market penetration has been about 100 years'. Imagine all the carbon that human activities will have placed in the atmosphere if it takes a century for non-carbon, renewable energy to achieve half of the global energy market. Worse yet, the other half will still be fossil fuels emitting carbon. These fossil-fuelled practices would likely result in the elimination of the Arctic ice cover which has reflected much of the sun's radiation back out into space and the thawing of permafrost letting loose the potent greenhouse gas methane hitherto trapped underground. Both are currently melting because of the fossil-fuelled warming that has already occurred. This could well lead to runaway global warming by nature's autonomous dynamics by the time clean renewable energy constitutes one-half of the global energy supply. Little wonder Rhodes (2018: O4) concluded that 'energy transitions take more time than a world faced with global warming may have'. The disjuncture between socioeconomic causal temporalities and socially acceptable mitigating temporalities renders management of the fossil-fuelled climate crisis extremely difficult for a global population increasing in number and energy consumption per person.

Because fossil-fuelled climate change creeps up slowly in most of the world, it is a threat more for the future than for the present. If fossil-fuelled climate change were to suddenly cause a rise of one metre in the level of oceans thereby immediately affecting polluters by flooding many of the world's cities, few would deny or be apathetic about it. But the ocean level rise is predicted by science to creep upward slowly and take a century or more to reach that height. Immediacy of consequences is not the physical property of the fossil-fuelled greenhouse effect. Hence, it can be pushed to the back of mind as near-term economic pursuits worsen the problem. Most of the world's population and countries perceive it as distant, leading to a tendency to discount it and shove aside costly and/or inconvenient changes in social practices. Nevertheless, in the Arctic, temperatures are increasing at twice the rate as in the rest of the world, causing near-term harm. At the Arctic Council whose members include the USA, Russia, Sweden, Norway, Denmark, Finland, Canada, Iceland, and the Inuit Circumpolar Council, the Swedish Foreign Affairs Minister stated that a 'climate crisis in the Arctic is not a future scenario, it is happening as we speak' (Dickson 2019: A7). The Inuit representative added that 'Inuit are feeling the effects of climate change every day, … our people are witnessing the adverse impacts of climate change. What about us and our reality?'

Globalization has tightly coupled not only societies and industries but also social dynamics with worldwide biophysical dynamics. In small scale, complex, tightly coupled systems manipulating nature's dynamics, such as nuclear reactors, the accumulation of minor inadvertent errors has led to disasters (Perrow 1984). There is no reason or evidence to conclude disasters could not occur on a global scale, namely because of a failure of foresight resulting in the incubation of dangerous fossil-fuelled climate change in a tightly coupled complex world.

Neither of the benign labels 'global warming' (which sounds nice for cold countries) nor 'climate change' (which implies it could change for the better) does justice to the gravity of the problem. When a Canadian province fought a national carbon tax in its supreme court, that court approved the tax and concluded that climate change constitutes an emergency. Branding the problem as 'global warming' and 'climate change' has failed to incite mitigation. After almost a half-century of accelerating atmospheric carbon buildup since those terms were introduced, it is now appropriate to redefine the problem as a 'fossil-fuelled climate crisis'.[4] The Oxford dictionary defines 'crisis' as a time of intense difficulty or danger, when important decisions must be made, and a turning point when an important change takes place leading to recovery or demise. All this is applicable to the situation in which societies now find themselves because of their ongoing combustion of fossil fuels. Admittedly, it is not a disaster everywhere right now, but it is a crisis in the sense that if preventive measures are not taken promptly then far worse consequences will unfold in slow-onset fashion. Thus, it is a creeping crisis similar to that caused in New Orleans prior to its disaster by shortsighted economic development in flood-prone and hurricane-prone areas in a below sea level city surrounded by the Gulf of Mexico, the Mississippi River, and Lake Pontchartrain. The disaster there was scientifically predicted in advance, only the timing was unknown, but canal construction and wetlands destruction continued thereby exacerbating vulnerability (Freudenburg et al. 2009), and disaster occurred in 2005 when Hurricane Katrina struck. Global warming does not feel like a crisis to many people, but neither did the threat of a pandemic predicted by epidemiologists, so the lack of foresight, discounting danger, and failure to mitigate promptly when COVID-19 surfaced resulted in huge costs in lives and economic losses. In both pandemics and fossil-fuelled climate change, science is needed to explain the danger and causal linkages, and to make informed decisions determining whether demise or a turning point leading to recovery will occur.

The threat is one of the piling up an enormous environmental debt to be paid by future generations including our grandchildren. The environmental debt will belatedly be paid in disasters or costs of disaster preparedness, adaptation, and resilience. This generation, especially high emitters, insists on externalizing the costs of fossil fuels to the environmental commons. There is an increasing 'normalization of the environmental crisis .... [characterized by] the curious simultaneity of an unprecedented recognition of the urgency of radical ecological policy change ... and an equally unprecedented unwillingness and inability to perform such change' (Blühdorn 2011: 36). The crisis consists of a refusal to stop the environmental debt from accumulating, what Beck (1995: 48–49) refers to as 'the death reflex of normality', and the evidence indicates that it currently applies to fossil-fuelled normality. Thus, an increasing number of scholars who have studied fossil-fuelled climate change are referring to it as a crisis (Speth 2009, 2012; Flannery 2009, 2015; Nordhaus 2013; Suzuki and Hanington 2017). It is a material crisis of the beneficial biospheric habitat for humanity, but it is also a crisis of culture and imagination, of faith in the market, in production science and technological innovation, and on the other hand faith in impact science and socioeconomic innovation. 'I would call it a crisis of belief' (Safran Foer 2019b: R14).

Fossil Fuels

Fossil fuels have been central to modern societies and their development. They power automobiles, trucks, planes, boats of all sorts, heating and air conditioning, factories, social media servers, and much more. The military is a fossil-fuel glutton and thereby a mammoth greenhouse gas-polluter. The US Department of Defense is the world's single largest institutional petroleum consumer and the single biggest consumer of energy in the USA. As a result, its CO_2 emissions surpassed those of many industrialized countries, for example in 2017, 59 million tonnes for the American military compared to 48 million tonnes for all of Sweden and 34 million tonnes for all of Switzerland (McCarthy 2019). Mechanized food production for a large and still growing population depends on fossil fuels, which have liberated vast proportions of the population from toiling in the fields like their great-grandparents. Oil 'packs a huge amount of energy by volume and weight: three large spoonfuls of crude oil contain about the same amount of energy as eight hours of human manual labour, and when we fill our car with gas, we're pouring into the tank the energy equivalent of about two years of human manual labour. Oil is also versatile, convenient, and still relatively cheap. No other substance or fuel comes close to matching its properties' (Homer-Dixon 2006: 82–83). Low-carbon renewable energy forms a tiny proportion of the world's energy compared to fossil fuels, which currently provide 81% and has only decreased marginally (World Bank 2015), leaving emissions greatly exceeding carbon withdrawals thereby worsening global warming. In some social practices the fossil fuel combustion is visible, but in many others it is not. Social media, cloud data storage, and smartphones depend on servers which consume enormous amounts of electricity for functioning and cooling, most of it supplied by combusting fossil fuels. 'With so much of the world now dependent upon cloud computing, social media, online streaming, and the like we have all become cheap electricity addicts' (Carolan 2014: 151). Users are unaware of the massive upstream carbon emissions involved in social media. Even cryptocurrencies like bitcoin, which appear to be created from nothing are constructed by combusting fossil fuels to produce electricity needed by computers for their complex computations. Most of the activities and conveniences people need, enjoy, or desire are powered by fossil fuels. They are the inanimate sources of energy that drive modern societies. Desrochers and Szurmak (2018) describe how important fossil fuels have been, but their tribute ignores how threatening this transfer of carbon from safe storage underground to a long-lasting danger in the sky has become. The multiplicity of fossil-fuelled social practices means that there are many dimensions to global warming. That will require many solutions, but the overall danger has been well documented by science. Mitigating fossil-fuelled climate change is much more difficult than solving urban water and air pollution and the like. This book's focus is on the role of fossil fuels in the climate crisis because they are so central to modern societies and such a big part of the problem.

The problem is particularly difficult to deal with because fossil-fuelled global warming is cumulative and largely invisible to the senses. The increase of greenhouse gases in the atmosphere can't be seen with the naked eye, nor can a global temperature increase be felt, unlike a local increase. In fact, a 2 °C increase seems insignificant. The climate seems normal because it is often confused with weather whose fluctuations have been experienced previously. Natural science is needed to know that the carbon content and temperature of the global atmosphere are increasing significantly, dangerously, and rapidly. In this sense, fossil-fuelled climate change is similar to the depletion of the ozone layer by CFCs, which could not be seen and was only made visible by science. This comparison shows that problems can be solved if there is political will by decision-makers and the population. But anthropogenic climate change is much more difficult to remedy because it involves an enormous number of social practices based on fossil-fuel energy that has propelled society's development. Unlike CFCs, there are no easy cost-effective substitutes for flexible, energy-laden fossil fuels used in so many sectors. Alternatives for fossil fuels require replacing not only those fuels but also much of the present physical and social infrastructure that has enabled their use. Moreover, the most severe consequences of global warming foreseen by science are distant in space and time; hence carbon polluters are not dissuaded from polluting by experiencing immediate harm. The threat of fossil-fuelled climate change is actualized in an insidious, creeping fashion, which reduces the conditions for creating strong social movements to combat it. The metaphor of a frog in water slowly brought to a boil is apt.

This book deals with only fossil-fuelled climate change, but climate change is a particularly important cause of environmental degradation affecting all the others. Fossil fuels power deforestation, overfishing, mining, agriculture, urbanization, plastic pollution of oceans, etc. (IPBES 2019). It is essentially a socioeconomic problem in its causes, consequences, and hopefully solutions. Societies have become dependent on machines that use greenhouse-gas emitting fossil fuels as their source of primary energy. This creates structured, locked-in, path-dependent (Nye 1998) predispositions, habitus, and perceived entitlements to emit carbon into the atmosphere. Dependence on fossil fuels is fostered by powerful fossil fuel oligopolies, related vested interests, consumer predispositions and habits, identities, values, reluctance to change, etc. Removal of that dependence requires not only technological innovations of low-carbon alternative energy sources, but also improved political and media institutions, environmental education, enhanced future orientation, and foresight. Companies justify their exploitation of fossil fuels by claiming they are satisfying demand by consumers. They are, however, also stimulating demand, not only by advertising but also by increasing supply thereby lowering price and increasing consumption, innovating new fossil-fuelled products, etc. The need to bring emissions into balance with carbon withdrawals from the atmosphere implies that most of the fossil fuels will have to be left safely in the ground unless inexpensive carbon capture and storage or geoengineering solutions are implemented. Thus, fossil fuel industries, their workers, and especially countries whose economies depend on extraction and export of fossil fuels feel threatened by the conclusions of science.

Berners-Lee and Clark (2013: 2) argue that societies have been unable to change the emissions trajectory, the problem is urgent, technological solutions are a long way off, the choice is between taking unimaginable risks and leaving fossil fuels in the ground, so it is 'critical that we get a proper understanding of the core barriers that are holding us back'. Boström, Davidson, and Lockie (2018: abstract) conclude that it is crucial that 'individual and collective actors, lay persons and experts, develop the reflexive capability to promote change, and counteract structural and cultural forces that prevent change'. That is the goal of this book, which seeks to move the investigation forward on the solid foundation of natural science conclusions.

The Autonomous Dynamics of Nature

The environment for humanity consists of complex systems of biophysical dynamics. Far from being constant or passive, they are actants that act and react, not intentionally but autonomously nevertheless. I (Murphy 2002) argued that modern societies are internalizing autonomous dynamics of nature as never before. The autonomous dynamics of nature are significant actants, whether they act independent of human social practices or are unleashed inadvertently by them. Nuclear reactors, satellites, planes, and the like consist of recombinations of nature's dynamics in the context of nature's broader dynamics. The autonomy of the actions of nature's dynamics becomes evident when reactors meltdown as in Chernobyl or are struck by tsunamis as in Fukushima, when satellites and electrical grids are disrupted by sun storms, when earthquakes and extreme weather devastate cities (Zebrowski 1997; Murphy 2009). As Adam (1995, 1998, 2000) argued, social constructions are superimposed upon the biophysical constructions of nature. Long-term sustainability or unsustainability is determined by whether material social constructions are or are not in harmony with the constructions of the natural world. The latter have sustained human societies and enabled their development during the Holocene. The issue is whether fossil-fuelled practices of humans will degrade nature's services in the long run. The focus of this social science analysis is on the interaction (i) of fossil-fuelled social practices at all levels with the dynamics of nature, (ii) of social constructions with nature's constructions, and (iii) of the actions of human actors having intended and unintended consequences with the actions of nature as an autonomous actant.

Environmental Social Closure

There is a serious weakness in conceptions of environmental impacts in the Anthropocene. They typically depict an undifferentiated humanity as the cause, and a homogeneous humanity suffering the effects. Hence, they fail to capture the socioeconomic dynamics that are the drivers of human impacts. This book attempts to correct those deficiencies by introducing the Weberian theoretical framework of social closure (Murphy 1988) to the environmental social sciences, and showing how it helps elucidate the socioeconomic drivers of environmental problems, victimization from them, and the resulting reaction. Closure refers to processes of monopolization of resources thereby closing off opportunities to others. The book analyses environmental social closure involving the appropriation of biophysical resources, including carbon sinks, by the present generation disproportionately benefiting some of its members, resulting in the risk of excluding latecomers from such benefits, as well as other species. Latecomers consist of poor individuals, poor societies, and future generations. The global biophysical environment such as the atmosphere constitutes a commons shared by everyone, including future generations, and is a medium that carries social relations of monopolization and exclusion across space and between generations over time. Priority given to near-term economic benefits to the exclusion of long-term costs, which are discounted, results in social closure becoming embedded in culture, practices, and physical infrastructures. Reaction against such environmental closure is led by environmental movements, impact scientists, and social democratic governments. Reaction also consists of nature as an actant whose biophysical dynamics strike back against their manipulation by humans, for example fossil-fuel combustion produces global warming thereby unleashing more intense wildfires and hurricanes.

Is the fossil-fuelled climate crisis the result of the interests, values, and actions (or inaction) of decision-makers and the powerful in society, or of actions by the overall population steered by their interests and values? Some people and companies disproportionately appropriate not only benefits but also influence and decision-making, hence make a particularly heavy impact on the environment and on climate change.

Leadership is crucial in determining whether there will be successful mitigation or a failure of foresight. Nevertheless, the remainder of the population are not passive followers or cultural dopes (Lynch 2016). They too have interests, values, and impacts, especially because of their large numbers, and are active agents in their relations with nature. Hence, it is important to take into account not only the concentration of decision-making, influence, and benefits but also social practices at all levels of society. Rowe et al. (2016: 234–235) apply Gramsci's insight 'that elite power is maintained not only through coercion, but also through the everyday actions of people that confer consent upon hierarchical social orders, ... the carboniferous capitalism of today, while reinforced by a daunting nexus of corporate and state power, is reproduced daily by the everyday consent of popular publics. This consent manifests itself in consumer decisions, the driving of cars, voting patterns, hands-off approaches to pension and mutual fund investments, and political quiescence'. Redclift (2010: 132) counsels 'analysing how current behavior is tied into patterns and cycles of carbon dependence' and advocates taking the long view of society. The concentration of benefits and decision-making regarding fossil fuels and the disproportionality of carbon pollution practices are important, but so are the huge number of small polluters who need or enjoy fossil-fuelled polluting practices. It would be an oversimplification to reduce the problem to the practices of big polluters or to demand for fossil fuels by the enormous number of smaller polluters. Hence, this book will include both dimensions in terms of (a) the social closure theoretical framework and (b) the social practices paradigm. These overlap in the monopolizing practices of the powerful.

Social Practices

The anthropogenic effect on the environment is usually represented as the consequence of human activities. Social scientists (Shove 2012a, b; Shove et al. 2012; Shove and Spurling 2013) have replaced 'activities' with the concept of 'social practices' and use it to study climate change.

They argue that practices involve the integration of three elements—materials, meanings, and competences—into performances. Thus, the practice of intercontinental tourism consists of the integration of a plane combusting jet fuel, competent pilots and employees, and meanings that motivate flying. Important also are marketing practices of aviation companies, operation of airports as shopping centres, etc. Following Latour's (2000: 113) argument that artefacts 'are in large part the stuff out of which socialness is made', Shove et al. (2012: 9) emphasize 'the constitutive role of things and materials in everyday life. … [and] that practices are constituted through the actions of material entities as well as people'. Social practices impact the environment because materiality is an indispensable ingredient out of which socialness is made.

The focus on social practices that degrade the environment and result in global warming, compared to other practices that are more benign, as well as their constitutive elements is important. But two improvements are needed. First, although there had been a cultural turn in some social theory that resulted in evacuating materiality leaving a significant gap, this was not true of most of the environmental social sciences like environmental sociology. Researchers (see Catton and Dunlap 1980; Freudenburg and Gramling 1993; Freudenburg, Frickel, and Gramling 1995; Gramling and Freudenburg 1996; Dunlap and Catton 1994; Benton 1994, 2001; Murphy 1994, 1997, 2002; Dickens 2004; Dunlap 2010; Foster, Clark, and York 2010) explicitly and continually analysed interactions between the material and the sociocultural. Second, authors like Latour and Shove et al. who include materiality typically reduce it to artefacts, things, etc., and have only an abstract, unexplicated sense of how they act. They are shy about incorporating the most important elements of materiality, namely nature's dynamics and properties, its services and threats, into social theory. This must be explicitly integrated into a theory of social practices.

The Inuit Nobel Peace Prize nominee Sheila Watt-Cloutier (2019: O11) gives an example of social practices in their physical context, which is shifting because of global warming. 'As a young child growing up in Kuujjuaq, Que., I travelled only by dogsled for the first 10 years of my life. I was often snuggled into warm blankets and fur as my family set out on hunting and fishing trips. The vast Arctic sky surrounded us and the ice was strong beneath our feet – the foundation that carried us across the frozen land. … But, in my generation, the Arctic sea ice and snow, upon which we Inuit have depended for millenniums, is now diminishing'. This resulted from fossil-fuelled practices of people far away because there was no carbon pollution in traditional Inuit transportation practices. But in Watt-Cloutier's generation, the transportation practices of the Inuit themselves have changed to non-traditional, fossil-fuelled ones: snowmobiles, all-terrain vehicles, motorboats, planes, etc.

Social practices can be portrayed as purely cultural only by abstracting them out of their material context. Social media interaction appears to be only cultural, but it and data storage clouds depend on servers consuming massive amounts of electricity, much of it coming from fossil fuels. Nevertheless, social practices vary enormously in their interaction with broader dynamics of nature. Travelling in an automobile propelled by fossil fuels emitting CO_2 contributes to global warming whereas propelling oneself on a bicycle does not. Differences in social practices matter.

It is informative to juxtapose the travelling practices of the rich and powerful with those of a young climate activist. Sir Elton John flew Prince Harry and his wife from England to Nice on his private jet. Each way combusted 1868 litres of fuel, which emitted 4.7 tonnes of carbon dioxide into the atmosphere, for a round-trip total of 9.4 tonnes (Reality Check Team 2019). Specialists state this should be multiplied by 1.9 to reflect emissions at high altitudes. Sir Elton claimed he paid a certified company for carbon offsets, but did not say how much, and it is difficult, expensive, and dubious to offset almost 10 tonnes of emissions. He used security to justify this private jet flight, but strict security reasons do not apply to Sir Elton himself when he flies around in his private jet. Prince Harry doesn't show much restraint either, since he and Meghan flew in a private jet earlier to vacation in Ibiza, with round-trip emissions calculated to be 10.4 tonnes. This case is not presented to single out the celebrity musician nor the prince for hypocrisy, as global warming deniers do. For every private jet flight of the prince, there are thousands of such flights by the rich who discount jet-fuelled danger. The fossil-fuelled practices of the rich and the powerful winging about in private jets disproportionately cause global warming compared to flyers packed like sardines in economy class of commercial airlines, and more so compared to those who rarely fly.

The distinctive contribution of the teenage climate activist Greta Thunberg is that her social practices speak louder than words. When she travels in Europe, she does so in electrically powered trains, which underscores the global warming consequences of combusting gasoline or diesel to power vehicles and jet fuel for short-haul flights. When invited to a summit on global warming in New York, she accepted to cross the Atlantic in a solar-powered racing yacht built for speed rather than comfort, but was warned that the journey would take two weeks and the ride would be choppy. Nobody expects people, including Thunberg in the future, to cross oceans like that, but she called attention to fossil-fuelled practices like flying causing global warming. Thunberg accused world leaders of mouthing 'empty words', and called for action to cut carbon pollution and prevent climate change. These two cases demonstrate the significant consequences of flying for global warming, the differential environmental impact of the powerful who monopolize private jets compared to ordinary folk, the difficulty of finding low-carbon alternatives for travelling long distances, but nevertheless the need for action and not just words.

Shove et al. (2012: 162) succinctly make three important arguments: 'people are somehow captured by the arrangements they sustain and to which they devote finite amounts of time, attention and resources. … there are no reliable means of steering or governing transitions in practice: systemic forms of policy intervention only have effect when taken up in (and through) practice'. First, in North America, people are captured by low gasoline prices into buying gas-guzzling SUVs, crossovers, pickup trucks, and fossil-fuelled adult toys such as motorboats, four wheelers, snowmobiles, etc. Then they develop a sense of entitlement such that they resist carbon taxes. Second, if these authors are right, then whether a transition sufficient to mitigate anthropogenic climate change will be achieved must remain an open question because there are no reliable means of governing it in practice. The third point correctly concludes that policy discourse is vacuous unless implemented into changed practices, implying that the key variable is not so much policy construction as implementation.

If one extracts or/and combusts fossil fuels to make money with no thought of global warming, or with the intention of causing it (which is rare), the result is the same: global warming. The converse is also true, with Giddens (2009) giving an example. In the nineteen-seventies, Germany, France, and other European countries increased the tax on petrol (gasoline), hence its price. The motivation was to enhance each country's energy security and independence from unreliable Middle Eastern sources of oil after the formation of OPEC. Nevertheless, it decreased the use of petrol and provided monies to build efficient public transportation systems. This policy indirectly contributed to mitigating climate change. What is significant are social practices and their effects, regardless of intentions. The analytical focus should be on social practices because they are consequential for climate change.

Emissions-reducing technological improvements have great difficulty keeping up with the treadmill of social practices that increase emissions. There was a 40% increase in the carbon footprint of tourism between 2009 and 2013, much of it from aviation (IPBES 2019). According to the International Civil Aviation Organization (ICAO 2018), there was a passenger-kilometer increase of 7.1% in 2015 and 6.3% in 2016. Hence, the International Air Transport Association (IATA) will have extreme difficulty achieving its 50% net emissions reduction goal by 2050 even if flights remain at present levels, and moreso if there is a global increase in flying, which is likely. The fossil-fuelled global auto fleet is expected to increase by 80% by 2035, especially in Asia (Reguly 2018: 19). The very belief that fossil-fueled engines are becoming more efficient leads decision-makers and the population to use more of them, thereby undermining the goal of emissions reduction (York 2012; York and McGee 2016). Use of social media and data clouds has exploded, requiring energy-glutton data servers that are insidious because there are no visible exhaust pipes as in vehicles nor do they leave contrails visible in the sky like planes. As developing countries rise out of poverty, their populations engage in the same social practices that inhabitants of wealthy countries have long enjoyed: drive cars, eat more meat, see the world, use air conditioning and data servers, etc. Even if greenhouse-gas emissions per person remain constant, population growth, expected to climb from

7.7 to over 9 billion, would result in more emissions. The treadmill of carbon polluting practices is accelerating.

The Struggle Between Value Spheres

A century ago, Max Weber (1946: 147–148) argued that 'the various value spheres of the world stand in irreconcilable conflict with each other … here too, different gods struggle with one another, now and for all times to come'. There are many conflicting value spheres, which are relevant to the fossil-fuelled climate change threat: economy versus environment; near term versus long-term; free market versus government intervention; temporalities of climate change and temporalities of politics (Lockie and Wong 2018); faith in technological innovation to master nature versus hope that humans can master their own social practices. It is an open question whether those conflicting value spheres can be reconciled. The pair of warring gods consisting of near-term economic prosperity worshiped in modern societies versus long-term environmental sustainability is more complex than economy versus environment. Stern (2009) documents that future costs of fossil fuel global warming will far outweigh current economic benefits of fossil fuels. Similarly, the Fourth National Climate Assessment documented that climate change is engendering substantial net damage to the US economy during this century and will cause further damage (US Global Change Research Program 2018). The costly 2018 wildfires in California, in Australia in 2020, and the 2017 hurricanes striking wealthy Houston and poor Puerto Rico are foretastes of what is coming. Fossil-fuelled social practices by the present generation are accumulating expensive environmental debts to be paid by future generations. Near-term priorities are excluding concern for long-term needs, so the tension is between discounting future danger and foresight. Solutions require having the foresight to act to prevent adverse consequences predicted by science. This involves shifting to clean, renewable energy and cultural shifts to more inclusive, cosmopolitan, future orientations.

Political and business decision-makers claim the economy and the environment are being reconciled, especially by themselves, and many ordinary people act and vote as if they have been reconciled. It is necessary to examine material social practices to determine whether such discourse constitutes reconciliation or greenwashing. The pursuit of profit and affluence through market dynamics have resulted in priority given to current economic growth at any environmental cost. Is mitigating emissions compatible with present economic growth, and how could this be done? Can economic growth be based on clean renewable energy, or must fossil-fuel consumption be reduced, and will that entail major sacrifices?

Near-term economic goals can be reconciled with long-term sustainability, but it is a difficult, enduring task that involves bringing social practices into harmony with dynamics of nature. The most sustainably prosperous societies (Switzerland, Sweden, South Korea, Germany, and Japan) do not have a drop of oil and typically develop their human capital to add value to a minimum of raw materials, produce a diverse range of goods, and develop their intellectual property and service sectors. Economies based on exporting crude oil (Saudi Arabia, Russia, Venezuela, and Nigeria) suffer boom-and-bust cycles of the market. To reduce their contribution to global warming, they become dependent on innovating technological solutions, which must be technically effective, cost-effective, and scalable, and are nowhere to be found. Therefore clean energy movements seeking to displace carbon polluting fossil fuels threaten them. Countries exporting crude fossil fuels typically become leaders of the laggards concerning global warming.

A Foreseeable Threat with Unforeseeable Specific Harms

The overall trajectory of fossil-fuelled global warming is foreseeable by science if present trends continue. Hyper-carbon societies and a hyper-carbon world are emerging, thereby brownfielding the atmosphere and the oceans. Claims of unforeseeability of the overall trajectory are a refuge for those who do not want to foresee and discount danger. However, the specific long-term impacts, locations, timing, etc., are difficult to foresee and in many cases unforeseeable. For example, global warming is causing Greenland's and Antarctica's glaciers to melt into the oceans, which may cause temporary cooling in some regions. The diversity and unforeseeability of harms of global warming has correctly led to the more general concept of climate change. The principal unknown is whether fossil-fuelled global warming will tip the planet into a new irreversible state less beneficial as a habitat for humanity. This book investigates how societies are responding to their foreseeable creation of brownfields in the sky having unforeseeable damages (Lockie and Wong 2017), that is, to their present incubation of the possible unsustainability of their beneficial habitat.

Biophysical uncertainties pale in comparison to uncertainties about the socioeconomic response. 'Today, the largest uncertainty in projecting future climate conditions is the level of greenhouse gas-emissions going forward. Future global greenhouse gas-emissions levels and resulting impacts depend on economic, political, and demographic factors that can be difficult to predict with confidence far into the future' (US Global Change Research Program 2018: Chapter 1 Overview). Mitigating anthropogenic climate change will be expensive and likely require changes of fossil-fuelled practices. It requires a better balance between near-term economic goals and minimizing long-term harm to the human-supporting environment and human health, but achieving that balance is socially problematic. It could spark serious tensions, likely by exacerbating pre-existing social divisions. Environmental problems could be mitigated if there is sufficient understanding and willingness to improve social practices causing those problems. But such willingness is in short supply, which could lead to a 'failure of foresight' and the 'incubation' of a slow-onset global calamity that researchers (Turner and Pidgeon 1978) have long documented in small scale 'man-made disasters'. If danger is discounted for slow-onset threats to gain near-term economic benefits, then serious harms may not be experienced until it is too late to avoid tipping environments and perhaps the whole planet into a new less beneficial state. This is what Giddens (2009) labelled as his paradox.

For many global environmental problems—fossil-fuelled climate change, degradation of oceans, biodiversity loss, etc.,—there are enormous time lags and/or spatial distances between causal social practices and consequences, and issues of scale whereby any one cause can be dismissed as minor. Land, water, and the atmosphere on our planet are huge, so it takes enormous accumulation of pollution and much time to degrade them globally. But precisely because they are huge, it would take much time, effort, and expense to rectify anthropogenic global environmental problems. Natural disasters and anthropogenic disasters are merging empirically. Fossil fuels are combusted and forests are cut down, resulting in global warming. This makes extreme weather (wildfires, droughts, hurricanes and floods, etc.), which has always occurred, more intense and frequent, and will eventually result in the rise of ocean levels. Despite the empirical merging, it is important to distinguish those causes, and especially to avoid depicting anthropogenic unleashing of nature's forces as natural occurrences. The distinction is necessary for scholarly rigour and appropriate preparedness and prevention. Disasters can be viewed as 'focusing events' (Birkland 1998) and prompts to preventive or/and preparatory action leading to greater sustainability, but they are often dismissed and written off as Acts of God or Mother Nature to avoid annoying conclusions of human responsibility (Zebrowski 1997).

The Need to Draw on the Natural Science of Fossil-Fuelled Climate Change

Admittedly, some risks are only sociocultural scares with improbable or negligible impact. It is important to avoid conflating all claims of hazards, such as unfounded apocalyptic climate predictions centuries ago with contemporary scientifically documented fossil-fuelled global warming. If conflated, then the distinction between high-impact probable hazards as opposed to imaginary or improbable threats is blurred, and the sense of urgency is diluted for real threats being caused now. This also incites the response that there are so many difficult threats we might as well discount danger and enjoy what we are doing because it won't last. A judicious assessment of threats is necessary to prioritize them, and identify and implement means to mitigate them promptly. How can probable high-impact threats be distinguished from all the alarmist claims continually being socially constructed, such as vaccinations causing autism? The best available evidence and understanding must be used, and this comes principally from natural science for a global, invisible, slow-onset biophysical threat like fossil-fuelled climate change. Hence, this social science analysis of the human-made climate crisis builds upon the conclusions of natural science (see Clark 2011; Wong and Lockie 2018). Because scientific consensus is based on confirmed theory, rigorous methodology, empirical evidence, and peer-reviewed analysis, it is the most reliable in a world filled with a cacophony of voices about threats.

Appearances to the eye are often deceiving. The Earth seems flat and the Sun appears to go around the Earth from east to west. These common sense presumptions were believed for most of the existence of humanity until science provided evidence that the Earth is round and rotates around the Sun. Science faced stiff opposition when it confronted cherished beliefs with its evidence-based theory, but eventually prevailed. Hopefully, that will also be the outcome for the fossil-fuelled climate crisis. Avoidance of misleading conclusions and deceptive wishful thinking requires a natural science understanding of the issues to build valid social science analyses.

It is important to avoid conflating 'solutionism' with practical solutions and improved practices. 'Solutionism' consists of changing discourse without changing practices and improving policies without implementing them, which goes with worsening of problems. It most misleadingly involves putative solutions that fail to address problems effectively. Hence, the best available understanding, concepts, evidence, and conclusions of natural science must be used to learn whether the problem is being lessened or worsened, and to carry out a nuanced social science analysis of why this is so. A clear understanding of the depth of the challenges is needed to lay the foundation to mitigate the threat. Earth scientists have documented that fossil-fuelled climate change unleashes self-reinforcing feedbacks of the Earth system resulting in the likelihood of crossing thresholds to much higher global average temperatures and sea levels than at any time in the Holocene. 'If the threshold is crossed, the resulting trajectory would likely cause serious disruptions to ecosystems, society, and economies. Collective human action is required to steer the Earth System away from a potential threshold and stabilize it in a habitable interglacial-like state. Such action entails stewardship of the entire Earth System – biosphere, climate, and societies – and could include decarbonzation of the global economy, enhancement of biosphere carbon sinks, behavioural changes, technological innovations, new governance arrangements, and transformed social values' (Steffen et al. 2018: abstract). They conclude that addressing these issues 'requires a deep integration of knowledge from biogeophysical Earth System science with that from the social sciences and humanities on the development and functioning of human societies' (Steffen et al. 2018: 8252).

Humanity has enjoyed amazing progress in health, education, and affluence since the development of science. Many challenges have been successfully met. The single best overall indicator is arguably the rise in life expectancy. Population increased from one billion around 1860 to over 7.6 billion in 2017 and appears to be going up to 11.8 billion by 2100, yet the proportion of people living in extreme poverty is decreasing and affluence is increasing. Water and air quality in European cities is better today than at the time of the industrial revolution. Natural disasters in countries like China and Japan killed people in greatest numbers after population increase but before science developed. Now applied science coupled with foresight has made infrastructures more robust and societies more resilient resulting in fewer fatalities (Zebrowski 1997; Murphy 2010). Depletion of the ozone layer by the innovation of CFCs was made visible by impact scientists, international negotiations led to the Montreal Protocol which restricted their use, and harm was diminished. In the 1970 and 1980s in Europe and North America, acid rain resulting from sulphur dioxide and nitrogen oxides produced by the combustion of fossil fuels was a major problem. Subsequent regulations and cap-and-trade systems successfully abated the problem. Environmental problems can be dealt with *if* they are understood, *if* there is willingness to pay the cost of prevention and adaptation, and *if* social practices and/or technologies are changed to meet the challenges. Those are three big '*if*'s for dealing with the fossil-fuelled climate crisis.

The Structure of the Book

After this introductory chapter, the book consists of two sections, each with five chapters. Part I analyses the fossil-fuelled climate crisis. Although other human actions contribute to global warming and climate change, the focus is on fossil fuels because they are the most significant single cause. They are also the most challenging since they are the energy source of most of what people have needed and enjoyed since the industrial revolution. Fossil-fuelled climate change is a physical problem, nevertheless it has socioeconomic causes and consequences. Part I doesn't shy away from explaining the depth of the problem. That is not intended to lead to despair and foster fatalism, but rather to provide a firm basis for solutions. Part II assesses solutions that have been proposed, of which there are many. It examines both their possibilities and weaknesses with eyes wide open in order to distinguish between hope and wishful thinking, and provide a solid foundation for mitigating the problem.

Chapter 2 presents a brief summary of current natural science understanding of the fossil-fuelled climate crisis. It underscores the crucial concept of net change in atmospheric carbon (if emissions exceed carbon withdrawals, then global warming is worsening) and the concept of a global carbon budget. The chapter assesses the scientific validity of the oft-used concept of greenhouse-gas emissions per GDP (intensity) as an indicator of improvement. It examines science's roadmap for limiting global warming to 2 °C and the possibilities and weaknesses of assuming improvements in technical efficiency will solve the problem. The chapter probes a paradox. Scientific knowledge of global warming has been available, increasing, and disseminated for a quarter-century, yet emissions continue to rise, as if the findings of impact science have had little effect. However, to legitimate the accelerating treadmill of fossil-fuelled social practices, faith in science to innovate just-in-time technological solutions abound. The chapter also assesses the concept of the Anthropocene for social science.

Chapter 3 uses the concept of social closure for the social science analysis of the climate crisis. It first examines whether recent market dynamics have resulted in greater concentration of economic power.

Then it analyses environmental social closure involving the appropriation of biophysical resources, particularly the atmosphere as a carbon dump, by the present generation, disproportionately by some companies and groups. It investigates whether emissions from fossil-fuelled practices degrading the environment are closing off resources and opportunities to latecomers, namely future generations and poor countries. Moreover it examines whether monopolization of resources by high consuming humans are excluding other species from needed resources and habitats, resulting in human-induced extinctions. It probes the purposeful reaction to environmental closure by impact scientists, environmental movements, and political actors. The chapter also studies nature's reaction as an actant biting back with autonomous dynamics: extreme weather, wildfires, sea level rise, etc. It examines environmental regulations as processes of demonopolization, and inquires whether deregulation involves practices of monopolization.

Chapter 4 probes how fossil fuel use by one group can close off resources and opportunities for other groups even on the opposite side of the planet or living centuries later. How can the environmental commons, namely the atmosphere, oceans, and land, act as a medium that carries social relations of monopolization and exclusion over space and time? To answer, it draws on the theory of externalities. Unlike wind and solar energy, fossil fuels have environmental costs, namely costs of wildfires, floods, biodiversity loss, etc., because of global warming. These are not included in their price. Their unpaid cost accumulates into an environmental debt, which will be paid belatedly by others. The chapter investigates the resistance of carbon polluters, big and small, to pay the full cost of fossil fuels upon use through carbon taxes, regulations, etc. It also examines whether supply or demand generates fossil-fuel use and emissions, and the effectiveness of additionality, offsetting, and incrementalism.

Chapter 5 presents a social science documentation of the road actually being travelled concerning society's response to the scientific findings of global warming, which is compared with the roadmap deemed necessary by science. It examines whether society is (i) heading towards transitioning out of fossil fuels, which requires leaving them stored safely in the ground if a safe technological solution is not implemented, or

(ii) is bedeviled by a Cassandra-like syndrome discounting scientific warnings and stuck in a path-dependent fossil-fuelled old normal characterized by failure of foresight. It investigates the accelerating treadmill of carbon polluting practices, market innovation as part of the problem, apathy, excuses, lucrative discounting of danger, top-down discounting and bottom-up discounting, backsliding, the culture of every man for himself, and free-ridership. It exposes dark sides of adaptation and resilience building. Thus it probes the depth of the fossil-fuelled climate crisis.

Chapter 6 documents a pattern that emerges and the phases or steps occurring as science reveals that near-term economic benefits from the exploitation of a dangerous resource are bringing long-term harm. It examines the extraction and combustion of one type of fossil fuel, and hypothesizes that the pattern is true for all fossil fuels. The fossil fuel investigated is oil from Canada's bituminous sands, often known as tar sands, which contain the world's third largest oil reserves. This analysis shows how challenging it is for states to transition away from fossil fuels, especially those that base their economies on fossil-fuel extraction. The chapter takes for comparison the steps travelled by that same country concerning its world-leading reserves of the valuable but dangerous resource of asbestos. The harm it caused and failure to find ways to make it safe prompted the innovation of technological alternatives. The massive asbestos reserves are being left safely underground, but only after a century-long struggle and enormous harm in many countries. Will the same outcome occur for bitumen and all fossil fuels, or will the steps be longer and more arduous, the consequences more severe, and the end-stage different because of the greater importance of fossil fuels to the modern economy?

Chapter 7 is the first chapter of Part II assessing possible solutions. It begins with an analysis of risk, its assessment, the actualization of risk into disaster, and uncertainty. Scientific assessments of risk cannot be communicated to non-scientists by simply presenting complex computer models. Therefore, the chapter next investigates staging strategies for the danger of climate change and its suitability for staging. Beck's staging framework is then turned right-side-up by analyzing the more significant staging of safety and of discounting danger. The chapter examines the uncertainty of whether there will be foresight under threats of fossil-fueled climate change.

Chapter 8 examines whether safe social practices are coming, and elucidates grounds for hope. It evaluates the theory of the preeminent social scientist of risk Ulrich Beck that anticipation of global catastrophe will result in a cosmopolitan inclusive worldview, emancipation from dangerous social practices, and a safe new normal. Empirically, it investigates social change after Hurricane Katrina catastrophe devastated New Orleans, which Beck used to support his theory, and the BP Deepwater Horizon blowout which gushed oil pollution for two months. The chapter then descends from Beck's high level of abstraction. It examines practical cases where foresight prevailed and the economy reconciled with the environment.

Chapter 9 investigates whether faith in mastering nature through future technological climate fixes (i) has collapsed when confronted with global environmental problems like anthropogenic climate change, or (ii) has taken a new form. It examines whether reluctance to modify fossil-fuelled practices incites reliance on future technological innovation conceptualized here as 'faith 2.0 in the mastery of nature'. The reformed faith believes in future technological innovations that will control nature's forces and impacts and create just-in-time mitigation, increased efficiencies, adaptation, robustness, and resiliency when the problem becomes sufficiently serious. The chapter investigates ways that belief in nature's mastery has legitimated present fossil-fuelled practices. It clearly distinguishes between (1) *hoping for* technological remedies, and (2) *relying on* such solutions because society refuses to change its fossil-fuelled practices.

Chapter 10 assesses technological climate fixes. It distinguishes two broad types. One would mitigate global warming while enabling fossil-fuel use to continue. Proposals assessed include carbon capture and storage (CCS), direct air capture (DAC) storing the carbon underground, innovations in cement, cultivating seaweed to draw down atmospheric carbon, and a geoengineered sunscreen in space. The other would leave fossil fuels safely underground with solar, wind, hydro, tidal, geothermal energy, and direct air capture of carbon producing energy. Although informed by technical knowledge, the objective of this social science assessment is to investigate their socioeconomic implementation and consequences. Thus, the chapter assesses the solution of economists of paying the full cost of fossil fuels upon use instead of belatedly, and exposes deficiencies in the iron law of climate change.

Chapter 11 gives suggestions to enhance foresight, many of which are derived from the analysis in the previous chapters. These include communicating climate science in words and illustrations understandable to non-scientists, making the danger concrete, focussing on social practices and timeliness, valuing nature's services, combatting fallacies that carbon taxes are job killers, and promoting inclusion and equality of opportunity particularly for future generations. The chapter also assesses proposed solutions of divestment, lawsuits, reducing consumption practices especially of fossil fuels, moral suasion impelling action, and restraining population growth by educating girls. It evaluates proposals to transcend capitalism, examines governance and in particular social democracy. The book ends by examining three possible outcomes of the fossil-fuelled climate crisis with different energy futures and relates the outcome to the central issue of whether foresight or discounting danger will prevail.

Notes

1. American President Trump called the COVID-19 virus an 'invisible enemy'. Greenhouse gases like carbon dioxide are also invisible enemies, but they are much worse. Carbon dioxide remains in the atmosphere for a century and accumulates with intensifying effect. Imagine if the virus were to accumulate for a century with increasingly disastrous consequences. The virus appeared as I was at the end of writing this book, so I will leave to others to explore the interesting parallels between this global pandemic and global warming.
2. There are other greenhouse gases in addition to long-lasting carbon dioxide (CO_2) and potent methane (CH_4) but these two are the main ones. When 'carbon' is used in the book, it refers to greenhouse gases.
3. The more general term 'anthropogenic climate change' would in addition deal with other greenhouse gases like HCFCs and other sources such as those from ruminants like cattle and sheep, as well as land use changes like

deforestation and certain agricultural practices (see IPCC 2019). Nevertheless, the combustion of fossil fuels is by far the biggest contributor of human activities to global warming. It is also the most socially problematic to deal with because of the centrality of fossil fuels to modern economies.

4. Nevertheless, the terms 'global warming' and 'climate change' will also be used in this book because they are the terms currently most often used.

References

Adam, Barbara. 1995. *Timewatch: The Social Analysis of Time*. Cambridge, UK: Polity Press.

Adam, Barbara. 1998. *Timescapes of Modernity: The Environment and Invisible Hazards*. London: Routledge.

Adam, Barbara. 2000. The Media Timescapes of BSE News. In *Environmental Risks and the Media*, ed. S. Allan, B. Adam, and C. Carter, 117–129. London: Routledge.

Beck, U. 1995. *Ecological Politics in an Age of Risk*. Cambridge: Polity Press.

Benton, T. 1994. Biology and Social Theory in the Environmental Debate. In *Social Theory and the Global Environment*, ed. M. Redclift and T. Benton, 28–50. London: Routledge.

Benton, Ted. 2001. Why Are Sociologists Naturephobes? In *After Postmodernism*, ed. J. Lopes and G. Potter. London: Athlone.

Berners-Lee, M., and D. Clark. 2013. *The Burning Question*. London: Profile.

Birkland, T.A. 1998. Focusing Events, Mobilization, and Agenda Setting. *Journal of Public Policy* 18 (1): 53–74.

Blühdorn, I. 2011. The Politics Of Unsustainability: COP15, Post-ecologism, and the Ecological Paradox. *Organization & Environment* 24 (1): 34–53.

Boström, Magnus, Debra J. Davidson, and Stewart Lockie. 2018. Conclusions: A Proposal for a Brave New World of Conceptual Reflexivity. In *Environment and Society: Concepts and Challenges*, ed. Magnus Boström and Debra J. Davidson, 351–374. Cham, Switzerland: Palgrave Macmillan.

Carolan, Michael. 2014. *Cheaponomics: The High Cost of Low Prices*. Abingdon, UK: Earthscan.

Catton, W., and R. Dunlap. 1980. A New Ecological Paradigm for Post-exuberant Sociology. *American Behavioral Scientist* 24: 15–47. https://doi.org/10.1177/000276428112400103.

Clark, N. 2011. *Inhuman Nature*. London: Sage.
Desrochers, Pierre, and Joanna Szurmak. 2018. *Population Bombed: Exploding the Link Between Population and Climate Change*. Toronto: The Global Warming Policy Foundation.
Dickens, P. 2004. *Society & Nature*. Cambridge, UK: Polity Press.
Dickson, Janice. 2019. Chrystia Freeland Says It's a "Disappointment" Arctic Council Could Not Issue Joint Communique. *Globe and Mail*, 7 May: A7.
Dunlap, R. 2010. The Maturation and Diversification of Environmental Sociology. In *The International Handbook of Environmental Sociology*, 2nd ed, ed. M. Redclift and G. Woodgate, 15–32. London: Edgar Elgar.
Dunlap, R., and W. Catton. 1994. Struggling with Human Exemptionalism. *The American Sociologist* 25 (1): 5–30.
Flannery, Tim. 2009. *Now or Never*. Toronto: HarperCollins.
Flannery, Tim. 2015. *Atmosphere of Hope: Searching for Solutions to the Climate Crisis*. New York: Atlantic Monthly Press.
Foster, John Bellamy, Brett Clark, and Richard York. 2010. *The Ecological Rift: Capitalism's War on the Earth*. New York: Monthly Review Press.
Freudenburg, W. 2006. Environmental Degradation, Disproportionality, and the Double Diversion. *Rural Sociology* 71 (1): 3–32.
Freudenburg, William R., and Robert Gramling. 1993. Socioenvironmental Factors and Development Policy. *Sociological Forum* 8 (3): 341–364.
Freudenburg, William R., Scott Frickel, and Robert Gramling. 1995. Beyond the Nature/Society Divide. *Sociological Forum* 10 (3): 361–392.
Freudenburg, William, Robert Gramling, Shirley Laska, and Kai Erikson. 2009. *Catastrophe in the Making*. Washington: Island Press.
Giddens, Anthony. 2009. *The Politics of Climate Change*. Cambridge: Polity Press.
Global Carbon Project. 2018. *Global Carbon Budget Summary Highlights*. http://www.globalcarbonproject.org/carbonbudget/18/highlights.htm. Accessed 6 December 2018.
Gramling, Robert, and William R. Freudenburg. 1996. Crude, Coppertone, and the Coast. *Society and Natural Resources* 9: 483–506.
Homer-Dixon, Thomas. 2006. *The Upside of Down*. Toronto: Random House.
ICAO. 2018. Traffic Growth and Airline Profitability Were Highlights of Air Transport in 2016. *Uniting Aviation: A United Nations Specialized Agency*. https://www.icao.int/Newsroom/Pages/traffic-growth-and-air line-profitability-were-highlights-of-air-transport-in-2016.aspx. Accessed 29 January 2018.

IIPBES (Intergovernmental Science-Policy Platform on Biodiversity and Ecosystem Services). 2019. *Global Assessment Report on Biodiversity and Ecosystem Services*. Bonn: IPBES. https://ipbes.net/sites/default/files/2020-02/ipbes_global_assessment_report_summary_for_policymakers_en.pdf. Accessed 10 April 2020.

IPCC. 2018. Intergovernmental Panel on Climate Change. *Global Warming of 1.5 °C*. http://report.ipcc.ch/sr15/pdf/sr15_spm_final.pdf. Accessed 8 October 2018.

IPCC. 2019. *The Ocean and Cryosphere in a Changing Planet*. https://report.ipcc.ch/srocc/pdf/SROCC_SPM_Approved.pdf. Accessed 25 September 2019.

Latour, Bruno. 2000. When Things Strike Back. *British Journal of Sociology* 51: 107–123.

Lockie, Stewart, and Catherineg Wong. 2017. Risk, Sustainability and Time: Sociological Perspectives. In *Social Science and Sustainability*, ed. Heinz Schandl and Iain Walker, 187–198. Melbourne: CSIRO Publishing.

Lockie, Stewart, and Catherine Wong. 2018. Conflicting Temporalities of Social and Environmental Change. In *Environment and Society: Concepts and Challenges*, ed. Magnus Boström and Debra J. Davidson, 327–350. London: Palgrave Macmillan.

Lynch, Michael. 2016. Cultural Dopes. https://doi.org/10.1002/9781405165518.wbeos0712. Accessed 31 January 2020.

McCarthy, Niall. 2019. Report: The U.S. Military Emits More CO_2 Than Many Industrial Nations. *Forbes*, 13 June. https://www.forbes.com/sites/niallmccarthy/2019/06/13/report-the-u-s-military-emits-more-co2-than-many-industrialized-nations-infographic/#52f5901f4372. Accessed 10 April 2020.

Murphy, Raymond. 1988. *Social Closure: The Theory of Monopolisation and Exclusion*. Oxford: Oxford University Press.

Murphy, Raymond. 1994. *Rationality and Nature*. Boulder: Westview.

Murphy, R. 1997. *Sociology and Nature: Social Action in Context*. Boulder: Westview.

Murphy, Raymond. 2002. The Internalisation of Autonomous Nature into Society. *The Sociological Review* 50: 313–333.

Murphy, Raymond. 2009. *Leadership in Disaster: Learning for a Future with Global Climate Change*. Montreal: McGill-Queens University Press.

Murphy, Raymond. 2010. Environmental Hazards and Human Disasters. In *The International Handbook of Environmental Sociology*, 2nd ed, ed. Michael Redclift and Graham Woodgate, 276–291. Cheltenham, UK: Edward Elgar.

Nordhaus, William. 2013. *The Climate Casino: Risk, Uncertainty, and Economics for a Warming World*. New Haven: Yale University Press.
Nye, D. 1998. *Consuming Power*. Cambridge, MA: MIT Press.
Perrow, C. 1984. *Normal Accidents*. New York: Basic Books.
Pielke, Roger Jr. 2010. *The Climate Fix*. New York: Basic Books.
Reality Check Team. 2019. Prince Harry and Private Jets: What's the Carbon Footprint? *BBC News*, 20 August. https://www.bbc.com/news/uk-49408915. Accessed 21 August 2019.
Redclift, Michael. 2010. The Transition Out of Carbon Dependence: The Crises of Environment and Markets. In *The International Handbook of Environmental Sociology*, 2nd ed, ed. Michael Redclift and Graham Woodgate, 121–135. Cheltenham UK: Edward Elgar.
Reguly, Eric. 2018. Bet Big on Big Oil. *Report on Business*, February: 19.
Rhodes, Richard. 2018. How Will the World Overcome Its Largest-Ever Energy Crisis? Slowly. *The Globe and Mail*, 16 June: O1, O4.
Rowe, James, Jessica Dempsey, and Peter Gibbs. 2016. The Power of Fossil Fuel Divestment (and Its Secret). In *A World to Win: Contemporary Social Movements and Counter-Hegemony*, ed. William R. Carroll and Kanchan Sarker, 233–249. Winnipeg: ARP Books.
Safran Foer, Jonathan. 2019a. *We are the Weather: Saving the Planet Begins at Breakfast*. New York: Farrar, Strauss, and Giroux.
Safran Foer, Jonathan. 2019b. 'A Crisis of Belief'. *The Globe and Mail*, 28 September: R14.
Shove, E. 2012a. Putting Practice into Policy: Reconfiguring Questions of Consumption and Climate Change. *Contemporary Social Science* 9: 1–15.
Shove, E. 2012b. Energy Transitions in Practice: The Case of Global Indoor Climate Change. In *Governing the Energy Transition: Reality, Illusion or Necessity?*, ed. G. Verbong and D. Loorbach. London: Routledge.
Shove, E., and N. Spurling (eds.). 2013. *Sustainable Practices: Social theory and Climate Change*, 208. Routledge: London.
Shove, E., M. Pantzar, and M. Watson. 2012. *The Dynamics of Social Practice: Everyday Life and How It Changes*, 240. London: Sage.
Speth, James Gustave. 2009. *The Bridge at the Edge of the World: Capitalism, the Environment, and Crossing from Crisis to Sustainability*. New Haven: Yale University Press.
Speth, James Gustave. 2012. *America the Possible: Manifest for a New Economy*. New Haven: Yale University Press.

Steffen, Will, Johan Rockström, Katherine Richardson, Timothy Lenton, Carl Folke, Diana Liverman, Colin Summerhayes, Anthony Barnosky, Sarah Cornell, Michel Crucifix, Jonathan Donges, Ingo Fetzer, Steven Lade, Marten Scheffer, Ricarda Winkelmann, and Hans Joachim Schellnhuber. 2018. Trajectories of the Earth System in the Anthropocene. *PNAS Proceedings of the National Academy of Sciences of the United States of America* 115 (33): 8252–8259. https://doi.org/10.1073/pnas.1810141115

Stern, Nicholas. 2009. *A Blueprint for a Safer Planet*. London: Random House.

Suzuki, David, and Ian Hanington. 2017. *Just Cool It: The Climate Crisis and What We Can Do*. Vancouver and Berkeley: Greystone Books.

Turner, B., and N. Pidgeon. 1978. *Man-Made Disasters*. London: Wykeham.

UNEP (United Nations Environmental Programme). 2018. *Emissions Gap Report 2018 Executive Summary*. Nairobi: UNEP. https://wedocs.unep.org/bitstream/handle/20.500.11822/26879/EGR2018_ESEN.pdf?sequence=10. Accessed 28 November 2018.

U.S. Global Change Research Program. 2018. *Fourth National Climate Assessment Vol. II: Impacts, Risks, and Adaptation in the United States*. Washington. https://nca2018.globalchange.gov/. Accessed 26 November 2018.

Victor, David. 2011. *Global Warming Gridlock: Creating More Effective Strategies for Protecting the Planet*. Cambridge: Cambridge University Press.

Watt-Cloutier, Shiela. 2019. If We Protect the Arctic, We Save the Planet. *Globe and Mail*, 5 October: O11.

Weber, Max. 1946 (1958). *From Max Weber: Essays in Sociology*. H.H. Gerth and C. Wright Mills (eds.). New York: Oxford University Press.

Wong, C., and S. Lockie. 2018. Sociology, Risk and the Environment: A Material-Semiotic Approach. *Journal of Risk Research* 21 (9): 1077–1092.

World Bank. 2015. Fossil Fuel Energy Consumption (% of Total). World Bank. https://data.worldbank.org/indicator/eg.use.comm.fo.zs. Accessed 1 April 2019.

York, R. 2012. Do Alternative Energy Sources Displace Fossil Fuels? *Nature Climate Change* 2 (6): 441–443.

York, R., and J.A. McGee. 2016. Understanding the Jevons Paradox. *Environmental Sociology* 2 (1): 77–87. https://doi.org/10.1080/23251042.2015.1106060.

Zebrowski Jr., Ernest. 1997. *Perils of a Restless Planet: Scientific Perspectives on Natural Disasters*. Cambridge: Cambridge University Press.

Part I

Analysing the Problem

2

Cooperation Between Natural Science and Social Science

The ingenuity of humans to manipulate dynamics of nature to add value to biophysical resources has resulted in unparalleled comforts, convenience, consumption, and life expectancy. This has produced a population surge. That very success of human rationality has, however, caused unmatched degradation of the global environment and is unleashing powerful and threatening new forces of nature, which constitute inadvertent and paradoxically irrational consequences (Murphy 1994). The most insidious impacts are those where causes are invisible and harm slow onset, which lull populations into believing all is normal even as abnormality creeps forward. Arguably the most serious and pervasive danger consists of extracting and combusting fossil fuels, which causes long-lasting carbon pollution of the atmosphere, a greenhouse effect, global warming, and climate change.

Anthropogenic global warming has been socially caused by giving priority to near-term economic interests at any environmental cost. It can either be suffered or mitigated by social action. Hence social science is needed to analyse causes, consequences, and possible resolution. Nevertheless, the social action consists of interaction with properties and dynamics of nature, including unleashing its threatening forces. The

R. Murphy, *The Fossil-Fuelled Climate Crisis*,
https://doi.org/10.1007/978-3-030-53325-0_2

social science analysis of such action must be grounded on a solid foundation of natural science knowledge to understand the depth of the problem. Natural science is necessary to make visible global atmospheric warming and distinguish climate change from weather change. The best available evidence is needed to transform aspirations for solutions into reality rather than underestimating problems and falling into wishful thinking or greenwashing. A cooperative relationship between social science and natural science is needed.

The Natural Science of Anthropogenic Climate Change

Science is admittedly fallible, and the evidence and understanding could change in the future. Nevertheless, natural science is the unsurpassed source of knowledge of such a biophysical problem where the connections between cause and effects extend across centuries and oceans, and are largely invisible to the senses. Even the admitted fallibility of science is a positive feature in that it involves non-dogmatic, self-corrective features. A brief summary of current natural science understanding of human-made climate change is warranted because there are misunderstandings in society which creeped into some social science.

Industrialization led to the massive combustion of coal, and later modernization was powered by combusting oil and natural gas. Fossil-fuel combustion and land use changes such as deforestation emit large amounts of carbon dioxide, which is the principal greenhouse gas. Fossil-fuel extraction and transport release methane. Nitrous oxide is emitted mainly through agricultural activities. Carbon dioxide, methane, nitrous oxide, and a few other gases cause a greenhouse effect trapping heat in the atmosphere.[1] Properties of specific greenhouse gases (GHG) vary. Methane is a potent greenhouse gas, but is removed from the atmosphere by nature's chemical reactions in about 12 years. Carbon dioxide is more complicated, with about two-thirds remaining in the atmosphere between 20 and 200 years, and the remainder removed by slower processes that take a hundred thousand years. Emissions accumulate carbon in the atmosphere; therefore it is crucial to reduce emissions promptly. Vegetation growth absorbs carbon dioxide from the atmosphere, as do oceans and soil. Land use changes, particularly deforestation and some types of agriculture (IPCC 2019), diminish nature's capacity to withdraw atmospheric carbon dioxide. The US Environmental Protection Agency presented the evolution of greenhouse gases in the atmosphere in Fig. 2.1 and concluded as follows. 'This graph shows the increase in greenhouse gas (GHG) concentrations in the atmosphere over the last 2000 years. Increases in concentrations of these gases since 1750 are due to human activities in the industrial era. Concentration units are parts per million (ppm) or parts per billion (ppb), indicating the number of molecules of the greenhouse gas per million or billion molecules of air. … Carbon dioxide is the primary greenhouse gas that is contributing to recent climate change' (EPA 2017a).

Given the complexities and interactions, and because Earth is a huge planet, it took evidence, increased understanding, and time to convince scientists that human activities are producing planetary warming. Early on, they raised the question everyone is now asking: how can actions of humans change the Earth's climate. Being closest to the evidence, they were first to answer affirmatively. In 1971, Rasool and Schneider (1971) ran a computer model that omitted to include the stratosphere, which resulted in the inference that cooling by aerosols outweighed warming

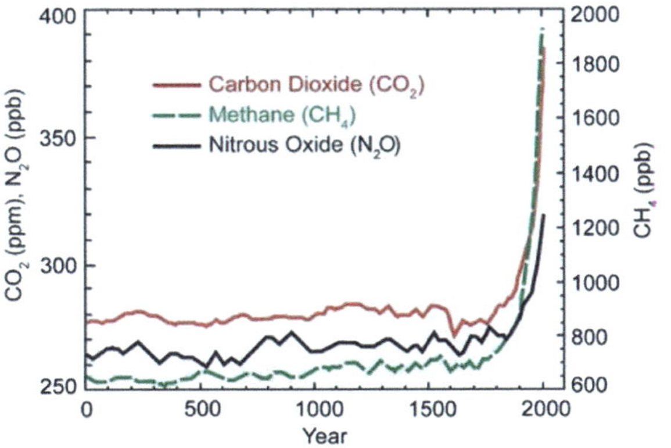

Fig. 2.1 Evolution of heat-trapping gases in the atmosphere that result in the greenhouse effect (*Source* EPA 2017b)

by carbon dioxide. After scientific debate, Schneider (2009) corrected his model, redid the calculations, and became a leading scientist investigating how greenhouse gases result in global warming. Similarly, Heal (2017: 31) in the late 1970s didn't believe humans could change the climate. But when he checked the theories and calculations, he understood the threat. 'What convinced me was that burning a ton of coal releases about two and a half tons of CO_2. A large power station can burn 10,000 tons of coal daily, releasing 25,000 tons of CO_2 daily, or more than 7 million tons annually. This amounts to billions of tons of greenhouse-gas emissions each year when you take into account that there are thousands of coal power stations around the world. Throw in other types of fossil fuel power stations as well as cars, planes, boats and trains, and it's not hard to see that this all adds up to a huge amount of CO_2, quite sufficient to change the composition of the atmosphere'. He then became a leading environmental economist. This evolution is similar to initial disbelief about smoking cigarettes slowly causing lung cancer. As scientific evidence mounted, doctors were closest to the evidence and were first to stop smoking.

Impact science is needed to understand how common social practices, like driving a car 100 miles by combusting 5 gallons of gasoline, which weigh about 31 pounds, produce 100 pounds of CO_2 when combusted. The remaining 69 pounds come from oxygen in the air. Combustion enables each carbon atom to combine with two oxygen atoms, which have greater atomic weight, from surrounding air thereby producing CO_2 much heavier than the original gasoline.[2] Non-scientists need to learn how much greenhouse gases their common fossil-fuelled social practices pump into the atmosphere. The manufacture of concrete, used extensively in buildings, bridges, roads, dams, pipes, etc., results in eight per cent of global emissions of carbon dioxide. Concrete is bound together by cement, which is manufactured by combusting fossil fuels to very high temperatures causing calcium carbonate to break down into the binding agent lime and emitted carbon dioxide. Hence cement is a double jeopardy carbon polluter, with carbon dioxide emitted both in combusting fossil fuels and the ensuing thermal chemical decomposition of calcium carbonate, such that the weight of carbon dioxide emitted is equal to the weight of the cement. Given all the cement used in the world, this weight of CO_2 emitted is enormous.

NASA (2018) found 'the planet's average surface temperature has risen about 1.62 degrees Fahrenheit (0.9 degrees Celsius) since the late 19th century, a change driven largely by increased carbon dioxide and other human-made emissions into the atmosphere. Most of the warming occurred in the past 35 years'.[3] Over the last 136 years, 17 of the 18 warmest years occurred since 2001. Global warming is speeding up. CO_2 emissions increased from 2 billion tons annually in 1900 to 35 billion tons in 2010, a 2.6% annual increase which is compounded and cumulative (Nordhaus 2013: 21 Fig. 2). Scientists at the American Environmental Protection Agency (EPA 2017a) documented that 'worldwide, net emissions of greenhouse gases from human activities increased by 35 percent from 1990 to 2010. … Concentrations of carbon dioxide and other greenhouse gases in the atmosphere have increased since the beginning of the industrial era. Almost all of this increase is attributable to human activities. Historical measurements show that the current global atmospheric concentrations of carbon dioxide are unprecedented compared with the past 800,000 years, even after accounting for natural fluctuations'. Human activities have loaded the atmosphere with an unparalleled amount of greenhouse gases, and despite talk about transitioning to low-carbon economies, the trajectory continues towards an atmosphere even more carbonized. The carbon dioxide content of the atmosphere prior to the industrial revolution was 280 parts per million (ppm), by 2008 it had risen to 387 ppm, humans are currently adding 3 ppm each year and this rate of added carbon dioxide is increasing.

The 2015 Paris goal of limiting atmospheric temperature increase to 2 °C implies limiting its carbon dioxide content to 450 ppm (Dyer 2008: 14, 231, 177), hence societies are rapidly reaching their hoped-for limit. The IPCC (2018: SPM-4) calculated that 'global warming is likely to reach 1.5 °C between 2030 and 2052 if it continues to increase at the current rate' and rise beyond 1.5 °C after that. To limit the increase to 1.5 °C, net global human-caused emissions of carbon dioxide would have to decrease from their 2010 level by 45% by 2030, and achieve net zero around 2050. This is so socioeconomically challenging most leaders have weakened the goal to an increase of 2 °C.

Nevertheless, 'the hard choices have to be made now. We are poised to set in motion irreversible alterations to the climate and the natural world around us. Because climate changes slowly in response to CO_2 concentrations, the effects of what we do in the next twenty-five years will play out over the next hundred in terms of mass extinctions, sharp rises in sea level, decreases in food production, and possible other trends not yet anticipated' (Heal 2017: 42). The World Meteorological Organization (2019) found that 'greenhouse gas concentrations in the atmosphere have also increased to record levels, locking in the warming trend for generations to come'. It documented that those concentrations are accelerating and are on track to reach or exceed 410 ppm by the end of 2019, and caused sea level rise, shrinking ice in the Arctic, Antarctic, and Greenland, increased the heat content and acidity of the oceans, and extreme weather like wildfires which release massive amounts of carbon dioxide into the atmosphere. The IPCC (2019: SPM-8 A2) similarly found 'it is virtually certain that the global ocean has warmed unabated since 1970 and has taken up more than 90% of the excess heat in the climate system (high confidence). Since 1993, the rate of ocean warming has more than doubled (likely). Marine heatwaves have very likely doubled in frequency since 1982 and are increasing in intensity (very high confidence). By absorbing more CO_2, the ocean has undergone increasing surface acidification (virtually certain). A loss of oxygen has occurred from the surface to 1000 m (medium confidence)'. The World Meteorological Organization (2019) concluded that 'to stop a global temperature increase of more than 2 degrees Celsius above pre-industrial levels, the level of ambition needs to be tripled. And to limit the increase to 1.5 degrees, it needs to be multiplied by five'. The emissions gap report (UNEP 2019) found that, even with current promises to cut emissions, temperature will increase by 3.2 °C this century, resulting in widespread destructive consequences, and the promises are not being implemented.

Aren't fossil fuels being replaced by low-carbon renewable energy? The International Energy Agency (2018) documented that strong global economic growth in 2017 resulted in a 2.1% increase in energy consumption, and that 70% of the increase occurred for oil, natural gas, and coal, with all types of renewable energy only accounting for 30%.

Combustion of coal increased by 1% and improvements in energy efficiency slowed. The result was (i) a 1.4% rise in energy-related carbon dioxide emissions to a historical high of 32.5 gigatonnes in 2017 and (ii) fossil fuels accounting for 81% of total energy consumption which, despite fluctuations due to economic growth or recessions and increases in low-carbon energy, has not decreased over the last three decades. By 2016, 96 million barrels of oil were being pulled out of the ground and combusted globally each day, which equals 7.5 billion litres daily. And the rate is increasing. Little wonder this massive transfer of carbon from ground to sky is changing the climate.

Scientific projections are based on empirical data and confirmed knowledge. 'Sea level continues to rise at an increasing rate. Extreme sea level events that are historically rare (once per century in the recent past) are projected to occur frequently (at least once per year) at many locations by 2050 in all RCP scenarios, especially in tropical regions' (IPCC 2019: SPM-22 B3). IPCC's expected scenario assumes rapid economic growth in China, India, and other developing countries, moderate global population growth, substantial implementation of non-fossil-fuel energy, and greater fossil fuels efficiencies. Despite these improvements, the carbon dioxide content of the atmosphere will be almost 700 ppm in 2100. That is not far away: the increase will occur in the lifetime of infants born now. This results in an estimated atmospheric temperature increase of 2.8 °C above its temperature in 1990, which was already above the pre-industrial level. This scenario is dangerous. If population growth is higher and if efficiencies, use of non-fossil-fuel energy, reduction of fossil-fuel emissions are less than hoped for, then the carbon dioxide content could hit 800 ppm in 2100. This is exceedingly threatening. Paleontologists documented there were five mass extinctions (Kolbert 2014). Only one was caused by an asteroid striking Earth, namely when dinosaurs disappeared. The others involved high levels of carbon dioxide in the atmosphere causing runaway chains of positive feedback loops destroying habitats of most species. 'That last greenhouse extinction occurred when there were only about eight hundred parts per million of carbon dioxide in the atmosphere, a level we might well achieve in this century on a "business-as-usual basis"' (Dyer 2008:

231). The long view is so menacing for the habitat of humanity that it is excluded from consideration by all but natural scientists.

Emissions and Withdrawal: The Net Change in Atmospheric Carbon

NASA (2007) documented that capping greenhouse-gas emissions at today's rates would not stop global warming because even with that cap, humans would be emitting carbon into the atmosphere faster than natural processes put it back underground, and this has been occurring since the industrial revolution. It allows anthropogenic climate change to be aggravated, and is worse if the cap is allowed to increase to promote economic growth, as in China and Alberta where the latter's cap of 100 megatonnes of emissions represents a 50% increase of its carbon pollution from 70 megatonnes. Even decreases in emissions per unit of GDP or per barrel of oil that leave emissions higher than withdrawal rates exacerbates global warming. Reducing emissions would only slow the increase of global warming, unless emissions fall below the amount of carbon that plants, the ocean, and rocks absorb from the atmosphere.

Global warming is determined by the net change in atmospheric carbon: the difference between the amount of carbon going into the atmosphere by nature's processes (forest fires) and anthropogenic emissions (fossil fuels) compared to the amount taken out by nature's processes (growth of forests, absorption by oceans) and humanly initiated processes (afforestation, direct air capture). Anthropogenic climate change can be worsening even as clean, renewable energy is developing and emissions are decreasing if they are still above withdrawal rates. It is aggravated not only by fossil-fuel combustion resulting in emissions but also by land use changes, such as some agricultural practices and deforestation (IPCC 2019), that reduce the capacity of nature's dynamics to provide the free service of withdrawing carbon from the atmosphere. IPCC (2019: A3) estimated that, from 2007 to 2016, land absorbed 29% of carbon dioxide equivalent emissions, called a sink, but added that 'the persistence of the sink is uncertain due to climate change'. Global warming is transforming parts of the huge boreal forests of Canada,

Alaska, and Russia from carbon sink to the source of carbon emissions: 'Climate warming and drying has led to more severe and frequent forest fires, which threaten to shift the carbon balance of the boreal ecosystem from net accumulation to net loss, resulting in a positive climate feedback' (Walker et al. 2019: abstract). The 2019 huge fires in the Amazon, many set by ranchers and farmers to clear land but made worse by hot, dry conditions caused by global warming, are changing large tracts of the Amazon from carbon sink to carbon source. This decreases services forests render in withdrawing atmospheric carbon and storing it safely in trees. Oceans are becoming more acidic as they absorb atmospheric carbon, which diminishes their capacity to absorb more carbon and act as a future carbon sink.

The concept 'global carbon budget' (Berners-Lee and Clark 2013) compares atmospheric carbon content with a specific target, such as the carbon that can be emitted yet remain under the 2 degrees temperature increase mandated in the Paris Agreement. The concepts net change in atmospheric carbon and global carbon budget are the most valid concepts to indicate whether climate change is lessening or worsening. They take into account greenhouse-gas emissions and human activities that weaken nature's carbon withdrawal capacity, such as deforestation (IPCC 2019). Even decreases of total emissions give misleading indications of improvement if excess of emissions over withdrawal continues to increase atmospheric carbon and worsen the problem.

Figure 2.2, taken from the Carbon Dioxide Information Analysis Center (CDIAC 2018) at Berkley, illustrates the dynamic of a global carbon dioxide budget. Between 2006 and 2015, fossil fuels were taken from underground geological reservoirs and combusted, resulting in 34.1 gigatonnes of carbon dioxide emitted into the atmosphere yearly. Land use changes (e.g. deforestation) resulted in 3.5 gigatonnes more emissions per year. The oceans absorbed 9.7 gigatonnes of carbon dioxide each year and the land, such as forests and soil, absorbed 11.5 gigatonnes. The difference resulted in the carbon dioxide content of the atmosphere growing by 16.4 gigatonnes yearly, worsening the greenhouse effect.

Figure 2.3, also taken from CDIAC (2018), shows where emissions originated and where carbon dioxide went from 1880 to 2015. Since 1900, fossil-fuel combustion constituted the principal source, and after

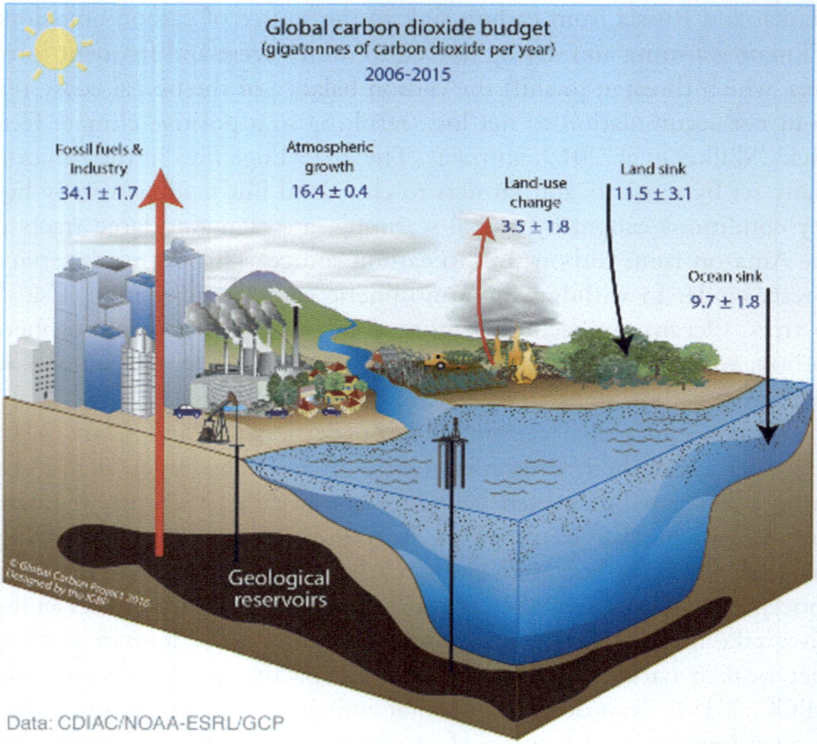

Fig. 2.2 Global carbon dioxide budget (*Source* CDIAC 2018)

1945 fossil-fuel emissions have risen steeply. That is why this book focusses on fossil-fuelled climate change. Land use changes resulted in some increase, but the curve is irregular and much flatter. Concerning carbon withdrawal, land absorbed increasing but fluctuating amounts, oceans absorbed increasing amounts, and the biggest increase in carbon dioxide absorption has been in the atmosphere. Remember that carbon dioxide remains in the atmosphere a century causing global warming before descending to the land and oceans.

It is true that 'three-fifths of countries in the EPI [Environmental Performance Index] have declining CO_2 intensities, while 85–90% of countries have declining intensities for methane, nitrous oxide, and black

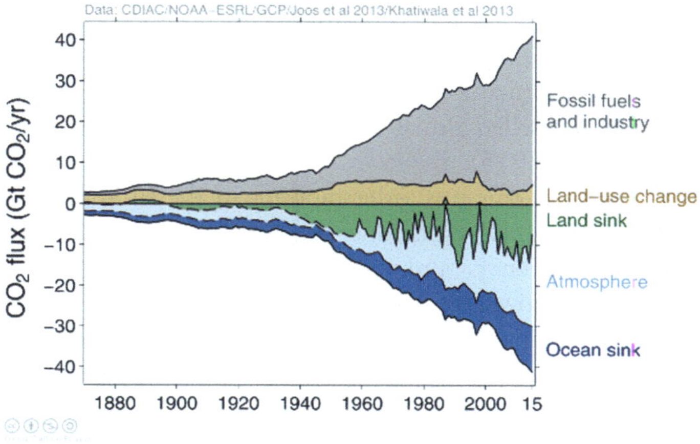

Fig. 2.3 Source of carbon dioxide emissions, and sinks for carbon dioxide (*Source* CDIAC 2018)

carbon' (Yale University 2019: Executive summary). Pielke (2010: 3) claims this shows 'the world has been decarbonizing for more than a century'. This has also been called dematerialization and decoupling economic growth from emissions growth. Pinker's (2018) climate change optimism is based on the indicator that the amount of carbon dioxide emitted per dollar of GDP has been reduced by 44%. Fossil-fuelled climate change lends itself to these misleading indicators, in some cases used naively by social scientists and the population but in others promoted by interest groups, falsely showing that the problem is being solved when it is being made worse. Such claims are based on intensity indicators of emissions per unit of economic activity as measured by GDP. But why then have emissions of carbon dioxide, methane, nitrous oxide (used in fertilizers), and black carbon increased, global warming intensified, and climate change worsened during that century? The deficiency of this way of thinking consists of the erroneous equation of decreasing carbon intensity of economic activity with mitigation of global warming, and ignores the crucial comparison documented by impact science of emissions with carbon withdrawals from the atmosphere, carbon-in versus carbon-out. Although the economy has superficially been decarbonizing for a century in terms of somewhat less emissions per GDP, the absolute amount of emissions has been increasing because of economic growth and far exceeds carbon withdrawal, so the world has been carbonizing for a century in its atmosphere and oceans, and therefore fossil-fuelled global warming and climate change have been worsening (Murphy 2015).

Figure 2.4 corrects this misunderstanding in the social sciences. The difference between GHG emissions per GDP (intensity), which indicates economic efficiency, and overall GHG emissions, which cause global warming, is shown for Canada. The yellow line shows that, from 1990 to 2017, Canada succeeded in reducing its emissions for every unit of economic activity. But the blue line shows that nevertheless Canada's emissions increased. Since it is overall emissions that causes global warming, Canada's contribution increased and worsened it. Hence Canada's environmental performance deteriorated. The yellow line of emissions intensity is used by the fossil-fuel industry to legitimate their carbon pollution of the atmosphere, which is instead accurately indicated by the blue line. The difference results from economic growth, which increases emissions even when emissions per GDP decrease. To mitigate global warming, technological innovation of emissions reduction per GDP would have to be faster than economic growth. This did not occur. York (2010, 2012a, 2012b) documented that the world's highest emitting countries—USA, China, Russia, India, and Japan—decreased

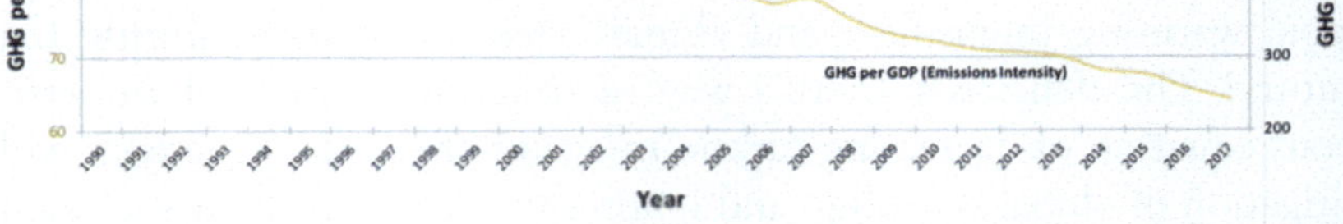

Fig. 2.4 Canadian greenhouse-gas emissions and indexed trend emission intensity (excluding Land Use, Land Use Change and Forestry) (*Source* Government of Canada 2019)

their greenhouse-gas emissions per unit of GDP between 1980 and 2005, but increased their total emissions because of economic growth, which is typical for almost all countries. Social science conceptions of decarbonization, dematerialization, decoupling, ecological modernization, etc., using intensity indicators have been misleading.

Pielke (2010: 230) states: 'If there is a single variable that will serve as a measure of progress toward emissions reduction or carbon-intensity goals, it will be the proportion of global energy consumption that comes from carbon-neutral (or even negative) sources'. He asserts that the proportion has been well under ten per cent and will need to exceed ninety per cent to stabilize global warming at low levels. Even if the proportion of carbon-neutral global energy were to increase to twenty per cent, a worsening of the carbon content of the atmosphere and of global warming would result if economic growth produces more emissions than withdrawals. That could even be true at ninety per cent carbon-neutral energy if there is enormous global economic growth without much growth in carbon withdrawal. Social science analysis of fossil-fuelled climate change needs to build upon natural science understanding of emissions compared to carbon withdrawals. If carbon emissions exceed withdrawals, then global warming worsens because of carbon's physical properties of (i) causing a greenhouse effect when in the atmosphere, and (ii) remaining there for 100+ years and accumulating. Both carbon emissions and withdrawals can have natural causes (forest fires and forest growth, respectively) and human causes (fossil-fuel combustion and purposive afforestation, respectively), but so far anthropogenic withdrawals are infinitesimal compared to anthropogenic emissions.

Limiting Anthropogenic Global Warming to 2 °C

There is massive scientific consensus that fossil-fuel combustion, land modification (e.g. deforestation), and other human activities have increased global surface temperatures since the pre-industrial period. The precise amount depends on what is taken as the pre-industrial period. Different years have been used as reference points yielding variable amounts of increase, and even the IPCC has left 'pre-industrial period' undefined. Hawkins, Ortega, and Suckling (2017) argue that 1720–1800 is the most suitable starting period. Their assessment shows that 'this preindustrial period was likely 0.55 °C–0.80 °C cooler than 1986–2005 and that 2015 was likely the first year in which global average temperature was more than 1 °C above preindustrial levels'. Some studies use the Kyoto reference year of 1990 as the benchmark rather than the pre-industrial period. Since greenhouse gases are accumulating in the atmosphere and increasing the temperature, the end date also determines the exact figure of global warming. Despite these technical differences yielding slightly different figures, there is vast agreement fossil-fuelled climate change is occurring and accelerating.

Superficially 1 °C seems trivial, but it nevertheless consists of global warming that has already melted the Arctic ice cover, glaciers, permafrost, destroyed coral, etc. The change from reflective white ice on the Arctic Ocean to dark water absorbs more of the sun's radiation. Melting permafrost releases the powerful greenhouse-gas methane which had been trapped safely in the ground. Both unleash nature's own runaway processes of global warming which will be exceedingly difficult to reverse. Steffen et al. (2018) documented the trajectory that fossil-fuelled social practices are on, which are resulting in 'the risk that self-reinforcing feedbacks [of nature's dynamics] could push the Earth System toward a planetary threshold that, if crossed, could prevent stabilization of the climate at intermediate temperature rises and cause continued warming on a "Hothouse Earth" pathway even as human emissions are reduced. Crossing the threshold would lead to a much higher global average temperature than any interglacial in the past 1.2 million years and to sea levels significantly higher than at any time in the Holocene', namely the last 12,000 years which have been so favourable to human development. Bringing emissions in line with carbon withdrawals is urgent because if the threshold is crossed, our planet could enter into an irreversible state much less advantageous to humans.

Despite the urgency and gravity of fossil-fuelled global warming, so little progress has been made to bring emissions in line with carbon withdrawal that decision-makers have given up on renaturing to the pre-industrial global temperature, and instead aim to limit global warming to 1.5 °C–2 °C. Preventive action would have been required globally at the Rio Earth Summit in 1992 to halt fossil-fuelled global warming. The question now is whether further social practices will make it bad or very bad, leading to either limited or extensive degradation of the present human-supporting habitat.

No evidence exists that cost-effective technology will be promptly innovated to massively withdraw and safely store carbon dioxide from the atmosphere. Hence 'to have at least a 50 percent chance of keeping the world below two degrees of warming, global emissions must steadily decline to zero and ultimately become negative (i.e. removing more CO_2 from the air than is added) in the second half of the century. … there is no way around the need to reduce emissions to near-zero or below' (Harvey and Orbis 2018: 290). But annual emissions are currently about 35 gigatonnes or 70 trillion pounds of carbon dioxide and increasing (1 gigatonne = 1 billion tonnes; 1 tonne = 2000 pounds; hence 35 gigatonnes = 35 × 1,000,000,000 × 2000 = 70 trillion pounds). Emissions need to be reduced quickly to avoid tipping the planet into runaway global warming. If global warming had not caused drought in California and Australia, their wildfires would have been easily and inexpensively extinguished. Like the Notre Dame Cathedral fire and the corona (COVID-19) virus, if caught early, fossil-fuelled global warming would have been relatively easy and inexpensive to mitigate, but left to intensify, it is costly and hard to stop destructive dynamics of nature that have been unleashed.

In 2013, Berners-Lee and Clark (2013: xii, xiv) stated 'scientists estimate that humans can pour roughly 565 more gigatonnes of carbon dioxide into the atmosphere by mid-century and still have some reasonable hope of staying below two degrees. … [but that] the amount of carbon already contained in the proven coal and oil and gas reserves' is 2795 gigatonnes, namely five times more, which fossil-fuel companies and states plan to combust. Therefore 80% of proven reserves will have to be kept underground between now and 2050, and exploration halted. McGlade and Ekins (2015: abstract) distinguished types of fossil fuels and calculated that 'globally, a third of oil reserves, half of gas reserves and over 80% of current coal reserves should remain unused from 2010 to 2050 in order to meet the target of 2 °C. … [concluding that] policy makers' instincts to exploit rapidly and completely their territorial fossil fuels are, in aggregate, inconsistent with their commitments to this temperature limit'. Biello (2015) established that American, Russian, and Chinese coal reserves must not be exploited, nor should Middle Eastern natural gas reserves, nor Arctic oil, and that almost three-quarters of the oil in Canada's bituminous sands will have to remain in the sand. 'At the rate we're burning fossil fuels, we'll have used up the entire carbon budget by 2028 – just over halfway into the budget period' (Flannery 2015: 105–106). The International Energy Agency (IEA 2018) argued that carbon prices will have to rise substantially to reduce demand to meet the 2 °C limit, but if that is not done, global temperature increases will result in far more expensive impacts later. Necessary restraint confronts desires for fossil-fuelled economic growth by both companies and fossil fuel extracting countries.

Fossil fuels are not created equal. The combustion of coal emits more greenhouse gases than any of the others. Heavy oil and oil extracted from tar sands have higher emissions intensity per barrel than oil from wells. Natural gas emits less than the others but is still a carbon-emitting fossil fuel. Emissions could be significantly reduced while generating the same electricity by globally replacing thermal coal-fired electricity plants with renewable energy or nuclear energy, and even natural gas. Substituting natural gas for coal was the principal way the USA reduced emissions under the Obama Administration. But there are upward trends too. Since the 2000s, oil companies have been increasingly exploiting heavy crude oil, bitumen (tar sands), deepwater drilling, Arctic oil reserves, etc., and other remote locations (Hughes 2009; Davidson and Andrews 2013). These deposits require more energy to extract, upgrade, and transport, almost invariably supplied by fossil fuels, and hence emit more greenhouse gases per useful barrel. To attain a balance between emissions and withdrawal, the most energy and economic benefit must be obtained for the least emissions until carbon capture and storage or carbon removal technologies are implemented. Much of the most carbon-polluting fossil fuels, such as coal and unconventional heavy oil, will have to remain in the ground where they are safely stored to limit warming to 2 °C.[4] The less polluting fossil fuels, like natural gas and conventional oil from wells, will have to be used first, judiciously and slowly, not rapidly.

Although different types of oil cause similar carbon pollution when combusted, they differ significantly in greenhouse-gas emissions to extract, upgrade, and transport. A Stanford University study (Masnadi et al. 2018) analysed 2015 data from 9000 oil fields in 90 countries, which accounted for 98% of global oil production, concerning extraction, and transportation of crude to refineries. It found that oil fields with the highest carbon intensity had almost triple the emissions compared to those with the lowest. Saudi Arabian oil had relatively low-carbon intensity because less energy and therefore less emissions were needed to extract oil from its wells and little flaring is done, whereas Venezuelan and Albertan tar sands oil had among the world's highest emissions because their unconventional oil required more energy and emissions to extract. The study found that the carbon intensity of Alberta's oil was 70% higher than the global average. Another study estimated that each barrel of oil generates 18 kilograms of greenhouse gases as the world's average, but 44 kilograms for Alberta's oil sands (Graney 2020).

Still another study (IHS Energy 2014) yields similar conclusions. The findings are presented in Fig. 2.5, which shows in yellow that combustion emissions for all types of oil are similar, but extraction, upgrading, and transportation emissions, called 'well-to-retail pump' and shown in blue,

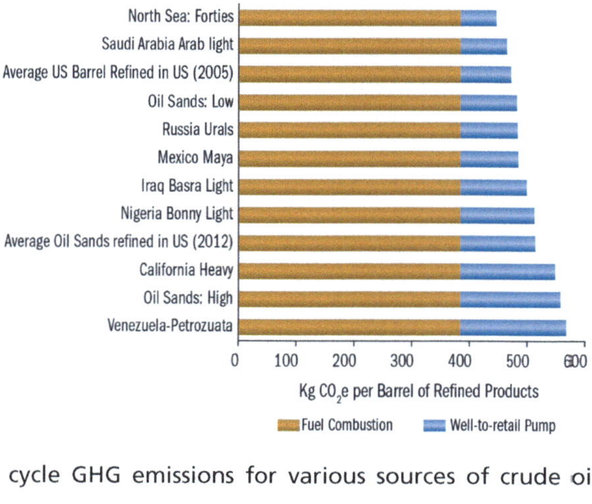

Fig. 2.5 Life cycle GHG emissions for various sources of crude oil (*Source* IHS Energy 2014)

vary. The graph demonstrates that combining emissions from combustion with emissions from extraction, upgrading, and transportation of oil masks important differences between extractions from different sources. Calling the total "well to retail pump" is a misrepresentation of life cycle GHG emissions because many sources are not wells, because there are a variety of engines combusting it instead of just vehicles receiving it from retail gas pumps, and because the carbon ends up in the atmosphere. It would be more accurate to portray carbon as going from "ground to sky", namely from safe storage in the ground to its threatening accumulation in the sky.

Table 2.1 gives a recalculation of Fig. 2.5's blue part, showing emissions involved in getting the oil out of the ground, upgrading it, and transporting it to be combusted. For consistency with the previous graph, the label 'well-to-retail pump' is retained with the above caveat.

The concept 'ethical oil' could be assessed in terms intrinsic to oil, namely the extent to which its level of emissions are less likely to harm future generations by causing global warming. Table 2.1 shows that North Sea and Saudi oil are more ethical than heavy oil from California, Alberta, and Venezuela. The least carbon-polluting types of oil should be used first to give time for technical and social innovations to decrease emissions. But that is socially problematic because types of

Table 2.1 GHG emissions for extracting, upgrading, and transporting various sources of crude oil (blue part 'Well-to-retail pump' of Fig. 2.5)

North Sea	60
Saudi Arabia	70
Average US Barrel Refined in USA (2005)	80
Alberta Oil Sands Low	90
Russia Urals	95
Mexico Maya	100
Iraq Basra Light	115
Nigeria Bonny Light	130
Average Alberta Oil Sands Refined in USA (2012)	135
California Heavy	160
Alberta Oil Sands High	170
Venezuela Petrozuata	180

(*Source* IHS Energy 2014)

oil vary by country, implying that countries having the most polluting reserves should dramatically decrease extraction until the least polluting sources—North Sea and Saudi oil—are exhausted. Imagine the difficulty of convincing countries with high-emissions oil to slow down extraction.

The Underestimation of Global Warming

Estimates of global warming by the IPCC are consensus driven, which means they require agreement by scientists from all countries along with input by their governments. Many scientists are concerned this approach has resulted in underestimation of fossil-fuelled global warming (Heal 2017), with research showing that estimates of emissions reported by industry and government using United Nations protocols and internationally recommended methods understate the problem. This was known for methane but now has been demonstrated for carbon dioxide. Even the best bottom-up estimates of absolute emissions and intensities for Alberta's oil sands, calculated by ground measurements and modelling, called Tier 3, underestimated emissions by 30% compared to actual top-down measurements done by aircraft flyovers of emitting facilities. The results indicate '64% higher annual GHG emissions from surface mining operations, and 30% higher overall OS [oil sands] GHG emissions (17Mt) compared to that reported by industry, despite emissions reporting which uses the most up to date and recommended bottom-up approaches' (Liggio et al. 2019: abstract). They infer that emissions globally may be universally underestimated. What the authors call 'the unaccounted emissions' may be one factor in explaining more-rapid-than-expected global warming, including rapid melting of the Arctic ice cover.

Wagner and Weitzman (2015) demonstrated that the possibility of catastrophic outcomes of fossil-fuelled global warming is not being taken into account, even in modelling by environmental economists, because the probability is relatively small. But it is not zero and low probability high-impact events occur all the time in nature (think of the COVID-19 pandemic). Hence in what is called the Dismal Theorem, they conclude that estimates of the threat are understated, rosy cost–benefit analyses are dubious, and that it is essential to take into account catastrophic, low probability consequences of global warming.

Most attention concerning greenhouse-gas emissions has been paid to the combustion of fossil fuels for the production of electricity and transportation. However the fastest growing sector of oil consumption, especially high-emissions heavy oil, involves petrochemicals: fertilizers, plastics for packaging, digital devices, car parts, clothing, etc. Petrochemical refineries are particularly profitable and being built rapidly. The executive director of the International Energy Agency stated: 'Petrochemicals are one of the key blind spots in the global energy [and emissions] debate, especially given the influence they will exert on future energy [and emissions] trends' (quoted in Willis 2019: B5).

What Is Needed to Limit Global Warming to 2 °C?

The president of the International Institute for Sustainable Development (Vaughan 2014: 4) concluded that a '40 per cent to over 60 per cent decrease in greenhouse gas emissions in the near term and net zero emissions must be achieved between 2050 and 2100 to keep under the 2 degree Celsius cap'. Flannery (2015: 202) quoted a United Nations study showing ways how the 15 highest emitting countries, which total 70% of global emissions, could reduce their emissions in half while tripling their economic output. In an article in *Science*, Rockström et al. (2017) present the following schedule, which they call a roadmap, of specifically what is needed and when it is needed to limit warming to 2 °C.

2017–2020
Global carbon emissions need to peak by 2020 and decline thereafter. Emissions trading needs to increase to a price of $50/ton, much more than triple what California emissions traded at in 2017. The Paris agreement needs to be enhanced.

2020–2030
Short-haul air traffic needs to be replaced by rapid rail, carbon taxes imposed on air transport and shipping, development of alternate aircraft

propulsion must begin, and technical removal of CO_2 from the atmosphere must be scaled up from zero at present.

2030–2040
By 2040 oil will need to exit the global energy mix. Road transportation, shipping, and aircraft will have to become carbon neutral. Emissions-free concrete and steel need to become omnipresent.

This roadmap estimates what is needed and when it is necessary. Jaccard (2018: A13) is somewhat less stringent arguing that oil demand needs to be reduced about 33% by 2050 to meet the 2 °C target. Nevertheless, there is broad scientific consensus oil demand needs to be dramatically reduced soon. These changes will be difficult, annoying, and expensive. To pay for them, Rockström et al. (2017) suggest appropriate international corporate taxes and inheritance taxes on historical wealth generated by fossil fuels to retroactively include previously externalized costs in their price. Such political-economic innovations do not currently exist and would be fiercely opposed in a tax reticent world, which underscores that fossil-fuelled climate change is a social problem, not just a physical one. This roadmap demonstrates the depth and urgency of the problem, and the socioeconomic difficulty of dealing with it. If decision-makers and populations refuse to take a road like this towards emancipation from excessive carbon emissions compared to withdrawals, science forecasts fossil-fuelled social practices will increase global warming more than 2 °C compared to the pre-industrial period, with all the dangers that entails. The 2017–2020 specification in the roadmap provides a valuable near-term reference marker to make visible whether humanity globally and individual countries are on the road to limiting global warming to 2 °C or on the path to making it much worse.

Building sufficient windmills, solar panels, geothermal equipment, batteries to store energy, etc., requires materials. It has been estimated that three billion tons of minerals and metals will be needed by 2050 to build this low-carbon energy infrastructure to limit global warming to 2 °C: graphite, lithium, cobalt, copper, aluminum, nickel, molybdenum, and chromium (World Bank Group 2020). These are found only where nature's processes put them, in some cases in only one country such as the war-torn Democratic Republic of Congo for cobalt and China for graphite. The minerals extraction industry is itself a significant carbon emitter, but the report estimated that its emissions would only be 6% of fossil-fuel emissions, a significant reduction. Nevertheless this would be equal to the 2018 emissions of the USA and China. Recycling would have to be scaled up greatly, but even then, additional materials would have to be dug out of the ground. Aluminum is widely used in low-carbon energy production, but requires much energy for its production, which would have to be low carbon. The demand for energy by a 7.7 billion human population has a huge impact on the planet's climate, but it can be made less damaging.

The Anthropocene

Science and technology, greater organizational efficiencies, market capitalism, and modern legal systems, what Weber (1930, 1978; see also Murphy 1994) called rationalization, have led to much greater consumption, conveniences, and life expectancy. This resulted in rapid population growth, increased extraction of raw materials transformed into commodities, and a globalization of pollution into the atmosphere and oceans. Fossil fuels have been the inanimate energy source for all this. Although there has been a long slow increase of the impact of human activities throughout evolution, the inflection point to an exponential rise is very recent in geological terms. The increasing impact of human activities on the biosphere of planet Earth has led to debate among scientists (Crutzen and Stoermer 2000; Steffen et al. 2011, 2015; Finney and Edwards 2016) concerning whether the present epoch should be characterized as the Anthropocene, that is, fundamentally different from the Holocene which lasted 12,000 years and was propitious for the development of humanity and societies. 'The Anthropocene is a proposed new geological epoch (1) based on the observation that human impacts on essential planetary processes have become so profound (2) that they have driven the Earth out of the Holocene epoch in which agriculture, sedentary communities, and eventually, socially and technologically complex human societies developed' (Steffen et al. 2018). When pressed to give a starting point to the Anthropocene, 1945 is usually the designated year. Since 1945 was also the first time in human history that nature's forces were reconfigured into atomic bombs and deployed to kill hundreds of thousands of civilians in cities, this should alert us that the development of human rationality and technology is not only a major accomplishment with beneficial consequences but also brings significant risks of harmful and irrational consequences.

Agreed that social practices are now having an impact on our planet unlike any other epoch in its existence. But notes of caution are needed before welcoming the concept "Anthropocene" into the social sciences (for others see Lidskog and Waterton 2016). Acceptance of the postulate that we are now in the Anthropocene does not give warrant to popular conceptions of the mastery of nature by human reason, the 'ultimate resource' (Simon 1981, 1996), nor does it support reliance on the premise that technological innovation will always give humans the capacity to adapt to anything nature throws at us as a result of human activities, such as global warming. Neither does it support social science narratives in the 1990s about the social construction of nature (Eder 1996; Evernden 1993). Even scientific proponents of the concept Anthropocene see humanity as at most a force equal in impact to other processes of nature. Nature's dynamics are still there, with human activities superimposed upon them (Adam 1995, 1998, 2000; Adam, Beck, and Loon 2000). Natural scientists must avoid naturalizing our present epoch, and social scientists must resist sociologizing it.

Far from entering the driver's seat of planetary change, it is equally possible that the enormous consequences of social practices are tipping the planet into new dynamics of nature's driverless transformations beyond human control. Even if humanity were to become the driver, it does not imply that the cliff ahead has been eliminated. The fact that social practices are causing global warming and climate change (IPCC 2013, 2018), degradation of oceans, biodiversity loss, etc., implies that this could result in nature's forces becoming more threatening by unleashing more powerful cyclones, flooding, droughts, wildfires, ocean level rise, dead zones in the ocean, etc. The Anthropocene could paradoxically usher in a subsequent biophysical epoch where nature's autonomous dynamics would be less propitious for sustaining human life and prosperity than in the Holocene. That is the concern of many scientists who argue that sustainability in the Anthropocene requires that humans modify their impacts on their biophysical environment.

The Anthropocene could be short if social practices trigger the emergence of stronger forces of nature which overpower human control. It could even be a mirage—more apparent than real and always receding from control—which is why specialists hesitate to label the period 'the Anthropocene'. Although it is improbable that the impact of social practices could throw the planet back to a Paleocene–Eocene Thermal Maximum, their interaction with the global forces of nature results in uncertainty rather than predictability. The interaction leads to not only nature's future dynamics that we know we do not know (known unknowns) but also to other forces of nature that we can't even image at present and don't even know that we don't know (unknown unknowns).[5] This uncertainty concerning never before experienced powerful forces of nature is another reason why it is folly to presume that market-based technological innovation will always be able to master nature's dynamics.

Limitations of Science

Undeniably science has limitations. Its applications enable the extraction of bountiful fossil fuels in deep oceans, the Arctic, shale, tar sands, etc., but have not enabled emissions-free combustion. It can document the carbon buildup in the atmosphere, the resulting greenhouse effect, and the occurrence of global warming due to the combustion of fossil fuels, but because planetary dynamics consist of a very complex multicausal system, science cannot specify the timing, location, and detailed impacts. For hurricanes, science can predict days in advance their trajectory and force reasonably well, but applied science is incapable of stopping or even weakening them. Hence the only action to decrease fatalities is to get out of harm's way by evacuation. Because applied science has not yielded a cost-effective way to combust fossil fuels without emitting greenhouse gases, getting out of harm's way for the threat caused by humans combusting fossil fuels would involve moderating the social practices that emit greenhouse gases and cause harm. For an intense wildfire or hurricane or more frequent sets of them, at present science can only conclude that such consequences fit what one would expect from the scientific understanding of an increasingly carbon-laden, greenhouse-engendering atmosphere. Such answers are not very satisfying, however, for decision-makers and populations who frame questions in terms of whether this specific wildfire or hurricane is or is not caused by global warming. Hence the temptation is great to push the scientific understanding of global warming to the back of the mind and shelve annoying and costly but necessary measures to mitigate the problem. Impact natural science has given convincing evidence and a logical explanation of where the climate is going because of fossil-fuelled global warming, but there remain many uncertainties concerning the details of the threats. The direction of travel is foreseeable but the specifics of the destinations are unforeseeable. This creates problems for adaptation and constructing resilience as a response because there is much uncertainty concerning what society will need to adapt to and bounce back from. Worse still, it is not sure that adaptation and resilience will be possible if global warming tips the climate into a state significantly less beneficial to humans, perhaps irreversibly so. Prevention is admittedly difficult and bothersome, but it may turn out to be more feasible than adaptation and resilience.

Science as Both Feckless and All-Powerful to Master Nature

Earth system scientists have studied trajectories between different global biophysical epochs separated by thresholds determined by feedbacks, interactions, and non-linear processes. Their conclusions indicate the gravity, urgency, and scope of the fossil-fuelled climate crisis. 'Social and technological trends and decisions occurring over the next decade or two could significantly influence the trajectory of the Earth System for tens to hundreds of thousands of years and potentially lead to conditions that resemble planetary states that were last seen several millions of years ago, conditions that would be inhospitable to current human societies and to many other contemporary species' (Steffen et al. 2018). This warning refers to discounting danger and it clearly places responsibility on social and technological trends (read fossil-fuelled ones) and decisions. It is a conclusion based on a peer-reviewed scientific study published in a renowned scientific journal presenting the best available evidence and understanding. There is no excuse for the population and especially leaders to ignore that conclusion because it and links to the study appeared in quality newspapers, television channels, and social media. The outcome could be otherwise if societies reduce their emissions to match rates of greenhouse gas withdrawal from the atmosphere, which means having the foresight to change their fossil-fuelled social practices. However, this is not being done.

Science has been portrayed as a powerful cultural force in modern societies, but in the case of the fossil-fuelled climate crisis, science seems feckless when it comes to inciting societies to change their harmful social practices. Hydrocarbons continue to be extracted in huge amounts from safe storage in the ground and combusted to accumulate in the atmosphere. Blühdorn (2011: 36) notes the strange combination of recognition of the environmental crisis and need for urgent change documented by science but the unwillingness of the population and leaders to do that change. The warnings of science are ignored when they bring troubling news that fossil-fuelled social practices threaten to incubate a global catastrophe and when it counsels changing fossil-fuelled social practices. Nordhaus (2013: 307) concludes that 'when science collides with deep convictions (such as those on religion or politics), conviction often trumps science, even for those who are highly educated'. Scientific predictions of the incubation of climate catastrophe have not resulted in emancipation of societies from dangerous fossil fuels. Instead, the impotence of predictions of environmental catastrophe when confronting near-term economic interests have led to a type of anticipation in which little that is adequate is being done to diminish the threats.

Nevertheless, to legitimate maintaining current fossil-fuelled practices, science is assumed capable of empowering societies to adapt just in time to their unleashing of dangerous new forces of nature in the future and resiliently solving environmental problems on demand. This belief will be examined in Chapter 9. These are two social sides of science: influence where its applications bring benefits but lack of influence when indicating the danger of fossil-fuelled climate change. A biophysical transformation of climate and the environment is occurring as a result of fossil-fuelled social practices, but a sociocultural, politico-economic, and technological transformation appropriate for dealing with its scale and urgency is lagging far behind. The focus of social science analysis must include the investigation of why the scientifically documented urgent problem of climate change is being socially defined as a non-problem (Freudenburg 2006) or a non-urgent problem that can be discounted, shelved in the back of the mind, and not incite preventive action.

A Social Science Analysis Needed

Although the impact applied natural sciences have documented what is necessary and provided a roadmap to mitigate fossil-fuelled climate change, they run up against production scientists, the power of the fossil-fuel industry, other vested interests, path dependency, habitus, and sense of entitlement to fossil-fuelled social practices. It cannot be presumed that impact science will win this confrontation. Needs are not always met. Nevertheless, it is important to keep in mind the best available understanding of problems in order to deal with them. This is especially true for anthropogenic global warming. Hardly a day goes by without the media reporting on new wind farms in China and Texas, the reduction in price of solar panels, electric vehicles, etc. There is much less reporting, if any, on whether carbon emissions are exceeding withdrawal rates, on the increase in carbon dioxide in the atmosphere, and on the rise in global atmospheric temperature. Valuable as renewable energy is, it does not reduce global warming unless it replaces the combustion of fossil fuels and reduces emissions to the level of carbon withdrawals. The upsurge in renewable energy is so far going hand in hand with the intensification of atmospheric carbon pollution. Subtraction of fossil fuels, not just addition of renewable energy, is urgently needed.

The conclusions of impact science have important socioeconomic implications. Fossil-fuelled climate change threatens to be very costly and impose involuntary economic sacrifices, lower standards of living, and downward intergenerational mobility for future generations as a result of a degraded natural environment with more dangerous forces of nature. Unless science is totally wrong or unless a safe, cost-effective, technological breakthrough is implemented promptly and globally to bring emissions in line with carbon withdrawal, which shows no sign of appearing, the pace of extraction and combustion of fossil fuels will need to slow down drastically to mitigate global warming, and the most polluting types of fossil fuels will have to be left safely in the ground. The future will bring a discontinuity from present fossil-fuelled normality. Either fossil-fuelled normality will be (i) disrupted by the unintended harmful perverse consequences of fossil-fuel combustion unleashing dangerous new forces of nature through global warming, or (ii) disrupted by the socially purposive emancipation from fossil fuels thereby intentionally changing to low-carbon polluting societies. Destruction of present fossil-fuelled normality is coming, but we don't know when. The issue is whether it will be harmful destruction, injurious to societies and their members, or creative destruction leading to safer, more sustainable relations with the natural world.

The ideal solution based on the evidence from impact science would consist of an immediate leap to an economy having dramatically reduced emissions and/or radically enhanced carbon withdrawals. With current technology, the optimal way to do this is to combust fossil fuels only for functions for which there is no clean alternative, for example aviation combined with restraint in flying, and switch to non-carbon-polluting energy for other functions, such as electricity generation. On the resource side, it would be atmospherically optimal to combust only those fossil fuels whose use contributes least to emissions and leave others in the ground at least until carbon capture and storage and atmospheric carbon removal technologies are perfected and economically feasible. As shown in Fig. 2.1 and Table 2.1, this implies using natural gas first, as well as North Sea oil and Saudi oil, but leaving coal in the ground and also keeping heavy oil, bituminous sands oil, etc., underground for the near term. Since deposits of these resources are territorially based, this provides a technical climatological answer to the question of whose resources should be used first and whose should be left in the ground at present.

However, this answer is socioeconomically challenging because of the tension between goals of near-term, local, path-dependent economic prosperity, and those of a long-term clean, global, sustainable economy, and environment. Populations and principal decision-makers insist on retaining their sense of entitlement to inexpensive fossil-fuel energy and resulting carbon-polluting practices. Leaders have been elected who roll back global warming mitigation measures already taken. Social action and practices driven by interests, predispositions, and values determine the road actually being travelled now and in the future, the direction along this road, and the speed. Hence they determine whether impact natural science and its recommendations are being acted upon or ignored and whether global warming is being mitigated or worsened.

Only emphasizing these scientific conclusions may not necessarily be the best way to accomplish emission-reduction goals, and may even cause citizens deeply attached to fossil-fuel normality to give up hope and trigger what Beck (1995: 48–49) referred to as a 'death reflex of normality'. In the USA, the main determinant of whether one accepts the science of fossil-fuelled climate change and pricing carbon pollution to remedy it is political affiliation: Democrats accept it and Republicans don't (Dunlap, McCright, and Yarosh 2016). This division has grown over time even among the educated. It may be more strategic to proceed indirectly by underscoring near-term benefits (Pielke 2010) such as opportunities and jobs in the renewable energy sector, or energy security (Giddens 2009). Indirect approaches may be more effective in inciting social change. But strategies of what is socially acceptable in a particular conjuncture must not be mistaken for what is needed, which has to be kept in mind. Basing social science and policy analyses on the best available scientific evidence should not be portrayed as deploying scare tactics. Even if evidence is scary, social scientists, decision-makers, and populations need to face up to it and not denigrate it as scaremongering.

There is a need for a social science analysis of why so little is being done, even as carbon pollution accumulates in the atmosphere worsening the greenhouse effect, despite the compelling science of the problem. Natural science conceptions typically talk of the human impact on our planet, depicting an undifferentiated humanity as the cause, and a homogeneous humanity as suffering the effects. Hence they fail to capture the socioeconomic dynamics that are drivers of human impacts in the Anthropocene. In the next chapter, that oversimplified conception will be corrected by using the social closure theoretical framework to explain how humanity is differentiated and to help elucidate those socioeconomic drivers.

Notes

1. A simplified explanation of the greenhouse effect is given by Heal (2017: 205 fn. 2). 'CO_2 blocks heat leaving the earth but not coming in because they are at different wavelengths: incoming heat is ultraviolet and outgoing heat, because it has lost energy, is mainly infrared. CO_2 is opaque to infrared but not to ultraviolet'.
2. Gasoline is about 87% carbon and 13% hydrogen by weight, so 5 gallons (31 pounds) contains about 27 pounds of carbon. When gasoline is combusted, the carbon and hydrogen separate, with the carbon combining with oxygen to form carbon dioxide and the hydrogen combining with oxygen to form water. A carbon atom has an atomic weight of 12, an oxygen atom 16; therefore each carbon dioxide molecule (CO_2) has an atomic weight of 44 ($= 12 + 2 \times 16$). Hence 5 gallons of gasoline containing 27 pounds of carbon produces about 100 ($= 44/12 \times 27$) pounds of carbon dioxide. See https://www.fueleconomy.gov/feg/contentIncludes/co2_inc.htm.
3. The specific figure of temperature rise varies according to the chosen starting date and end date and whether a year is chosen (e.g. 1880) or a period (1880–1920).
4. Unconventional fossil fuels refer to heavy oil, shale gas, tar sands oil, kerogen, coal-bed methane, etc.
5. Rumsfeld (2011) became notorious for making these distinctions, which are nevertheless important.

References

Adam, Barbara. 1995. *Timewatch: The Social Analysis of Time*. Cambridge, UK: Polity Press.

Adam, Barbara. 1998. *Timescapes of Modernity: The Environment and Invisible Hazards*. London: Routledge.

Adam, Barbara. 2000. The Media Timescapes of BSE News. In *Environmental Risks and the Media*, ed. S. Allan, B. Adam, and C. Carter, 117–129. London: Routledge.

Adam, Barbara, Ulrich Beck, and Joost van Loon (eds.). 2000. *The Risk Society and Beyond: Critical Issues for Social Theory*. London: Sage.

Beck, U. 1995. *Ecological Politics in an Age of Risk*. Cambridge: Polity.

Berners-Lee, M., and D. Clark. 2013. *The Burning Question*. London: Profile.

Biello, D. 2015. Where in the World Are the Fossil Fuels That Cannot Be Burned to Restrain Global Warming. *Scientific American*, 7 January.

Blühdorn, I. 2011. The Politics of Unsustainability: COP15, Post-Ecologism, and the Ecological Paradox. *Organization & Environment* 24 (1): 34–53.

CDIAC. 2018. *Carbon Dioxide Information Analysis Center*. Berkeley: U.S. Department of Energy. cdiac.Ess-dive.lbl.gov.

Crutzen, P.J., and E.F. Stoermer. 2000. The "Anthropocene". *Global Change Newsletter* No. 41: 17–18.

Davidson, D., and J. Andrews. 2013. Not All About Consumption. *Science* 339 (6125): 1286–1287.

Dunlap, R., A. McCright, and J. Yarosh. 2016. The Political Divide on Climate Change: Partisan Polarization Widens in the U.S. *Environment Science and Policy for Sustainable Development* 58 (5): 4–23.

Dyer, Gwynne. 2008. *Climate Wars: The Fight for Survival as the World Overheats*. Toronto: Random House.

Eder, K. 1996. *The Social Construction of Nature*. London: Sage.

Evernden, N. 1993. *The Social Creation of Nature*. New Haven: Yale University Press.

EPA United States Environmental Protection Agency. 2017a. *Climate Change Indicators in the United States 2016: Greenhouse Gases*. www.epa.gov/climate-indicators/greenhouse-gases. Accessed 3 November 2017.

EPA. 2017b. Causes of Climate Change. *United States Environmental Protection Agency*. https://19january2017snapshot.epa.gov/climate-change-science/causes-climate-change_.html. Accessed 17 April 2020.

Finney, S.C., and L.E. Edwards. 2016. The "Anthropocene" Epoch: Scientific Decision or Political Statement? *GSA Today [Geological Society of America]* 26 (3): 4–10.

Flannery, Tim. 2015. *Atmosphere of Hope: Searching for Solutions to the Climate Crisis*. New York: Atlantic Monthly Press.

Freudenburg, W. 2006. Environmental Degradation, Disproportionality, and the Double Diversion. *Rural Sociology* 71 (1): 3–32.

Giddens, Anthony. 2009. *The Politics of Climate Change*. Cambridge: Polity Press.

Government of Canada. 2019. *Greenhouse Gas Sources and Sinks: Executive Summary 2019*. Ottawa: Government of Canada. https://www.canada.ca/en/environment-climate-change/services/climate-change/greenhouse-gas-emissions/sources-sinks-executive-summary-2019.html. Accessed 19 April 2019.

Graney, Emma. 2020. Newfoundland Warns of Oil Exodus Without Federal Aid. *Globe and Mail*, 27 May: B1, B4.

Harvey, Hal, and Robbie Orbis. 2018. *Designing Climate Solutions: A Policy Guide for Low-Carbon Energy*. Washington: Island Press.

Hawkins, Ed, Pablo Ortega, and Emma Suckling. 2017. Estimating Changes in Global Temperature Since the Preindustrial Period. *AMS American Meteorological Society Journals Online*. https://journals.ametsoc.org/doi/full/10.1175/BAMS-D-16-0007.1. Accessed 31 May 2018.

Heal, Geoffrey. 2017. *Endangered Economies: How the Neglect of Nature Threatens Our Prosperity*. New York: Columbia University Press.

Hughes, J.D. 2009. The Energy Issue. In *Carbon Shift*, ed. T. Homer-Dixon, 58–95. Toronto: Random House.

IEA International Energy Agency. 2018. Global Energy Demand Grew by 2.1% in 2017, and Carbon Emissions Rose for the First Time Since 2014. *International Energy Agency*. 22 March. Paris: IEA. https://www.iea.org/newsroom/news/2018/march/global-energy-demand-grew-by-21-in-2017-and-carbon-emissions-rose-for-the-firs.html. Accessed 23 March 2018.

IHS Energy. 2014. *Comparing GHG Intensity of the Oil Sands and the Average US* Crude Oil, May. Natural Resources Canada. http://www.nrcan.gc.ca/energy/publications/18731. Accessed 20 February 2017.

IPCC Intergovernmental Panel on Climate Change. 2013. *Fifth Assessment Report Climate Change 2013*. http://www.climatechange2013.org/images/uploads/WGIAR5-SPM_Approved27Sep2013.pdf. Accessed 28 September 2013.

IPCC. 2018. *Global Warming of 1.5 °C*. http://report.ipcc.ch/sr15/pdf/sr15_spm_final.pdf. Accessed 8 October 2018.

IPCC. 2019. *The Ocean and Cryosphere in a Changing Planet*. https://report.ipcc.ch/srocc/pdf/SROCC_SPM_Approved.pdf. Accessed 25 September 2019.

Jaccard, Mark. 2018. Divisive Carbon Taxes Are Much Ado About Nothing. *The Globe and Mail*, 15 December: O4.

Kolbert, Elizabeth. 2014. *The Sixth Extinction*. New York: Henry Holt and Company.

Liggio, John, Shao-Mend Li, Ralf M. Staebler, Katherine Hayden, Andrea Darlington, Richard L. Mittermeier, Jason O'Brien, Robert McLaren, Mengistu Wolde, Doug Worthy, and Felix Vogel. 2019. Measured Canadian Oil Sands CO_2 Emissions Are Higher Than Estimates Made Using Internationally Recommended Methods. *Nature Communications* 10: 1863. https://doi.org/10.1038/s41467-019-09714-9; www.nature.com/naturecommunications. Accessed 23 April 2019.

Lidskog, R., and C. Waterton. 2016. Anthropocene: A Cautious Welcome from Environmental Sociology. *Environmental Sociology* 2 (4): 395–406.

Masnadi, M., et al. 2018. Global Carbon Intensity of Crude Oil Production. *Science* 361 (6405): 851–853. https://doi.org/10.1126/science.aar/6859.

McGlade, C., and P. Ekins. 2015. The Geographical Distribution of Fossil Fuels Unused When Limiting Global Warming to 2 °C. *Nature* 517: 187–190.

Murphy, Raymond. 1994. *Rationality and Nature*. Boulder: Westview.

Murphy, Raymond. 2015. The Emerging Hypercarbon Reality, Technological and Post-carbon Utopias, and Social Innovation to Low Carbon Societies. *Current Sociology* 63 (3): 317–338.

NASA Earth Observatory. 2007. If We Stabilized Greenhouse Gas Emissions at Today's Rates, Would Global Warming Stop? http://earthobservatory.nasa.gov/blogs/climateqa/stabilize-gg-emissions-effects/. Accessed 28 December 2015.

NASA. 2018. Long-Term Warming Trend Continued in 2017: NASA and NOAA. *Global Climate Change Vital Signs of the Planet*. https://climate.nasa.gov/news/2671/long-term-warming-trend-continued-in-2017-nasa-noaa/. Accessed 28 April 2019.

Nordhaus, William. 2013. *The Climate Casino: Risk, Uncertainty, and Economics for a Warming World*. New Haven: Yale University Press.

Pielke, Roger Jr. 2010. *The Climate Fix*. New York: Basic Books.

Pinker, Steven. 2018. Doomsday Is (Not) Coming. *Globe and Mail*, 24 February: O1, O6–O7.

Rasool, S. Ichtiaque, and Stephen Schneider. 1971. Atmospheric Carbon Dioxide and Aerosols: Effects of Large Increases on Global Climate. *Science* 173: 138–141.

Rockström, Johan, Owen Gaffney, Joeri Rogeli, Malte Meinshausen, Nebojsa Nakicenovic, and Hans Schellnhuber. 2017. A roadmap for Rapid Decarbonisation. *Science* 355 (6331): 1269–1271.

Rumsfeld, Donald. 2011. *Known and Unknown: A Memoir*. New York: Penguin.

Schneider, S. 2009. *Science as a Contact Sport*. Washington: National Geographic.

Simon, Julian. 1981. *The Ultimate Resource*. Princeton: Princeton University Press.

Simon, Julian. 1996. *The Ultimate Resource 2*. Princeton: Princeton University Press.

Steffen, W., J. Grinevald, P. Crutzen, and J. McNeill. 2011. The Anthropocene: Conceptual and Historical Perspectives. *Philosophical Transactions of the Royal Society A* 369 (1938): 842–867.

Steffen, W., W. Broadgate, L. Deutsch, O. Gaffney, and C. Ludwig. 2015. The Trajectory of the Anthropocene: The Great Acceleration. *The Anthropocene Review* 2 (1): 81–98.

Steffen, Will, Johan Rockström, Katherine Richardson, Timothy Lenton, Carl Folke, Diana Liverman, Colin Summerhayes, Anthony Barnosky, Sarah Cornell, Michel Crucifix, Jonathan Donges, Ingo Fetzer, Steven Lade, Marten Scheffer, Ricarda Winkelmann, and Hans Joachim Schellnhuber. 2018. Trajectories of the Earth System in the Anthropocene. *PNAS Proceedings of the National Academy of Sciences of the United States of America* 115 (33) (August 6): 8252–8259. https://doi.org/10.1073/pnas.1810141115. Accessed 7 August 2018.

UNEP United Nations Environmental Programme. 2019. *Emissions Gap Report 2019 Executive Summary*. Nairobi: UNEP. https://wedocs.unep.org/bitstream/handle/20.500.11822/30798/EGR19ESEN.pdf?sequence=13. Accessed 26 November 2019.

Vaughan, Scott. 2014. The Challenge of Extreme Events and Their Impacts. *International Institute for Sustainable Development Speech*. Canadian Climate Forum's Symposium on Extreme Weather and Adaptation 23 April: 4.

Wagner, Gernot, and Martin Weitzman. 2015. *Climate Shock: The Economic Consequences of a Hotter Planet*. Princeton: Princeton University Press.

Walker, X.J. et al. 2019. Increasing Wildfires Threaten Historic Carbon Sink of Boreal Forest Soils. *Nature* 572 (7770) (August): 520–523.

Weber, Max. (1904–1905) 1930. *The Protestant Ethic and the Spirit of Capitalism*, trans. Talcott Parsons. London: Unwin.

Weber, Max. (1922) 1978. *Economy and Society*, ed. Guenther Roth and Claus Wittich. Berkeley: University of California Press.

Willis, Andrew. 2019. The Future Is Plastics: Murray Edwards, Li Ka-Shing Add to Oil Patch Holdings as Others Flee. *Globe and Mail*, 7 July: B1, B5.

World Bank Group. 2020. Minerals for Climate Action: The Mineral Intensity of the Clean Energy Transition. *Climate Smart Mining Facility*. Washington: The World Bank. http://pubdocs.worldbank.org/en/961711588875536384/Minerals-for-Climate-Action-The-Mineral-Intensity-of-the-Clean-Energy-Transition.pdf. Accessed 15 May 2020.

World Meteorological Organization WMO. 2019. *Global Climate in 2015–2019: Climate Change Accelerates*. https://public.wmo.int/en/media/press-release/global-climate-2015-2019-climate-change-accelerates. Accessed 23 September 2019.

York, Richard. 2010. Three Lessons from Trends in CO_2 Emissions and Energy Use in the United States. *Society and Natural Resources* 23 (12): 1244–1252.

York, Richard. 2012a. Asymmetric Effects of Economic Growth and Decline on CO_2 Emissions. *Nature Climate Change* 2 (11): 762–764.

York, Richard. 2012b. Do Alternative Energy Sources Displace Fossil Fuels? *Nature Climate Change* 2 (6): 441–443.

3

Social Closure in the Anthropocene: The Environment as a Medium for Monopolization and Exclusion

The objective of this chapter is to use the social closure theoretical framework to analyse the socioeconomic dynamics underlying what Ciplet, Roberts, and Kahn (2015) aptly refer to as the remaking of inequality and power in a warming world. It investigates environmental social closure based on monopolising resources and waste sinks, resulting in the exclusion of latecomers from opportunities. This emergent dimension of social closure will be examined to gain insight into the fossil-fuelled climate crisis. The chapter also aims to show how biophysical dynamics of the atmosphere, which is the ultimate commons for humanity across space and time, carry social relations of closure.

What Is Social Closure?

The concept social closure was proposed originally by Weber (1978) and then elaborated by Parkin (1972, 1974a, 1974b, 1979, 1980, 1982) and Murphy (1988). Collins' (1968, 1971, 1975, 1979, 1980, 1981, 1986) work has also been integrated into the closure framework. It continues to

R. Murphy, *The Fossil-Fuelled Climate Crisis*,
https://doi.org/10.1007/978-3-030-53325-0_3

be theoretically elaborated (Mackert 2004; Swartz 2008; Daly and Silver 2008) and deployed in empirical studies (Koch 2003; Minefee et al. 2015; Masuku and Rama 2020; Mauer 2020; Viola 2020).[1] Mackert (2004 and especially 2012) provides an exhaustive overview of research using the concept. Both his overview and my own search lead to the conclusion it has not been used in the environmental social sciences. I will endeavour to show its value to analyse fossil-fuelled climate change.

Closure refers to processes of subordination whereby one group monopolizes resources thereby closing off opportunities to other groups: 'material monopolies provide the most effective motives for the exclusiveness of a status group' (Weber 1978: 935). The concept enabled Weber to analyse monopolization of opportunities by property classes in the market and by status groups based on race, gender, religion, ethnicity, etc., in terms of an overarching coherent framework. It constitutes an enlargement of focus from monopolizing one particular resource (means of production) and one set of rules of closure (private property laws) to other forms of monopolization and exclusion based on gender, race, religion, credentials, and the Communist Party. Groups can monopolize opportunities only if they have the power to do so. There is not usually an absolute monopoly, but rather a process of monopolization and corresponding practices of exclusion. In the USA, Barak Obama became President and Nancy Pelosi Speaker, but such exceptions do not disprove prevalent dynamics of racial and gender monopolization and exclusion. Notwithstanding cases of rags to riches in market competition, which are as spectacular as they are rare, resources and opportunities tend to be concentrated, even intergenerationally. Dynamics of market closure lead to sequential oligopolies in particular economic sectors.[2]

Parkin distinguished collectivist exclusionary criteria such as race, gender, etc., which transmit advantage to descendants of the group within which one was born, from individualist criteria such as property and credentials, which are designed to legitimate advantage and are somewhat less efficient at transmitting it to future generations. The closure framework has been used to analyse monopolization and exclusion among races, ethnic groups, genders, and religions, for example

Koch's (2003) study of Northern Ireland. Collins focussed on positional property and political labour to impose one's definition of reality to legitimate exclusion. In his books *Conflict Sociology* and *The Credential Society*, he examined credentials as the basis for monopolizing organizational positions.

The focus is on objective relations of appropriation and exclusion. Their relationship to intentions and cultural struggle to either legitimate or oppose closure is subject to investigation. All forms of exclusion have potential to provoke reactions, which Parkin called usurpation, consisting of attempts to bite into advantages and power of privileged groups. Monopolization results in inequality of opportunity and exclusion, which can incite reactions and movements aiming to foster inclusiveness and greater equality of opportunity. It is hard to imagine how persistent slavery was as a practice of social closure two centuries ago, but it was eliminated. Aristocratic monopolization, and exclusion of commoners based on birth, was deeply embedded in the past, but it too was challenged and abolished. Apartheid has been overthrown in South Africa, but racial exclusion still exists there and elsewhere in lesser forms. Monopolization by the Communist Party has been toppled in Eastern Europe, the Soviet Union no longer exists, but that form of closure persists in North Korea. China has changed to a paired mode of power monopolization based on the Communist Party and market-based private property. In terms of economics, the closure framework and its focus on practices of monopolization and exclusion is related to the institutional economics and countervailing power analysis of Galbraith (1967) and to the work of Stiglitz (2012, 2015a, 2015b) and Spence (2019). Monopolization based on private property in the market has become stronger worldwide with the advent of globalization, neoliberalism, mergers, acquisitions, and international strategies of tax evasion, which have led to greater concentration of corporate power and wealth. Surprisingly the closure perspective has been rarely used to analyse monopolization and exclusion in the market, an omission this chapter seeks to correct as a secondary objective.

Market Dynamics of Monopolization and Exclusion from the 1980s to the Present

Since social closure based on market dynamics is so persistent, has become more predominant over the past three decades, and since the modern form if rooted in manipulating nature's dynamics yet threatened by them, the evidence about market monopolization and exclusion generally must be examined before analysing its relationship to fossil-fuelled climate change. Economists (Piketty 2014; Krugman 2007; Stiglitz 2012, 2015a, 2015b; Akerlof and Shiller 2015; Zucman 2015), sociologists (Carolan 2011, 2014), political scientists (Freeland 2012), law professors (Wu 2018) and journalists (McQuaig and Brooks 2010) have provided documentation of income and wealth disparities and the processes underlying them. This typically means practices of monopolization and exclusion. Note that monopolization is examined here as a process (a verb) and a set of practices. In some cases market monopolization results in a monopoly (a noun), with Microsoft being an approximation in its market niche. In other cases, it produces a duopoly (Boeing and Airbus). But in most cases, practices of market monopolization lead to more diffuse inequality (the 1% and the 99%).

Piketty (2014) argues as follows. Initial immigration to America led to less inequality there than in Europe, but by the early twentieth Century inequality was high in both. Later two world wars and the great depression reduced inequality especially in Europe. After World War II the expansion of education and the welfare state also reduced inequality. But since 1980 processes of divergence–Piketty's term for the concentration of income and wealth by a pecuniary elite–dominate. This was especially true in English-speaking countries: the USA, Britain, Canada, and Australia. 'The growth of capital's share accelerated with the victories of Margaret Thatcher in England in 1979 and Ronald Reagan in the United States in 1980, marking the beginning of a conservative revolution' (Piketty 2014: 42). He (Piketty 2014: 335) argues that 'the decrease in the top marginal income tax rate led to an explosion of very high incomes, which then increased the political influence of the beneficiaries of the change in the tax laws, who had an interest in keeping top tax rates low or even decreasing them further and who could use their windfall to finance political parties, pressure groups, and think tanks'.

Piketty documented two main dynamics of monopolization that led to greater inequality. The first consists of inheritance and accumulation of wealth: when the top tax rate was decreased and new forms of tax evasion were developed, the return to capital invested grew faster than wages, which stagnated. 'The entrepreneur inevitably tends to become a rentier' (Piketty 2014: 571) who hires talented professionals to monopolize wealth, avoid taxes and diminish risk. People without wealth (the vast majority) are excluded from such opportunities. The second involves the transformation of managers into compensation superstars who set their own remuneration, which has exploded. He shows that, although some managers have done remarkable work, the evidence is weak that remuneration for compensation superstars as a whole is meritocratic, but the claim of merit is nevertheless used to justify their compensation. Even some big institutional investors are questioning the remuneration of managers. Superstar monopolization of compensation resulted in greater inequality: 'in all the wealthy countries, including continental Europe and Japan, the top thousandth enjoyed spectacular increases in purchasing power in 1990-2010, while the average person's purchasing power stagnated' (Piketty 2014: 320).

Piketty (2014: 300, 257) documents that the top 1% in the USA appropriated 20% of income, the richest 10% owned 72% of its wealth, and the bottom 50% only 2%. In 2010–2011 in France, the richest 10% owned 62% of total wealth whereas the lowest 50% owned only 4%, and this likely underestimates inequality because of hidden wealth. Monopolization becomes more pronounced higher up the wealth and income curves. The two types of appropriation are mutually supportive: superstar managers become rentiers with their wealth and pass it on to their offspring. Duff (2017: B4) confirms that tax deductions for stock options and capital gains are monopolized by the wealthy, which enable them to pay lower rates of income taxes.[3] In the USA, the best universities and superior health care are private and costly whereas public universities and health institutions are criticized as unwarranted 'entitlements' and underfunded because of resistance of the wealthy to paying taxes.

Hence, the best universities and hospitals tend to be monopolized by the wealthy, and the majority of the population excluded. The offspring of the wealthy inherit the fortune, attend the most expensive schools providing quality education and insertion into advantageous networks, and receive the best health care money can buy. Such opportunities, even health care in the USA, are closed off to others by lack of money and networks. Oxfam documents that, since 2015, the richest 1% has owned more wealth than the rest of the world's population, eight men now own the same amount of wealth as the poorest half of the world, over the next 20 years 500 people will hand over $2.1 trillion to their heirs–a sum larger than the GDP of India, and a Fortune-100 CEO earns as much in a year as 10,000 people working in Bangladesh's garment factories (Hardoon 2017: 2). Monopolization of resources and opportunities persists even as large numbers of people are lifted out of extreme poverty into less poverty, especially in China.

There are differences in talent, effort, choice of activity, and therefore rewards, but monopolistic structures amplify benefits for the wealthy and prevent others from receiving them. Inequality of opportunity is at the core of social closure. Layered upon the objective facts of monopolization and exclusion is discursive struggle to either legitimate or undermine them. Despite claims of individual meritocratic accomplishment, achievement often involves differential insertion into networks and inheritance from one generation to another, with all the inequalities of opportunity that entails. Markets free of government regulations claim to increase liberty and freedom but in fact diminish them. Winners in market competition accumulate resources, so the next round of competition begins on an unequal footing. Winners can even curtail competition, which can only be prevented by government antitrust legislation. Free market theories reduce liberty by handcuffing the only institutions capable of controlling monopolization by billionaires and their corporations and by underfunding public institutions capable of enhancing equality of opportunity and thereby the liberty and freedom of the rest of the population. People are not free if they are unable to access quality education necessary to develop their talents or if they are sick and do not have adequate health care. Stiglitz (2015a, 2015b) and

Jackson (2017) following Galbraith (1967) decades earlier, argue that market power of big corporations and the wealthy has to be held in check by the countervailing power of democratic governments and trade unions.

Recognition that power wielded in market competition unconstrained by government regulation leads to oligopolies, dominance of the interests of large shareholders and top managers over those of consumers, workers, and small shareholders, and to extreme inequalities is not limited to left-wing economists. It is found also in analyses in serious business journals. Reguly (2017a: B2) documents that in the USA mergers turned eight major airlines into four, a single airline controls a majority of seats at 40 of America's 100 largest airports and one or two airlines at 93 of the top 100 airports, hence they enjoy 'oligopoly profits'. This monopolization has worsened since the year 2000. Airlines 'are not the only examples of highly concentrated corporate power – monopolies, duopolies and oligopolies – that have snuffed out competition, raised prices and harmed innovation' (Reguly 2017a: B2). Furthermore, they 'exert enormous lobbying pressure on even the strongest governments to bend rules in their favour' (Reguly 2017b: 19). Globally, industries as diverse as pesticides and seeds, baby formula, and cable TV are dominated by four or fewer companies. Google (Alphabet) has a market share of eighty per cent (Mortished (2017: B4). Amazon dominates its sector. Taplin (2017) documents this winner-takes-all economy. Money is spent on share buy-backs to benefit the biggest shareholders rather than investing in research, reducing prices for consumers, or increasing jobs and salaries for workers. Dow Chemical is merging with DuPont, ChemChina purchased Syngenta, and Bayer is taking over Monsanto (Reguly 2017c). Reguly argues that market concentration became extreme in the Reagan era under the influence of Milton Friedman when the deregulation cult terminated the government's trust-busting role. He contends oligopolies and duopolies eventually get diluted by new start-up companies with innovative ideas and competitive prices, but that takes decades and results in new oligopolies, hence competition can only be maintained by bringing back government's trust-busting role. Crowley (2017: B4) argues that 'the growth strategy of many companies now is to improve their profitability by buying up their rivals, and their market share, and thereby insulating themselves from the market discipline that puts limits on the prices they can charge'. The company Glencore controls more than half the global market for copper and zinc, the global tea trade is dominated by three companies, and 81% of the American beef market is controlled by four processing companies. Highly profitable mega-companies then shelter profits from taxes by buying political influence and lobbying. 'In the political climate of the United States today, monopoly per se is not a sin. On the contrary it is merely a measure of success. According to Peter Thiel, the PayPal co-founder, monopolies are the only businesses worth having' (Mortished 2017: B4). Moreover, ultra-rich severance packages for executives who failed to perform and astonishingly big rewards for nothing more than luck in the market refute myths that executive compensation is based on performance and that market competition renders it meritocratic (McGugan 2017a). Lazonick (2014) argues that the increasing practice of share buy-backs undermine investments in productivity, innovation and long-term health of a corporation, benefit only shareholders and top company executives, and result in 'profits without prosperity'. From his market competition perspective, Crowley contends monopolization can be countered by free-trade agreements between governments and action by lawmakers to curtail mergers and bust trusts to promote competition. Market competition freed from government regulation fosters monopolization and exclusion. It can only be overcome by government action to restore competition. If such action is precluded, then monopolization, exclusion, and inequality of opportunity proceed full bore.

Two interrelated developments modified dynamics of market closure in the past three decades. First technological innovations resulted in previous temporary oligopolies being undermined and new ones established. The development of the personal computer led to new hardware and software companies (Apple, Microsoft, etc.) usurping dominance by IBM's mainframe computers, to innovation of mobile phones (iPhones, Android phones, tablets), and to new monopolistic niches in specific areas. McGugan (2017c: 5) argues that Google and Facebook make fortunes by distributing work created by others and taking a freeride on their backs: 'The rise of the digital duopolists has devastated creative industries ranging from music to journalism'. McGugan (2017b) adds: 'Thanks to their ever-expanding utility, Google and Facebook are natural monopolies that seem destined to dominate their respective areas. At this point, no new rival could realistically challenge their supremacy'. They favour their own platform for advertising, and pit countries and regions against one another to evade taxes. Portraying them as 'natural' is an overstatement, but their monopolies are based on producing a useful service and on a network effect where their utility becomes greater the bigger their network. Srinivasan (2018: Abstract) argues that 'Facebook's ability to extract this qualitative exchange from consumers is merely this titan's form of monopoly rents. The history of early competition, Facebook's market entry, and Facebook's subsequent rise tells the story of Facebook's monopoly power. However, the history which elucidates this firm's dominance also presents a story of anticompetitive conduct. Facebook's pattern of false statements and misleading conduct induced consumers to trust and choose Facebook, to the detriment of market competitors and consumers' own welfare'. Thus she argues that the principles of antitrust should be used to induce competition. The co-founder of Facebook, Chris Hughes (2019) concludes that it 'has used its monopoly position to shut out competing companies or has copied their technology'. When users are annoyed with Facebook, they punish Facebook by moving to Instagram or WhatsApp, not realizing Facebook has bought those too. Thus Hughes argues regulations are needed and it's time to break-up Facebook through antitrust legislation like with AT&T's monopoly, which spurred innovation. Carolan (2014: 183) concludes that 'we are left living in a world where freedom for the pike is death for the minnow. … most socioeconomic "minnows" (the bottom 90 percent in the US and bottom 99.9 percent worldwide) exist in this state through no fault of their own'. Among his solutions are enhancing and enforcing antitrust laws.

However, antitrust legislators and regulators have great difficulty controlling such monopolistic companies because a break-up would diminish their utility. European lawsuits against monopolistic practices of Microsoft and Google have had limited success. The invisible hand of the market, far from enhancing competition, results in winners and losers, with the former accumulating resources to reinforce monopolistic practices in their particular domains, including mergers and acquisitions to diminish competition. Thus the principal barrier to market competition is the normal functioning of the market itself. The principal rewards go to the major shareholders and managers, often by way of dual-class shares to maintain control.

Increased efficiency in the flow of information, and in transportation of goods and people, has led to a second driver of modern forms of social closure, namely globalization. Owners have transferred production from developed countries (USA, Europe, Canada, etc.) where wages are high and labour and environmental regulation relatively strict to developing ones (China, India, Bangladesh, Mexico, etc.) where wages are low and regulations lax. This has been facilitated by free-trade agreements which enshrine shareholder property rights for international investment and enable transnational companies to sue governments. The result is increased profits for owners, a slight rise in earnings for the world's poorest, but decreased workers' wages in developed countries (Hammond 2017: B4). It has resulted in somewhat lower inter-country inequality but greater intra-country inequality. This has produced working-class downward mobility and resentment in developed countries, which has often been displaced towards immigrants in Europe and North America. If the broader population does not benefit through social welfare programmes and instead market winners monopolize benefits, Stiglitz (2017: A13) argues that 'Trumpian politicians may become a permanent feature of the landscape. … They [the Scandinavian countries] understood that the only sustainable prosperity is shared prosperity'. This requires political action to promote equal opportunity and include everyone as beneficiaries. Piketty advocates taxation, including international taxation, and quality public education to diminish inequality. Saunders (2017: F7) agrees these measures are needed to knock down walls making 'privilege a closed loop that excludes outsiders'.

Processes of Monopolization and Exclusion in the Anthropocene

Heal (2017: 160) demonstrates that GDP is not a good measure of well-being for two reasons. 'The first is that the measures offer no distinction between an increase in the nation's income that goes entirely to the rich (as has happened in the United States recently) and one that is distributed more uniformly. The second is that the measures don't in any way reflect damage to natural capital and, perversely, could even show gains from this'. This conclusion leads to the chapter's main objective: analysing emergent forms of closure characteristic of the Anthropocene, namely, ways social practices interact with planetary biophysical dynamics. 'An unquestioning society-wide commitment to economic growth at almost any cost; enormous investment in technologies designed with little regard for the environment; powerful corporate interests whose overriding objective is to grow by generating profit, including profit from avoiding the environmental costs they create; markets that systematically fail to recognize environmental costs unless corrected by governments; government that is subservient to corporate interests and the growth imperative; rampant consumerism spurred by a worshipping of novelty and by sophisticated advertising; economic activity so large in scale that its impacts alter the fundamental biophysical operations of the planet – all combine to deliver an ever-growing world economy that is undermining the planet's ability to sustain life' (Speth 2009: 7–8). Environmental social closure has distinctive features. It involves monopolizing biophysical resources and appropriating the atmosphere and oceans as pollution sinks in a first-come, first-served manner, whereby latecomers are threatened with exclusion. Benefits are disproportionally garnered by some, but consequences are suffered by others excluded from most benefits. Beck (1992, 1998, 2009) argued that there is a new power structure imbedded in the concept of global risk, namely organized irresponsibility: those who are making decisions are not accountable to those affected, and those affected have no participation in the decision-making.

Monopolization and Exclusion in Space

Monopolising raw materials has occurred for a long time. Rich nations typically extract raw materials in poor nations, then process those materials, and the processing makes them rich (Wallerstein 2011; Roberts and Parks 2007). This has intensified under globalization (i) with technologically enhanced capacity for extracting and transporting materials and communication and (ii) in terms of trade agreements. Hydraulic fracturing for oil and natural gas has been innovated, as has deepwater oil drilling and oil extraction from bituminous sands. Although it seems these have refuted theories of peak oil, likely that is not the case. A discovery of oil or gas may supply a large population for thirty years, but that is only one-third of a lifetime, hardly long enough to ensure sustainability. Oil and gas deposits in shale become depleted rapidly and new ones have to be found. There is an abundance of fossil fuels and other raw materials, but remaining deposits may be geographically or technologically hard to extract and require so much energy and cost to extract, upgrade, and transport that it would not be worthwhile. The affluent of this generation are picking low-hanging fruit, leaving difficult or impossible ones to latecomers whether poor individuals, poor countries, or future generations. This threatens to close off resources to them.

More novel and important (Jaccard 2009) is the appropriation of the atmosphere as a sink to dump greenhouse gases. It is a commons shared by all, including past, present, and future generations (Nordhaus 2013). Countries, companies, and individuals are taking as much of the atmospheric sink for their emissions as they want and can afford. Hence there is much inequity in its use to dispose of long-lasting carbon dioxide emissions. Since the Earth's atmosphere is finite, and since there are intensifying adverse consequences as it fills with greenhouse gases, unregulated dumping amounts to monopolising it as a pollution sink. A hundred companies account for more than seventy per cent of carbon dioxide emissions; the richest one per cent of the world's population emit more than the poorest fifty per cent; the wealthiest ten per cent emit half of humanity's emissions (Mason 2019: A11). Exploitation of fossil fuels has made billionaires of a few, provided some benefits to many (jobs, fossil-fueled combustion engines for travel, heating, and air conditioning), but are polluting the atmosphere needed by everyone. The carbon pollution spreads worldwide and is cumulative because of the physical properties of carbon dioxide which remains in the atmosphere for a century. Property laws grant ownership of fossil fuels without requiring responsibility for their pollution, thereby concentrating benefits with polluters and spreading danger to many, especially those living in distant locations excluded from the benefits. Thus Stern (2009: 13) argues that 'the poor countries are least responsible for the existing stock of greenhouse gases, yet they are hit earliest and hardest by climate change'. They have lowest emissions because inhabitants have no electricity, only rudimentary means of transportation, etc. (CDIAC 2018). Decisions to rely on fossil fuels in countries on one side of the planet (e.g. the USA and Canada) have threatening consequences for societies on the other side (Bangladesh, Chad, etc.) because the atmosphere is a medium which transports carbon combusted and emitted from one side to the other. Whereas emissions are 18 tonnes per-person per year in the USA, they are only 0.1 tonnes per person in Madagascar (Hawken 2017: 81). This constitutes monopolising the atmosphere as a carbon dump. Poverty results in exclusion from the pleasures of fossil-fuelled practices the affluent enjoy: 80 per cent of the world's population have never been on an airplane (Friesen 2019: A4). It is not the poor who are combusting jet fuel and thereby contributing to the climate crisis. The habitat of the Arctic Inuit is experiencing degradation earliest. 'In all four regions where the Inuit live – Northern Canada, Alaska, Greenland, and Siberia – every community is struggling to cope with extreme coastal erosion, melting permafrost and rapid runoff as temperatures rise' (Watt-Cloutier 2019: O11).

The drought and bush fires in coal-based Australia in 2020 during Melbourne's prized Australian Open tennis tournament gives a foretaste of monopolization and exclusion to come. The smoke gave the city the worse air quality in the world. A horse race was cancelled because thoroughbreds are so valuable. But lowly tournament qualifiers had to compete despite smoke that dried their throats and lungs, caused coughing fits, and forced one to withdraw. 'When the news is telling you to stay inside, to keep your pets inside, why are we exerting ourselves to the maximum? … There's everything at stake for these players in qualifying. This is their ticket to potentially break through into the top 100, make the main draw, be able to support themselves for the rest of the season. So of course they're going to play. … As the qualifiers were coughing and struggling outside, the top stars in the game – Federer, Novak Djokovic and others – were practising in air-filtered, air-conditioned comfort inside one of the three stadiums' (Myles 2020: B19).

Even safety from nature's forces is being monopolized. Louisiana has a prosperous oil industry, yet it also has one of the highest levels of inequality in the USA (McNichol et al. 2012). New Orleans is a city mostly below sea level surrounded by the Mississippi River, enormous Lake Pontchartrain, and the Gulf of Mexico, in an area prone to hurricanes and storm surges. The safe high ground of the city in the French Quarter and Garden District is monopolized by the wealthy, and closed off to the poor by housing prices in the market. The poor are housed in what is left, the vulnerable Lower Ninth Ward. Monopolization of safe areas by some and exclusion from safety for most became painfully visible when Hurricane Katrina struck in 2005 with catastrophic consequences for the excluded.

Some putative technological solutions to fossil-fuelled climate change increase the risk of monopolization: nuclear energy, carbon capture and sequestration, geoengineered sunscreens in space. 'These large-scale solutions also tend to concentrate power to the hands of governments and wealthy corporations as they require massive investment, central planning, and often intensive security measures' (Suzuki and Hanington 2017: 163, see also 222). The claim is made that the oppositional dynamics and equal voting rights for all regardless of knowledge lead democracy to be slow and ineffective for dealing with urgent, slow onset, invisible perils; hence it needs to be replaced by authoritarian monopolies by knowledgeable elites. This claim risks become more widespread as fossil-fuelled consequences becoming more severe.

Monopolization and Exclusion in Time

The gravest threat of fossil-fuelled climate change occurs over time. Appropriation of biophysical resources, including atmospheric sinks, by the present generation, with benefits disproportionately concentrated, results in the risk that future generations will be excluded from benefits, restricting their life chances: 'issues of climate justice include the excluded non-living generations, who are going to suffer most' (Beck 2015: 82). The atmospheric commons shared by present and future generations is a medium carrying social relations of monopolization and exclusion between risk makers and risk-takers over generations. For the present generation, principal monopolizers need to be differentiated from minor beneficiaries. Similarly for future generations, there will be significant differences between principal victims, namely the vulnerable poor and middle classes, compared to minor victims who are wealthy and have resources to protect themselves. Through the medium of the atmosphere, current social practices threaten to close off opportunities to future others. Priority given to near-term economic benefits to the exclusion of long-term danger constitutes a code of social closure and monopolization embedded in culture, social practices, and physical infrastructures. Pipelines need to transport oil for many decades to be profitable; hence decisions made today to build pipelines lock in emissions for the next half-century. Holmberg (2017) demonstrates that corporate short-termism, namely emphasis on short-term profits, is a driver of both social inequality and fossil-fuelled climate change.

As atmospheric greenhouse gases accumulate to dangerous levels and harm increases, safety will require that emissions decrease, more drastically the longer it takes. Either the atmosphere will have to be closed off to future emitters as a sink to ensure safety, or they will suffer the consequences. Latecomers, both the poor and future generations, will be excluded from using the atmosphere to dump greenhouse gases the way high polluters of the present generation are doing. Latecomers will bear the brunt of monopolization by high early emitters. McKibben (2019: O8) postulates an 'iron law of climate change [which] is that the less you did to cause it, the quicker and harder you're hit by its effects'. The less a country (think Madagascar) or a generation (think our grandchildren) did to cause fossil-fuelled global warming, the more their opportunities will be closed off in terms of nature's degraded services and the atmospheric carbon sink becoming full. There are market dynamics of monopolising both fossil-fuel benefits and security, and exclusion from well-being and safety.

Ethicists contend every person—past, present, and future—has the right to an equal amount of carbon they can emit into the atmosphere (Dyer 2008: 75, 174), hence its disproportionate use as a pollution dump resulting in exclusion of latecomers from using it, is unethical. Spokespersons from developing countries argue that countries which became wealthy by emitting carbon dioxide for centuries have a duty (i) to cut back their emissions more than recent emitter developing ones so that the latter won't be excluded from using the atmospheric carbon sink, and (ii) to compensate the latter by paying for their adaptation and transition to low-carbon energy. High carbon-polluting societies, companies, and individuals refuse this logic, a refusal understandable prior to the scientific documentation of harmful effects but unconvincing for subsequent emissions.

An important dimension consists of present generations' practice of monopolising what ecologists refer to as ecosystem services or what economists call natural capital, which consists of public common goods, such as the climate system, hydrological cycle, tropical forests, biodiversity, oxygen produced from photosynthesis, etc. It doesn't have market value, but it is essential to life and irreplaceable by technological innovations. Heal (2017: 171) concludes 'it is of immense value to human societies. We depend on it in many ways, and it provides services we could not replace. Yet we still deplete this natural capital, running it down so that future generations will inherit less than we have, and less than we inherited from our predecessors. … We are leaving our successors less and poorer natural capital – a world with a less stable and hospitable climate, fewer species, less water, and fewer of many other environmental assets. Perhaps this is condemning them to an impoverished lifestyle'.

The rapid pace of contemporary carbon-emitting fossil-fuel combustion results in the threat of diminishing supplies of usable, economic fossil fuels for future generations and of filling the atmosphere with carbon and acidifying oceans, thereby excluding future generations from benefitting from cost-free waste sinks that the present generation enjoys, and foisting onto the former dangers and costs of more frequent and intense hazards of nature like hurricanes, floods, and wildfires. To avoid overheating the atmosphere by more than 2 °C since pre-industrial time, humanity has a limited carbon dioxide budget to emit (Berners-Lee and Clark 2013). The more the present generation engages in fossil-fuelled practices and high emissions, the less of the budget is left for latecomers, especially future generations. The 1 °C increase already documented by 2015, with two-thirds occurring since 1975, indicates the present generation monopolized much of the carbon budget thereby closing off most of it to future generations. By refusing mitigation measures like a price on carbon pollution, the present generation is forcing latecomers (future generations including their grandchildren) to either suffer the consequences or pay a much higher price for carbon pollution. If the maximum emissions allowable by 2050 for a 2 °C temperature rise are all done by 2030, which appears likely, either no emissions will be allowed between 2030 and 2050 or the harm of a greater than 2 °C temperature rise will be suffered. Monopolising the atmospheric sink now means latecomers will be excluded from using it. Within the present generation, there are huge inequities in the carbon budget being spent, with giant corporate polluters and large individual ones using most. Mitigating fossil-fuelled climate change will require undermining monopolization of the limited carbon budget so that latecomers can have more equitable shares. The carbon budget should be seen in terms of struggle to maintain or usurp its monopolization. The longer the excess of emissions over carbon withdrawals festers, the worse consequences will be, and the sooner the atmosphere and oceans will have to be closed off as sinks.

Global warming will also result in future cohorts losing benefits of glaciers providing fresh water year-round, of Arctic ice cover reflecting the sun's radiation back into space, etc. Near-term economic interests are prevailing over long-term human interests, thereby risking the future to maintain present fossil-fuelled practices where monopolization of benefits is widespread. It constitutes a 'failure of foresight' that likely will lead, according to climate science, to the 'incubation' of future 'man-made disasters' (Turner and Pigeon 1978). Current fossil-fuelled practices—characterized by some groups monopolizing a disproportionate amount of wealth, depleting finite raw materials, and filling up limited pollution sinks in nature's commons—threaten to exclude latecomers and especially future generations from benefits presently enjoyed from self-sustaining dynamics of the natural world. 'Long-term' can be operationalized as the length of a human lifetime, about eighty-five years to 2105, which corresponds to the time frame when global warming is predicted by science to become severe. The fact children born today risk being gravely affected before they die demonstrates 'long term' is not far away.

Countries that became wealthy early disproportionately added to the stock of greenhouse gases in the atmosphere, whereas those developing now are increasing their current emissions but have relatively low cumulative emissions. 'China produces 22 per cent and the USA 18 per cent of current emissions, whereas the USA is responsible for 27 per cent and China for only 9 per cent of cumulative emissions' (Heal 2017: 76). This leads to inequities of binding targets for current emissions for all countries. A Brazilian official criticized this as like a person arriving late for a meal, takes only a coffee, but is expected to pay the full share of the bill.

Closing off Nature's Resources to Other Species

A growing population of high consuming humans, some more than others, is monopolising the biophysical resources of the planet, including those on land, lakes, rivers, oceans, as well as the atmosphere as a carbon pollution dump. This deprives other species of resources and habitats needed to survive, and led to about one million species currently threatened with extinction, many in a matter of decades, a far higher rate than in the past, and the rate is accelerating (IPBES 2019). Fossil-fuelled global warming is destroying habitats of species as disparate as polar bears and ocean coral. Deforestation of massive tracts of Alberta's Boreal forest caused by extraction of oil from its bituminous sands has devastated the habitat of its caribou herd, and decimated the herd. 'Even for global warming of 1.5 to 2 degrees, the majority of terrestrial species ranges are projected to shrink profoundly' (IPBES 2019). Present fossil-fuelled temperature rise is just a slight foretaste of planetary heating that will certainly increase by 2 °C and could rise by 5 degrees, yet already half of the species are failing to cope with it (Wiens 2016). Bar-On et al. (2018) estimate that:

- The biomass of all wild mammals combined is now 10 times less than the biomass of humans.
- The wild mammal biomass has been cut by 85%.
- Domesticated cattle and pigs outweigh wild mammals by 14 to 1.
- Domesticated foul (especially chickens) outweigh wild birds by 3 to 1.
- Total plant biomass (especially trees) has declined by half.

Monopolization by humans also involves exploiting bodies of other species for meat, eggs, milk, etc., and reconfiguring their lives in factory farms to produce food at least cost but maximum disruption for those species being exploited. The Anthropocene involves monopolisztion of the planet's resources by human agriculture, domestication of livestock, industrialization, deforestation, and fossil-fuelled practices. The issue is one of scale. Humans are predators, but the scale of human predation in the Anthropocene amounts to the monopolization of habitats and bodies of other species to an exceptional degree. The quantity of human monopolization of nature's resources has become a qualitative difference in kind. The Anthropocene is characterized by emerging relations of social closure between human populations distant in space, between generations distant in time, and between humans and other species.

Response by the Excluded and Backlash from the Manipulated

Is this new type of monopolization in the Anthropocene usurpation-free? How could future generations, not even born, undermine monopolization of biophysical resources they will need that are being depleted and pollution sinks being filled by the present generation: 'how to address and account norms of justice to subjects who do not live yet and therefore have no voice of their own in decision-making which affects their conditions of life dramatically' (Beck 2015: 82)? Similarly, the vulnerable poor in distant lands have little leverage to challenge the monopolization of nature's materials and its pollution sinks, especially in nature's commons such as oceans and atmosphere everyone shares. How could Madagascar, whose per capita emissions are infinitesimal, undermine American and Chinese monopolization of the carbon sink in the sky? Other species are helpless when confronting the enormous technological power of humans. McKibben's law of climate change—the less a person, company, country, or species did to cause it, the harder it is hit—seems iron clad. Where will usurpation of environmental social closure come from, if at all?

Purposeful Reaction

Intentional usurpation of social closure constitutes one reaction. Responses come from environmental, ecology, and conservation movements (Greenpeace, Earth First, Sierra Club, and environmental justice movements). The world wildlife and animal rights movements support species unable to speak for themselves, including species losing their habitats, animals in factory farms, and fish in aquaculture. Organizations like Oxfam advocate for future generations and poor countries. They demand action to limit degrading the global environment all humans need, including future generations, even if this means reducing privileges of monopolizers. A recent social movement, called Extinction Rebellion, responds specifically to the fossil-fuelled climate crisis. Environmental movements are similar to the feminist movement, labour movement, civil rights movement, etc., opposing specific forms of exclusion.

Impact scientists (Schneider 2009) have another crucial usurpationary role by making visible temporarily invisible dangers of present socioeconomic practices, both globally and in the future. This lays the foundation for changing those practices. Impact science was essential to revealing depletion of the ozone layer invisible to the naked eye, which led to the Montreal Protocol to eliminate CFCs. Similarly, impact science is needed to make visible slow-onset fossil-fuelled, global warming. It usurps privileges of chemical, coal, oil industries, and their consumers to pollute without paying. Impact science is transformative by adding an important dimension to scientific applications: knowledge of consequences of production rather than the only pursuit of profit, consumer goods, and conveniences. It shines light on harmful outcomes of technology and demonstrates that science is not so pure as once believed. It undermines vested interests and predispositions, thereby giving science a subversive role. Impact science constitutes an important component of the pursuit of knowledge when adverse consequences of production science are invisible or creeping from latency to being manifest. This makes science 'a contact sport' (Schneider 2009): impact scientists versus production scientists.

Political and business leaders with sufficient foresight to consider the long term also attempt to restrain fossil-fuelled pollution so that windfalls for the wealthy and minor benefits for the present generation will not damage the environment of the poor and future generations, including their own grandchildren. Populations in flooded and drought stricken lands are already being driven to migrate to less vulnerable, more prosperous locations where global environmental problems like climate change were largely caused by extensive rise of fossil-fuels. This migratory pressure will likely intensify as vulnerable environments are further degraded by fossil-fuelled global warming. These climate refugees challenge the ethics of exclusionary citizenship rights and privileges of closed borders.

Monopolization is often not resisted if there is upward mobility, especially of the intergenerational kind. Comparison with one's parents and the recent past can be more influential than comparison with the wealthy. If there is opposition, it cannot be presumed that the target of revolt will be the main monopolizers. Grievances can be hijacked by skilled demagogues gaining the confidence of the dissatisfied by promising near-term benefits but worsening long-term degradation of the environment future generations will need. Hence it could foment more revolt in the long run. The 2016 American election was instructive, where a bipolar reaction to social closure intensified by globalization, international trade agreements, automation, wage stagnation, and downward intergenerational mobility emerged in the form of diametrically opposed social movements led by plutodemocrat Donald J. Trump and democratic socialist Bernie Sanders. Fossil-fuelled environmental damage similarly threatens to produce downward intergenerational mobility, which in turn has the potential to incite demands for change, but its direction remains to be socially constructed.

The institution having most potential to mitigate monopolization and develop equitable sharing of benefits and risks consists of governance. Studies comparing environmental performances of societies documented that social democratic countries outperformed others (Yale University 2012, 2018; Germanwatch 2012, 2019). Social democracy deploys governments, trade unions, etc., to redistribute wealth more equitably within its borders thereby countering monopolization by the wealthy and reducing inequality of opportunity, and is typically more inclusive of the needs of future generations and poor countries because of environmental considerations.

Reaction by Nature's Dynamics Being Manipulated

Another response does not involve intentions. Since monopolization of environmental resources is based on manipulating nature's biophysical dynamics, the most consequential reaction will likely come from nature itself. Unlike the poor and future generations, it does not need other humans to act on its behalf. Nature's dynamics have the power to undermine the most powerful institutions. The earthquake and tsunami that devastated devoutly Catholic Lisbon in 1755 not only killed people and destroyed buildings but also shook the power of the Catholic Church: if they were God's punishment for sin, as Jesuits claimed, why did they destroy churches and spare brothels (Zebrowski 1997)? If this natural disaster was capable of undermining a religious monopoly, then an unnatural disaster (Abramowitz 2001) of nature's forces unleashed by economic pursuits has the potential to destabilize drivers of those pursuits. Nature is an actant whose dynamics strike back against their manipulation by humans (Tenner 1997; Clark 2011). The nature-is-like-putty assumption that humans can reconstruct nature without concern is a misrepresentation. Nature instead reacts to its manipulation by humans by biting back, rebounding, backlashing, boomeranging, recoiling, etc.

There are many examples of nature's dynamics striking back at their manipulation. The innovation of antibiotics caused the emergence of antibiotic-resistant bacteria. The social practice of smoking cigarettes causes runaway cancer cells in the lungs. Many apparently successful innovations later proved to be so harmful because of nature's backlash that they had to be abandoned: DDT, CFCs, and asbestos products. Airplanes and ocean tankers result in many benefits, but the former carry infectious diseases (COVID-19) and the latter transport ecosystem-damaging invasive species (zebra mussels). BP claimed the blowout protector on its oil-drilling rig was failsafe, yet it was overwhelmed by deepwater pressures of the Gulf of Mexico (Freudenburg and Gramling 2011). NASA stated the Challenger Space Shuttle was safe to launch, but physical forces on the O-ring led its rocket to explode (Vaughan 1996). Disaster researchers (Turner and Pidgeon 1978) have repeatedly found that nature's dynamics presumed harnessed can slip their leash. They investigated how the incubation of disaster was socially constructed by underestimating the power and autonomy of nature's forces.

Fossil fuels propel engines which shrink time and space, but also cause global warming, which increases the frequency and intensity of hurricanes that strike cities like New Orleans (Katrina) and New York (Sandy), of wildfires in Portugal, California, Australia, etc. Freudenburg et al. (2009) refer to this as catastrophes in the making, and fossil-fueled global warming could be the biggest of them all. Fossil-fuel combustion causes global warming which melts permafrost thereby letting loose previously stored greenhouse-gas methane and melts the Arctic ice cover letting in more of the sun's radiation previously reflected back into space. This constitutes a positive feedback loop whereby nature itself then intensifies global warming. First-order fossil-fuelled global warming threatens to cause second-order runaway global warming by nature's dynamics, unleashing more frequent and intense hurricanes, wildfires, droughts, floods, ocean level rise, etc. The Inuit Nobel Peace Prize nominee Sheila Watt-Cloutier (2019: O11) stated: 'In order to arrest this dangerous trajectory, the world has to take note of what is happening in the Arctic – because what happens in the Arctic doesn't stay in the Arctic. It's the planet's air conditioner, and as it melts, it causes havoc on the world'. Fossil-fuelled social practices threaten to cause a reaction by nature's forces that could undermine many human activities. Their enormous consequences may let loose more powerful forces of nature having the capacity to undercut those practices. There is a serious contradiction inherent in monopolising nature's resources and closing them off to other forms of life in that it threatens to undermine the very services that nature's species and its autonomous dynamics have provided free of charge for humans in the Holocene, which have enabled human development. Human monopolization of the planet threatens to backfire with a loss of ecosystem services for humans.

Because nature's forces let loose by human practices are so powerful and global, even the principal monopolizers and beneficiaries are threatened in the long run. Rising ocean levels because of fossil-fuelled global warming are causing storm surges that flood luxurious Miami Beach, where the porous soil would make a seawall ineffective. At the least, the backlash by nature's forces puts human innovation on a costly treadmill to keep up with nature's novel constructions. Paradoxically, it is not the failures of science and technology to manipulate nature's dynamics that unleashes nature's dangerous side effects, but instead their successes.

Usurpation by Environmental Regulation; Monopolization by Deregulation

Regulations, especially environmental regulations, are indirect ways of combatting monopolization. They entail a cost to the industry involved and its wealthy owners, but provide bigger benefits for society and its average citizens. In the USA the nonpartisan White House Office of Management and Budget (OMB 2018) monetized, in constant dollars, benefits, and costs of regulations from 2006 to 2016. It estimated that rules reducing toxic emissions from power plants cost that industry \$9.6 billion but brought air quality benefits worth between \$33 and \$90 billion. They were beneficial to ordinary citizens, especially those living near polluting power plants and highways, but at a cost to wealthy owners of those industries. They constitute usurpation of the industry's entitlement to pollute, which had been externalized and paid in terms of harm to the health of average Americans. Similarly, regulations increasing fuel economy standards cost industry between $0.8 and $1.1 billion but provided between $6.7 and $9.7 billion in benefits. The regulations consist of a downward redistribution of wealth. Moreover, environmental regulations shift costs to immediate pollution prevention from much higher but belated costs of pollution itself. The OMB (2018) also documented that environmental regulations caused no detectable harm for economic growth or national employment during that decade, although they may have incited shifts from one industry or location to another or had short-term consequences. The propaganda that regulations kill jobs was refuted; other factors are far more important.

Conversely, deregulation and failure to regulate pollution contribute to monopolising wealth by externalizing costs of pollution thereby increasing profitability. They involve an upward redistribution of wealth. Much like the wealthy and conservative politicians struggle against taxes and against antitrust legislation, they also fight environmental regulations because such regulations reduce their capacity to monopolize wealth and constitute usurpation of it. The first executive order of Donald Trump as American President was to deregulate the coal industry, allowing it to pollute profitably. This was followed by more environmental and socioeconomic deregulation measures and tax reductions, all of which disproportionately benefited the wealthy. Government regulations and taxation are weakened by the actions of the monopolizers. Frederick Koch Sr. built a fossil-fuel behemoth, which sons Charles and David inherited and developed into the second-largest privately held company in the USA with annual revenues of US$ 110-billion, making the Kochs the third richest family in America and Charles and David two of the wealthiest people in the world. The Koch brothers then used their profits to lobby to reduce taxation and deregulate fossil fuel pollution. Monopolization of wealth and power gives the monopolizers the capacity to influence governments to set rules in their interests, to abolish regulations such as antitrust and environmental protection rules that restrain their monopolistic and polluting practices, thereby creating slow-onset risks for everyone.

Environmental regulation promotes a more inclusive sharing of environmental benefits, even between generations and between those distant in space, thereby constituting an indirect form of undermining monopolization. Deregulation, on the contrary, constitutes a subtle type of monopolization by intensifying the externalization of environmental costs to be paid by victims. Placing a price on carbon pollution, either through a carbon tax or cap-and-trade, to include environmental and health costs in the price of fossil fuels reduces their profitability. However, it would bring long-term benefits to the population of mitigating global warming. Opposition to pricing carbon pollution is led by the fossil-fuel industry (e.g. Koch Brothers in the USA) and conservative politicians. Opposition to governments imposing a price on carbon pollution is leading some environmental economists to propose instead government regulations to phase out coal plants, enact strict rules to reduce emissions from gasoline vehicles, and implement upstream regulations decreasing emissions in the extraction, upgrading, and transport of fossil fuels (Jaccard 2018). If set at levels to mitigate global warming, these measures would result in keeping more fossil fuels safely in the ground and have long-term benefits for the population, but would eliminate the profitable coal industry and diminish the value of the oil industry. Such regulations would obstruct the path of least resistance to profitability so they incite deregulation political lobbies. Despite his market optimism, Rand (2018: B4) explains as follows how dominant fossil-fuel companies resist environmental regulations. 'If you make lots of money doing something, it's natural to want to keep doing the same thing. Dominant market players will try to defend and extend the status quo. That strategy works well - until it doesn't'. Struggles over environmental regulations involve dynamics of social closure, namely appropriating near-term economic opportunities versus inclusion in long-term enjoyment of environmental resources and the benefits and opportunities of a clean environment.

Big Win–Little Win or Big Win–Lose?

The social closure framework fits the empirical evidence of the concentration of wealth, specifically the monopolization of fossil-fuel wealth. It explains inequalities of opportunity, particularly emissions-generated global warming closing off life chances to latecomers (future generations, the poor) and other species. But if the most talented create wealth making everyone more affluent, and if they can technologically solve environmental problems, then it is a win–win outcome (Norberg 2003, 2016; Lomborg 2001, 2007), or more accurately big win–little win. The most talented cause a rising tide that lifts all boats and merit enormous rewards. What's wrong with that, and with the trickle-down theory describing it?

One criticism is that wealth appropriated by a tiny minority but somewhat beneficial to everyone is often temporary, followed by disastrous recessions if regulations are lax for risk makers, as occurred in the 1930s and 2008. Then everyone suffers, except the monopolizers if everybody becomes dependent on their institutions judged too big to fail and have to be subsidized to stay afloat. Moreover, win–win is often a mirage. The benefits of globalization since the 1980s were concentrated in the fortunes of the wealthiest shareholders whereas workers' wages in developed countries stagnated, which workers see as a loss. Inequality is corrosive to society; relative deprivation breeds resentment and conflict. Monopolization of resources and rewards produces major inequalities of opportunity, especially if differential resources are passed between generations (Corak 2013). These result in inequities from a moral perspective and wastage of human capital from an economic viewpoint. Attempts to justify the concentration of wealth by meritocratic discourse are refuted by objective differences in life chances. Some people are unusually talented, but there is little evidence that inequality overall is meritocratic.

Another criticism is becoming more significant. Big wins for present-day monopolizers and small wins for the rest of the population risk coming at the cost of big losses. These consist of exclusion of future generations from the benefits, opportunities, and services of the present bountiful environment in an intensifying dynamic of social closure. Although fossil fuels are abundant, it will not be worthwhile extracting them if the energy extracted barely exceeds the energy required to extract, upgrade, refine, and transport them. Hughes (2009) and Davidson and Andrews (2013) documented that the energy return on energy invested (EROI) is decreasing. The giant Ghawar oil field in Saudi Arabia came into production in 1951 yielding 100 barrels of oil for each energy-equivalent barrel input to extract it. After seventy years of extraction, it and other giant oil fields are in decline, with extraction propped up by technology that temporarily maintains extraction but hastens the arrival of steep decline. The energy yield for heavy oil, deepwater oil, oil from the Arctic and Siberia, etc., is dramatically lower than the initial Ghawar yield. Hydraulic fracking currently yields large amounts of tight oil and natural gas from shale, but these deposits become depleted rapidly and new drilling has to be continually done to compensate. Fossil fuels are on a drilling—depletion treadmill. Although appearing counterintuitive in an era of shale gas and oil abundance, the prediction of peak oil has not been refuted and has, according to estimates, only been postponed until after 2030. The likelihood of present high consumption closing off hitherto easily accessible oil to latecomers remains, for both the poor and future generations. The fossil-fuel industry has been exploiting the most accessible deposits, leaving future generations' only inaccessible ones both geographically and technologically.

Even more important is the availability of safe carbon sinks (Jaccard 2009). Presently, carbon pollution is treated as cost-free and danger-free dumped into the atmosphere and descending to oceans, but they remain finite as pollution sinks. They are presently monopolized by fossil-fuel companies and high consumers. Big market winners cause big pollution by their consumption and decisions as managers. Economists (Jaccard 2005, 2009; Nordhaus 2013) advocate either government regulations restricting carbon emissions or an escalating price on carbon to include pollution costs in the price of fossil fuels. These are necessary to reduce their consumption and leave raw materials and pollution sinks, which are bequeathed by nature's dynamics, to latecomers. But rare indeed are countries that have implemented a price on carbon pollution, especially its required escalating feature, and it continues to accumulate filling the atmosphere.

Sustaining the big win–little win treadmill of consumption would require certainty there are no biophysical limits to economic growth. Anything less would result in degrading the global environment thereby closing off opportunities to future humans, and possibly tipping humanity's only planet into a less advantageous state. The analogy of a rising tide lifting all boats should be replaced by a more appropriate one. Market-driven carbon emissions resulting in sea level rise risks leaving well-protected yachts unscathed but sending vulnerable tiny boats crashing against the rocks. Excessive fossil-fuelled practices in the near term are causing long-run harm by polluting the commons needed by all.

In principle, near-term economic interests and long-term environmental interests are reconcilable, and some countries, especially social democratic ones, are accomplishing reconciliation partially. These have regulations limiting monopolization and promoting inclusion, redistribution and equality of opportunity. They also have regulations to safeguard the biophysical commons needed by everyone. Inclusionary social democratic values and practices are extended to mitigating the closing off of biophysical resources and opportunities by one generation at the expense of future generations. However, these are exceptions. As the increase in emissions demonstrates, near-term fossil-fuel interests of extractors and consumers are globally trumping long-term environmental needs of future humans. This constitutes practices of monopolization and exclusion, namely social closure, whereby future dangers and needs are discounted to attain near-term economic goals. Near-term economic wins are pursued despite the risk of long-term loss of opportunities for latecomers, namely the poor and future generations.

Notes

1. The following summary is obviously too brief, so the reader is encouraged to consult Murphy (1988) and the aforementioned texts. They constitute a few examples of the construction and use of the closure theoretical framework, not an exhaustive list, since this study's objective is not an overview but rather to open up a new environmental dimension of that framework.

2. Monopolization and exclusion are not eliminated by market competition, but are more complex in terms of temporary, space specific, or technological monopolistic niches.
3. This shows that the example of multibillionaire investor Warren Buffet paying a lower rate of income tax than his secretary is not unusual.

References

Abramovitz, J. 2001. *Unnatural Disasters*. Washington: World Watch Paper.

Akerlof, George, and Robert Shiller. 2015. *Phishing for Phools: The Economics of Manipulation and Deception*. Princeton: Princeton University Press.

Bar-On, Yinon M., Rob Phillips, and Ron Milo. 2018. The Biomass Distribution on Earth. *Proceedings of the National Academy of Sciences of the United States of America (PNAS)*. http://www.pnas.org/content/early/2018/05/15/1711842115. Accessed 24 May 2018.

Beck, Ulrich. 1992. *Risk Society: Towards a New Modernity*. London: Sage.

Beck, Ulrich. 1998. *Democracy Without Enemies*. Cambridge: Polity Press.

Beck, Ulrich. 2009. *World at Risk*. Cambridge: Polity Press.

Beck, U. 2015. Emancipatory Catastrophism. *Current Sociology* 63 (1): 75–88.

Berners-Lee, M., and D. Clark. 2013. *The Burning Question*. London: Profile.

Carolan, Michael. 2011. *The Real Cost of Cheap Food*. Abingdon, UK: Earthscan.

Carolan, Michael. 2014. *Cheaponomics: The High Cost of Low Prices*. Abingdon, UK: Earthscan.

CDIAC. 2018. Carbon Dioxide Information Analysis Center. U.S. Department of Energy. Berkley. cdiac.Ess-dive.lbl.gov.

Ciplet, David, J. Timmons Roberts, and Mizan Khan. 2015. *Power in a Warming World: The New Global Politics of Climate Change and the Remaking of Environmental Inequality*. Cambridge, MA: MIT Press.

Clark, N. 2011. *Inhuman Nature*. London: Sage.

Collins, Randall. 1968. A Comparative Approach to Political Sociology. In *State and Society*, ed. Reinhard Bendix, 42–67. Boston: Little, Brown.

Collins, Randall. 1971. Functional and Conflict Theories of Educational Stratification. *American Sociological Review* 36: 1002–1019.

Collins, Randall. 1975. *Conflict Sociology: Toward an Explanatory Science*. New York: Academic Press.

Collins, Randall. 1979. *The Credential Society: A Historical Sociology of Education and Stratification*. New York: Academic Press.

Collins, Randall. 1980. Weber's Last Theory of Capitalism: A Systematization. *American Sociological Review* 45: 925–942.

Collins, Randall. 1981. *Sociology Since Midcentury*. New York: Academic Press.

Collins, Randall. 1986. *Weberian Sociological Theory*. Cambridge: Cambridge University Press.

Corak, Miles. 2013. Income Inequality, Equality of Opportunity, and Intergenerational Mobility. *Journal of Economic Perspectives* 27 (3): 79–102.

Crowley, Brian Lee. 2017. As Workers Get a Smaller Share of National Income, a Robust Competition Policy Could Be the Solution. *Global and Mail*, 7 February.

Daly, M., and H. Silver. 2008. Social Exclusion and Social Capital: A Comparison and Critique. *Theory & Society* 37: 537. https://doi.org/10.1007/s11186-008-9062-4.

Davidson, D., and J. Andrews. 2013. Not All About Consumption. *Science* 339 (6125): 1286–1287.

Duff, David. 2017. It's Time to Limit Tax Expenditures That Favour Our Highest Earners. *Globe and Mail*, 7 February.

Dyer, Gwynne. 2008. *Climate Wars: The Fight for Survival as the World Overheats*. Toronto: Random House.

Freeland, Chrystia. 2012. *Plutocrats: The Rise of the New Global Super-Rich and the Fall of Everyone Else*. New York: Penguin.

Freudenburg, W., and R. Gramling. 2011. *Blowout in the Gulf*. Cambridge, MA: MIT Press.

Freudenburg, William, Robert Gramling, Shirley Laska, and Kai Erikson. 2009. *Catastrophe in the Making*. Washington: Island Press.

Friesen, Joe. 2019. Academics Vow to Use Greener Transit. *Globe and Mail*, 16 July: A4.

Galbraith, John Kenneth. 1967. *The New Industrial State*. Boston: Houghton Mifflin.

Germanwatch. 2012. *CCPI the Climate Change Performance Index Results 2012*. http://germanwatch.org/klima/ccpi.pdf. Accessed 5 November 2012.

Germanwatch. 2019. *CCPI the Climate Change Performance Index Results 2019*. https://germanwatch.org/sites/germanwatch.org/files/CCPI-2019-Results-190614-WEB-A4.pdf. Accessed 22 September 2019.

Hammond, Andrew. 2017. BRICs Make the Case for Globalization. *Globe and Mail*, 4 August: B4.

Hardoon, Deborah. 2017. *An Economy for the 99 per cent*. Oxfam GB for Oxfam International under ISBN 978-0-85598-862-3. Oxford: Oxford International. https://www.oxfam.org/sites/www.oxfam.org/files/file_attachments/bp-economy-for-99-percent-160117-summ-en.pdf. Accessed 1 February 2017.

Hawken, Paul. 2017. *Drawdown: The Most Comprehensive Plan Ever Proposed to Reverse Global Warming*. New York: Penguin.

Heal, Geoffrey. 2017. *Endangered Economies: How the Neglect of Nature Threatens Our Prosperity*. New York: Columbia University Press.

Holmberg, Susan. 2017. *Boiling Points: The Inextricable Links between Inequality and Climate Change*. Roosevelt Institute. http://rooseveltinstitute.org/boiling-points/. Accessed 24 May 2017.

Hughes, Chris. 2019. It's Time to Break Up Facebook. *New York Times*, 9 May. https://www.nytimes.com/2019/05/09/opinion/sunday/chris-hughes-facebook-zuckerberg.html. Accessed 15 April 2019.

Hughes, J.D. 2009. The Energy Issue. In *Carbon Shift*, ed. T. Homer-Dixon, 58–95. Toronto: Random House.

IPBES. 2019. *Intergovernmental Science-Policy Platform on Biodiversity and Ecosystem Services*. Global Assessment Report on Biodiversity and Ecosystem Services. Bonn: IPBES. https://ipbes.net/sites/default/files/2020-02/ipbes_global_assessment_report_summary_for_policymakers_en.pdf. Accessed 10 April 2020.

Jaccard, M. 2005. *Sustainable Fossil Fuels*. Cambridge: Cambridge University Press.

Jaccard, M. 2009. Peak Oil and Market Feedbacks. In *Carbon Shift*, ed. T. Homer Dixon, 96–131. Toronto: Random House.

Jaccard, Mark. 2018. Divisive Carbon Taxes Are Much Ado About Nothing. *The Globe and Mail*, 15 December: O4.

Jackson, Andrew. 2017. The New Age of Corporate Concentration Requires More Than Just a Policy Fix. *The Globe and Mail*, 12 April: B4.

Koch, Max. 2003. Closure Theory and Citizenship: The Northern Ireland Experience. *Electronic Journal of Sociology*, ISSN 1198 3655. Accessed 28 November 2016.

Krugman, Paul. 2007. *The Conscience of a Liberal*. New York: W.W. Norton.

Lazonick, William. 2014. Profits Without Prosperity. *Harvard Business Review*, September. https://hbr.org/2014/09/profits-without-prosperity. Accessed 17 December 2017.

Lomborg, B. 2001. *The Skeptical Environmentalist*. Cambridge: Cambridge University Press.

Lomborg, B. 2007. *Cool It*. New York: Alfred A. Knopf.
Mackert, Jürgen (ed.). 2004. *Die Theorie sozialer Schliessung: Tradition, Analysen, Perspektiven*. Wiesbaden: VS Verlag für Sozialwissenschaften.
Mackert, Jürgen. 2012. *Social Closure. Oxford Bibliographies*. Oxford: Oxford University Press. http://www.oxfordbibliographies.com/view/document/obo-9780199756384/obo-9780199756384-0084.xml?rskey=qjC7fm&result=1&q=social+closure+theory#firstMatch. Accessed 21 November 2018.
Mason, Gary. 2019. How Many Pipelines Do We Really Need. *The Globe and Mail*, 15 November: A13.
Masuku S, Rama S. 2020. A Case Study of Government and Civil Societies'collaboration and Challenges in Securing the Rights of Congolese Refugees Living in Pietermaritzburg, KwaZulu-Natal, South Africa. *South African Journal of Science* 116 (1/2).
Mauer, K.W. 2020. Monopoly's Winners and Losers: Elwha River Dam Construction as Social Closure. *Journal of Environmental Studies and Sciences* 10: 137–147. https://doi.org/10.1007/s13412-020-00586-w.
McGugan, Ian. 2017a. Mayer's Yahoo Windfall Exposes a Convenient Cliché About Executive Compensation. *The Globe and Mail*, 14 June: B4.
McGugan, Ian. 2017b. Time to Push Back Against the Digital Monopolists. *Globe and Mail*, 29 July: B1, B5.
McGugan, Ian. 2017c. Dangerous Duopolists. *Globe and Mail: Report on Business*, September: 5–6.
McKibben, Bill. 2019. Our Climate Change Bill Is Coming Due. *Globe and Mail*, 28 December: O8.
McNichol, Elizabeth, Douglas Hall, David Cooper, and Vincent Palacios. 2012, Pulling Apart: A State-by-State Analysis of Income Trends. *Center on Budget and Policy Priorities*, November 15.
McQuaig, Linda, and Neil Brooks. 2010. *The Trouble with Billionaires*. Toronto: Viking Canada.
Minefee, I., V. Rabelo, O. Stewart, and N. Young. 2015. Repairing Leaks in the Pipeline: A Social Closure Perspective on Underrepresented Racial/Ethnic Minority Recruitment and Retention in Business Schools. *Academy of Management Learning and Education* 17 (1). https://doi.org/10.5465/amle.2015.0215.
Mortished, Carl. 2017. The Totalitarian Capitalism of Tech Giants Should Trump Your Fears of Populism. *The Globe and Mail*, 15 May: B4.
Murphy, Raymond. 1988. *Social Closure: The Theory of Monopolisation and Exclusion*. Oxford: Oxford University Press (Clarendon).

Myles, Stephanie. 2020. Qualifiers in Australia Face Extreme Conditions. *The Globe and Mail*, 16 January: B19.

Norberg, Johan. 2003. *In Defense of Global Capitalism.* Washington: Cato Institute.

Norberg, Johan. 2016. *Progress: Ten Reasons to Look Forward to the Future.* Kindle Edition. Oneworld Publications.

Nordhaus, William. 2013. *The Climate Casino: Risk, Uncertainty, and Economics for a Warming World*. New Haven: Yale University Press.

OMB Office of Management and Budget. 2018. *2017 Draft Report to Congress on the Benefits and Costs of Federal Regulations and Agency Compliance with the Unfunded Mandates Reform Act*. Washington: Office of the President of the United States. https://www.whitehouse.gov/wp-content/uploads/2017/12/draft_2017_cost_benefit_report.pdf. Accessed 19 March 2018.

Parkin, Frank. (1971) 1972. *Class Inequality and Political Order*. Frogmore: Paladin.

Parkin, Frank. 1974a. Strategies of Social Closure in the Maintenance of Inequality. Unpublished Paper Presented to the Eight World Congress of Sociology, Toronto.

Parkin, Frank. 1974b. Strategies of Social Closure in Class Formation. In *The Social Analysis of Class Structure*, ed. Frank Parkin. London: Tavistock.

Parkin, Frank. 1979. *Marxism and Class Theory*. London: Tavistock.

Parkin, Frank. 1980. Reply to Giddens. *Theory and Society* 9: 891–894.

Parkin, Frank. 1982. *Max Weber*. London: Tavistock.

Piketty, T. 2014. *Capital in the Twenty-First Century*. Cambridge, Mass.: Belknap Harvard.

Rand, Tom. 2018. A Sound Debate About Canada's Emissions, Brought on by Fury Over Trans Mountain. *Globe and Mail*, 3 May: B4.

Reguly, Eric. 2017a. It's Time to Call in the Oligopoly Busters. *Globe and Mail*, 15 April: pp. B1–B2.

Reguly, Eric. 2017b. Scary Monsters. *Report on Business*, July/August: 19.

Reguly, Eric. 2017c. Bayer, Monsanto Tie-Up Sows Seeds of Discord for Competition. *The Globe and Mail*, 2 September 2017: B1.

Roberts, J. Timmons, and Bradley Parks. 2007. *A Climate Of Injustice: Global Inequality, North-South Politics, and Climate Policy*. Cambridge, MA: MIT Press.

Saunders, Doug. 2017. Can We Ever Knock Down the Walls of Wealthy Ghettos? G*lobe and Mail*, 15 July: F7.

Schneider, S. 2009. *Science as a Contact Sport*. Washington: National Geographic.

Spence, Michael. 2019. Inequality Fuels Economic Failure—And Frustrations Are Spreading. *Globe and Mail*, 4 January: A11.
Speth, James Gustave. 2009. *The Bridge at the Edge of the World: Capitalism, the Environment, and Crossing from Crisis to Sustainability*. New Haven: Yale University Press.
Srinivasan, Dina. 2018. The Antitrust Case Against Facebook. *Berkeley Business Law Journal* 16 (1) https://ssrn.com/abstract=3247362. Accessed 15 April 2019.
Stern, Nicholas. 2009. *A Blueprint for a Safer Planet*. London: Random House.
Stiglitz, Joseph E. 2012. *The Price of Inequality: How Today's Divided Society Endangers Our Future*. New York: W.W. Norton. ISBN 9780393088694.
Stiglitz, Joseph E. 2015a. *The Great Divide: Unequal Societies and What We Can Do About Them*. New York: W.W. Norton. ISBN 9780393248579.
Stiglitz, Joseph. 2015b. *Rewriting the Rules: An Agenda for Growth and Shared Prosperity*. New York: W.W. Norton.
Stiglitz, Joseph. 2017. Globalism Needs a Bigger Boost Than the French Election'. *Globe and Mail*, 2 May: A13.
Suzuki, David, and Ian Hanington. 2017. *Just Cool It: The Climate Crisis and What We Can Do*. Vancouver and Berkeley: Greystone Books.
Swartz, David L. 2008. Social Closure in American Elite Higher Education. *Theoretical Sociology*, 37: 409. https://doi.org/10.1007/s11186-008-9064-2. Accessed 28 November 2016.
Taplin, Jonathan. 2017. *Move Fast and Break Things: How Facebook, Google and Amazon Cornered Culture and Undermined Democracy*. Hachette Audio.
Tenner, E. 1997. *Why Things Bite Back*. New York: Vintage.
Turner, B., and N. Pidgeon. 1978. *Man-Made Disasters*. London: Wykeham.
Vaughan, D. 1996. *The Challenger Launch Decision*. Chicago: University of Chicago Press.
Viola, Lora Anne. 2020. *The Closure of the International System: How Institutions Create Political Equalities and Hierarchies*. Cambridge: Cambridge University Press.
Wallerstein, Immanuel. 2011 (1980). *The Modern World System II*. Berkeley: University of California Press.
Watt-Cloutier, Shiela. 2019. If We Protect the Arctic, We Save the Planet. *Globe and Mail*, 5 October: O11.
Weber, Max. (1922) 1978. *Economy and Society*, ed. Guenther Roth and Claus Wittich. Berkeley: University of California Press.
Wiens, John. 2016. Climate Change and Extinction. *PLoS Biology*, December.

Wu, Tim. 2018. *The Curse of Bigness: Antitrust in the New Guilded Age*. New York: Columbia Global Reports.

Yale University. 2012. *EPI Environmental Performance Index 2012*. http://epi.yale.edu. Accessed 8 November 2012.

Yale University. 2018. *EPI Environmental Performance Index 2018*. https://epi.envirocenter.yale.edu/epi-country-report/CAN. Accessed 22 September 2019.

Zebrowski, Ernest, Jr. 1997. *Perils of a Restless Planet: Scientific Perspectives on Natural Disasters*. Cambridge: Cambridge University Press.

Zucman, Gabriel. 2015. *The Hidden Wealth of Nations: The Scourge of Tax Havens*. Chicago: University of Chicago Press (see Corporate Knights Spring 2016: 60–62).

4

Energy: Paying Its Full Cost, Belatedly or Upon Use?

In a market economy, price has a powerful influence on consumption: low price stimulates consumption; high price dampens it. This is true for all goods, and in particular fossil fuels. But what determines price? Is it only the cost of inputs (raw materials, labour, organization, etc.) and profits? Has something been excluded from the price? In a market free of government regulations, the cost of harm to health and the environment has not been included in the price of products. For a few goods, this changed after much conflict. Lawsuits forced asbestos companies to pay for the harm their products caused, which was so great the companies eventually went bankrupt. Cigarette companies similarly faced lawsuits, and taxes were imposed on cigarettes to pay for health care of victims. Acid rain was largely solved by mandating a charge for sulphur and nitrogen oxide emissions from industrial burning of coal, then capping and trading pollution allotments. This added to the cost of those industries, giving them incentives to introduce scrubbers and other technologies to reduce emissions. Some states are attempting to mitigate fossil-fuelled global warming by placing a price on carbon pollution, either through carbon taxes or cap-and-trade. Excellent research has been done on these efforts to make polluters pay some of the costs of

R. Murphy, *The Fossil-Fuelled Climate Crisis*,
https://doi.org/10.1007/978-3-030-53325-0_4

harm they cause rather than having only the victims of pollution pay. Before examining research on paying the full costs of fossil fuels, a brief history of their use is warranted because they are such an important energy source for modern societies.

Fossil Fuels: From the Underground Commons to the Atmospheric Commons

Fossil fuels originate in the ancient photosynthesis of energy from the sun into organisms which died, are buried, and decompose anaerobically over millions and hundreds of millions of years. They contain high proportions of carbon, and when burned, oxidize to form carbon dioxide and water releasing large amounts of energy. Coal, petroleum and natural gas constitute the main types of fossil fuels, with derivatives including kerosene and propane.

Before 1750, human and animal effort provided energy for labour, burning wood and peat supplied heat, and windmills and watermills gave energy for milling. Side effects of these energy practices were rare and the scale of their use was small, so there was little pollution of the atmosphere. Carbon emissions because of forest fires, etc., were balanced by the withdrawal of carbon through vegetation growth, soil and ocean dynamics, etc. Permafrost locked methane underground, as did shale for natural gas, and the ground for coal and oil. There was an equilibrium of carbon transferred from the ground to the sky compared to carbon transferred from sky to ground. For the atmosphere, carbon in equalled carbon out. As societies evolved, the burning of wood and peat, clearing of the land (deforestation) for agriculture thereby reducing its carbon absorbing capacity, etc., resulted in carbon deficits, but they were small. Little anthropogenic carbon deficit was left by one generation to the next to accumulate. The stability of the carbon content and temperature of the atmosphere was propitious for the development of human societies.

When coal and then petroleum powered steam engines and the industrial revolution, emissions began to systematically surpass withdrawals. The accumulation of carbon in the atmosphere increased dramatically. The famous hockey stick graph of a very slowly increasing line followed by an exponential increase accurately depicts this trajectory of carbon being anthropogenically transferred from safe storage in the ground to the sky where it causes global warming. This is shown in Fig. 2.1 of Chapter 2. The innovation of the internal combustion engine in vehicles heightened demand for gasoline and diesel, as did the invention of planes for jet fuel. Ships and trains became powered by fossil fuels. Another major deployment of fossil fuels came with the generation of electricity. It carries energy from primary sources, such as the combustion of coal or natural gas, nuclear, wind, solar hydro, etc., to where it will be used but is not a primary energy source itself. Coal and recently natural gas have been the cheapest and predominant primary sources of electrical energy. Electricity use is growing rapidly, and will continue its swift increase. Fossil fuels are also used as input for the petrochemical industry and plastics. Whereas renewables are used almost exclusively to generate electricity, fossil fuels also power aviation, shipping, vehicles, heating, etc. The pursuit of profit resulted in the discovery of fossil-fuel deposits and its extraction. By 2017, fossil fuels provided 85% of the world's primary energy sources: 34% petroleum, 28% coal, and 23% natural gas (Heal 2017; IEA 2018). This dwarfs renewables, which are growing rapidly, reaching 15% of primary energy in 2015, but so is the world's overall energy consumption, increasing 2.1% a year. The result is that renewables are adding energy but not replacing fossil fuels. Fossil fuels have been an amazing, even indispensable, source of energy. They have liberated humans from labour and raised living standards. Unfortunately, they also bring grave, slow-onset damage to the environment. Their combustion emits 21.3 billion tonnes of carbon dioxide a year, and natural processes only remove half this amount. Hence carbon dioxide is accumulating in the atmosphere causing a greenhouse effect, global warming and climate change with threatening consequences.

Externalities

Pigou was 'the presiding genius of Cambridge economics for several decades [1920s and 1930s], the formalization of the concept of external effects was his principal legacy' (Heal 2017: 16). Externalities mean that fossil fuels have effects that are costly, but those costs are not included in their price. In a major economics textbook, Baumol and Blinder (2010: 6) conclude that 'externalities' is one of the 10 great ideas of the discipline. For example, greenhouse gas emissions resulting from combustion of fossil fuels accumulate in the atmosphere causing global warming. This is melting Arctic ice upon which Inuit depend, producing drought in Africa, Australia, California, wildfires, floods and more intense hurricanes. This damage is extremely expensive, as is fighting fires and protecting against floods and extreme weather. 'Currently we [Canada] are spending about $1-billion on fire suppression every year. We are seeing up to a couple of thousand houses destroyed every year. When you look into the future with [fossil-fuelled] climate change, we are going to be seeing things like a doubling of area burned, more severe fires, faster-spreading fires' (Lynn Johnston quoted in Stueck 2019: A7). Eventually fossil-fuelled global warming will raise sea levels because land-based glaciers in Antarctica and Greenland are melting. Since the world's main cities are on coastlines, the resulting damage and costs of adaptation and making cities robust, resilient, or relocating them to higher ground will be enormous. Such costs are not included in the price of fossil fuels. They are paid not by polluters but belatedly by victims often distant in space or time from the pollution's source. Externalities are free for producers and consumers of fossil fuels, so they produce and consume without regard for the damage caused, resulting in incentives to pollute. Therefore 'externalities escape the control of the market mechanisms because no financial incentive motivates polluters to minimize the damage they do' (Baumol and Blinder 2010: 6). They constitute a failure of the market.

Fossil fuels are relatively cheap and profitable because harmful consequences of their combustion are someone else's problems. 'An externality arises when the action of one person [company, country] directly affects the prospects of another .... Emissions are clearly an externality and are thus a market failure .... There is a double inequity here: the poor countries are least responsible for the existing stock of greenhouse gases, yet they are hit earliest and hardest by climate change' (Stern 2009: 11–13). Atmospheric currents carry greenhouse-gas pollutants from high per capita emitting countries to low emitting, more vulnerable societies. The latter nations are largely excluded from benefits but suffer adverse consequences of fossil-fuel combustion, whose benefits tend to be monopolized by the former. This threatens to exclude poor countries from safe, clean environments providing nature's free services. Within countries, large differences exist between risk makers who disproportionately benefit from fossil-fuel combustion and vulnerable risk-takers who only receive marginal benefits but suffer major threats, as environmental injustice researchers (Bullard 2000, 2005; Agyeman et al. 2003; Roberts and Parks 2007; Miranda et al. 2011) have documented. Social justice researchers focus on externalities and the polluter pays principle. 'Having polluters pay is efficient, since it puts formerly externalized costs back on them, which should inspire their cleanup. … That polluters should pay the costs of dealing with their pollution reflects the most fundamental principles of justice and responsibility' (Kahn and Roberts 2013: 139–140).

The sociologist Fairbrother (2016: 376) argues that 'environmental problems are instances where negative externalities are imposed through the medium of the physical and natural worlds … An externalities-based view of environmental degradation emphasizes not just that pollution and resource use have costs, but that polluters and resource users over-engage in polluting and resource-using activities specifically *because* they burden others with some of the costs of those activities, rather than themselves'. Had polluters been forced to suffer consequences of their pollution, there would be less pollution. Polluters appropriate benefits of their fossil-fuelled practices, and slough off liabilities and risks to others. Fairbrother argues that monopolizing benefits and externalizing harm without appropriate compensation is theft. First-come, first-served practices close off limited resources and pollution sinks to latecomers, whether they be countries developing later, the poor in developed countries, or future generations. Profit is enhanced by freely dumping waste and pollution into 'land fill', 'ocean fill', and 'atmosphere fill'. Costs of commodities depend on whether externalities are included in prices or paid belatedly in terms of a polluted environment (i) particularly by the vulnerable poor and poor countries, or (ii) by future generations, because of the transmission of pollution over space and time in the shared atmosphere. Environmental and health costs are excluded from consideration in conventional accounting that cloisters benefits and harm in categories closed off from one another. Diamond (2019: 399–400) distinguishes 'the apparent cost' of fossil fuels consisting of the low per barrel cost of crude oil and per litre cost of gasoline, from the higher 'actual cost' that includes damage caused by fossil fuels: climate change; oil spills; air pollution; etc. The more the coal industry externalizes costs of harm caused by coal, the more it can lower coal's price and drive out clean competitors, such as wind, solar, and hydro industries. That makes polluting coal more profitable for coal barons like the Koch brothers, which enables them to financially support politicians who oppose regulating carbon pollution. If regulations forced the coal industry to install carbon capture and storage technology, direct air capture of its legacy emissions, or/and pay its environmental damage, thereby including its pollution costs in the price of coal, then coal would be less cheap, and wind, solar and hydro energy would outcompete it. Flannery (2015: 174) suggests a method to determine the cost of carbon externalities: 'the cost per tonne of CO_2 removed should act as a guide to the price polluters should pay to emit in the first place. Just as the size of a fine given, for example to the polluter of a lake, should reflect the clean-up cost'.

Heal (2017: Chapter 3), like Stern, depicts 'climate change – [as] the greatest external effect in human history'. Nordhaus (2013: 18) contends that global warming is 'the Goliath of all externalities' because it results from a huge number of activities, affects so many people, and the consequences are long-lasting: 'In the case of harmful externalities like CO_2, unregulated markets produce too much because markets do not put a price on the external damages from CO_2 emissions. The market price of jet fuel does not include the cost of the CO_2 emissions, and so we fly too much'. I would nuance the 'we' by adding: some people, companies, and countries far more than others. The Inuit Nobel Peace Prize nominee Sheila Watt-Cloutier (2019: O11) concludes that 'the human, environmental and economic costs are racking up around the world and need to be properly accounted for'. Unfortunately, that isn't happening. 'But for a tiny slice of emissions in Nordic countries, nearly the entirety of the globe's carbon pricing falls well below the U.S. Environmental Protection Agency's social cost [i.e. harmful effects] of carbon, about $41 per ton' (Harvey and Orbis 2018: 271).

Carolan rejects the concept 'externalities' of economists, claiming it leaves the impression that 'when a cost is externalized that means we no longer have to worry about it' (Carolan 2014: 8), and rejects their concept 'market failure' because it suggests 'markets *could* solve the world's ills if only we did a better job of pricing things' (Carolan 2014: 16). The argument of environmental economists is, however, that we do have to worry about externalities because they constitute costs that climate change victims have to pay, and the world's pollution ills would be mitigated by including them in fossil-fuel prices. Despite his criticism, Carolan recognizes the significance of costs not included in prices but calls them 'socialized': paid by others in society including future generations. Thus his book is entitled *Cheaponomics: The High Cost of Low Prices*: low prices of fossil fuels have high cost for global warming victims. Regardless of label, excluding expensive consequences of fossil fuels from their price sends misleading signals to consumers, companies, governments, and citizens resulting in incentives to consume more of them. That is especially true in North America where fossil fuels are cheap. The sense of entitlement which has developed to cheap fossil fuels is an inducement to drive and thereby emit greenhouse gases. Air fares that exclude harmful consequences of combusting jet fuel are incentives to fly, which amounts to discounting the danger of fossil-fuelled climate change.

Subsidizing Carbon Pollution

By refusing to include costs of pollution in prices of fossil fuels, carbon polluters are free-riders benefiting from unpaid costs of their fossil-fuelled practices that will be paid later as damage to the environment, to the health of others, and eventually closing off the atmosphere as a free pollution sink as it fills with carbon. Or it will have to be paid later through particularly high-carbon taxes needed to avoid further harm because carbon taxes were refused as emissions accumulated. Nordhaus (2013: 214) argues that 'those who burn fossil fuels are enjoying an economic subsidy – in effect they are grazing on the global commons and not paying for what they eat. … allowing firms to put carbon in the atmosphere cost-free is a similar valuable subsidy – it is the right to harm others'. Currently, polluters are subsidized by victims who will belatedly pay global warming costs.

Often one hears complaints about subsidies needed to start up emissions-free energy like wind or solar, but rarely is it understood that fossil fuels involve subsidies to users, with the externalized costs paid by victims in damages to their health or environment. Users appropriate the atmosphere as a pollution sink to dump greenhouse gases free of charge, thereby harming innocent bystanders distant in space and time. Global externalities are the intervening variables which carry a social relation between (a) companies, countries, and individuals that use carbon-emitting fossil fuels whose cost is only partially paid by the user and beneficiary, and (b) those who will have to pay the remainder of the cost by suffering the consequences of fossil-fuelled global warming, whether they be environmental or economic.

Although carbon pollution is invisible, it is not cost-free. It has a cost that someone has to pay, either users of fossil fuels or belatedly by victims. Excluding from the price of fossil fuels their cost of degrading the environment, which future generations, poorer countries, and other species need, constitutes a subsidy these victims are paying. This maintains the relatively low price of fossil fuels. Elucidating this shifts debate away from pious platitudes about climate change to analysis of who pays the pollution costs of fossil fuels. This generation of the affluent has hitherto enjoyed a discount for fossil fuels that will be paid later as costs of climate change. They won't like to hear they are passing costs they avoided onto their grandchildren. This is an uncomfortable truth for users and extractors of fossil fuels to hear, but shying away from it deepens the chasm between environmental discourse and environmental practices. Failure to understand how present fossil-fuel prices only cover part of their cost, with the rest being eventually paid by victims of fossil-fuelled global warming, is an important part of the failure to mitigate global warming.

In modern societies, everyone uses fossil fuels, but the difference between big and small users and beneficiaries must not be ignored. The bigger the carbon polluter, whether they be individuals, companies, or countries, the more they appropriate embedded subsidies based on the externalization of the costs of their carbon-polluting practices to innocent bystanders, such as poor countries and future generations. It is the oligopolistic industrialists, such as the Koch brothers in the USA, who became wealthy on the basis of unpaid externalized costs of fossil fuels. They lead the resistance to including the full cost of fossil fuels in their price.

The Full Cost of Fossil Fuels Left Unpaid and the Growing Carbon Debt

Whereas early industrial fossil-fuel combustion created local pollution which fouled nests of polluters, late modern fossil-fuel combustion emits enormous amounts of invisible greenhouse gases harming the vulnerable, even those distant in space and time and other species. The atmosphere and oceans are a commons shared by everyone, including future generations. They constitute a medium that carries social relations of unpaid costs between fossil-fuel users and victims of carbon pollution. That commons is being treated as a greenhouse-gas dump. In addition to consequences mentioned earlier, others include diminished services of nature such as glacier run off, weakened capacity of the Arctic Ocean to reflect sunlight, and of massive land-based glaciers in Antarctica and Greenland to hold water and limit ocean levels, lessened ability of permafrost to safely store methane, etc. Such effects amount to the high cost of cheap fossil fuels, to paraphrase Carolan (2014, 2018), where cheap price results in excessive consumption and the remaining cost is paid belatedly by victims of global warming.

The atmosphere and oceans are finite, so as they fill with carbon and consequences become more acute, costs of dealing with fossil-fuelled climate change become higher the longer carbon accumulates. If the full costs of fossil fuels are not paid upon use, this constitutes a deficit. Its accumulation, evidenced in terms of greenhouse gases in the atmosphere, amounts to a growing environmental debt owed by fossil-fuel users—whether they be companies, countries, or individuals in proportion to their fossil-fuel use—to victims (future generations, affected low users, etc.). Worse still, there is the equivalent of interest on the unpaid debt. 'Had serious climate action begun in 2010, the cuts required per year to meet the projected emissions levels for 2 °C and 1.5 °C would only have been 0.7 per cent and 3.3 per cent per year on average. However, since this did not happen, the required cuts in emissions are now 2.7 per cent per year from 2020 for the 2 °C goal and 7.6 per cent per year on average for the 1.5 °C goal. Evidently, greater cuts will be required the longer that action is delayed' (UNEP 2019: X.5). Refusal to reduce emissions now results in the next generation having the burden of making much greater reductions, with all the costs and changes in fossil-fuelled practices that involves. 'Fossil CO_2 emissions from energy use and industry, which dominate total GHG emissions, grew 2.0 per cent in 2018, reaching a record 37.5 GtCO2 per year. There is no sign of GHG emissions peaking in the next few years; every year of postponed peaking means that deeper and faster cuts will be required. By 2030, emissions would need to be 25 per cent and 55 per cent lower than in 2018 to put the world on the least-cost pathway to limiting global warming to below 2 °C and 1.5 °C respectively' (UNEP 2019: V.1).

There is general consensus among scientists that if carbon emissions had been reduced a decade ago, it would have been easier and less costly; if they had been reduced two decades ago, it would have been even easier and much less costly; and if the reduction is done a decade from now, it will be more difficult and expensive than doing it now. This indicates that a carbon debt is accumulating and growing because of fossil-fuel use. Fossil-fuel extraction will have to slow down and restraint practiced on using fossil fuels. And carbon capture and storage underground of some sort (afforestation, technical CCS, etc.) is needed. The debt takes many forms in nature. The capacity of oceans to absorb greenhouse gases decreases as more is absorbed in a process of saturation and diminishing returns. Global warming dries out forests, fosters insect infestations killing trees, and makes wildfires more likely, thereby emitting instead of absorbing carbon. It melts permafrost releasing the potent greenhouse-gas methane, and melts the reflective ice cover of the Arctic

Ocean thereby absorbing more of the sun's rays. The more emissions have accumulated, the more difficult and costly it is to scale up carbon capture and storage (CCS) and direct air capture (DAC).

Energy is not free. Every type of energy has a cost. Costs differ in kind. Some costs are paid in monetary terms; others are incurred in the form of harmful consequences of the energy used. Some are paid on use upfront, but others are deferred to be paid later. Some costs are fully paid by users of the energy; others are left in large part unpaid by users. The energy costs that are paid are usually those that make the energy ready for use, whereas those that remain unpaid to be paid later are typically those that take the form of charges for its harmful consequences. Some costs are paid by those who benefit from the energy, whereas other costs are paid by those who, for whatever reason, are excluded from the benefits of the energy used. The full costs of the energy used are paid eventually in one form or another. Will they be paid on use by the beneficiaries or deferred to be paid later in the form of harmful consequences for those who receive little or no benefits from its use? Energy is material, taken from nature and returned to nature in a more or less modified form after use, and energy is conserved. Energy can be taken from one place, where nature stored it safely, and transferred to another, where it has dangerous consequences. Fossil fuels are taken from underground, combusted with oxygen to form carbon dioxide, and emitted into the atmosphere where it causes a greenhouse effect and dangerous global warming for centuries. This can close off benefits from nature's services to others, which amounts to a cost to them.

Some types of energy have more pollution costs than others, for example, fossil fuels compared to solar and wind energy. If energy is carbon free (solar, wind), then there are no global warming externalities and the full cost is paid by the user upfront. Fossil fuels are more costly in the long run than low-carbon energy like wind and solar. Pielke (2010: 230) notes that carbon-neutral energy has been well under 10% of global consumption and this must rise to over 90% to stabilize the atmosphere's carbon content at a low level. '[A] political commitment to leaving fossil fuels in the ground will likely have to accompany innovation of alternative sources of energy' (Pielke 2010: 244). Prins et al. (2010: 24) argue that 'decarbonisation of energy supply is the only long-term approach that can deliver a radical acceleration of decarbonisation of economic activity'. If that is done, fossil fuels will remain safely stored underground and nature will heal itself through vegetation growth, oceans, soil, etc., absorbing carbon from the atmosphere, thereby reducing the carbon debt. Reducing inflow of carbon into the atmosphere decreases the carbon debt to be paid by future generations.

This involves the issue of how fossil-fuelled climate change is related to processes of market monopolization and exclusion, that is, social closure. Although the Anthropocene is usually portrayed in terms of human impact on the planet and its climate, humans are enormously differentiated according to their impact. The more a company, group, or individual appropriates wealth and power, the more it tends to use resources, especially fossil fuels, and impacts the biophysical environment, notably by emitting greenhouse gases into the atmosphere as a carbon sink. The social practices of wealthy individuals disproportionately cause unpaid damage to the atmosphere. Flying frequently in a private jet combusts much more jet fuel per capita and therefore greenhouse-gas emissions than not flying or rarely flying or doing it in economy class packed like sardines. Air conditioning, heating and lighting a huge mansion uses more electricity, often supplied by fossil fuels, than a bungalow or apartment. The resulting pollution contributes to closing off opportunities to latecomers and other species. Most important is the concentration of decision-making power and benefits concerning fossil-fuel extraction and combustion. The Koch brothers in the USA made a fortune from fossil fuels in companies that were among America's worst polluters by profiting from unpaid costs of carbon, then used it to finance American politicians who voted to eliminate environmental regulations that would decrease carbon pollution, such as placing a price on carbon emissions (Schulman 2014; Mayer 2016). They financed think tanks to undermine climate science and measures to address global warming that would diminish their fossil-fuel profits (Jaques, Dunlap, and Freeman 2008; Dunlap and Jaques 2013; Dunlap, McCright, and Yarosh 2016). The world's 3000 biggest corporations offloaded US$2.2 trillion in environmental damage in 2008 (UNEP 2010). Investors in carbon-polluting companies profit from externalities and the unpaid cost of their commodities. Carolan (2014: 4) argues that 'in a free market economy the incentives collectively encourage firms to externalize costs, as many as they can get away with' and that its proponents 'strive to make society pay for the costs of this economic system while concentrating its benefits within the hands of a few' (Carolan 2014: 3). Corporations are the principal beneficiaries of cost offloading, but fossil-fuel consumers are also beneficiaries of prices lower than the full cost of fossil fuels. Carolan (2014: 15) contends that 'the affordability of cheap goods and services is an illusion: those low retail prices are costing others – now and in the future – dearly'. His argument could be illustrated in terms of vehicle driving practices. Cheap gasoline is an incentive to drive gas-guzzling vehicles, resulting in more emissions and costly global warming, drought, wildfires, floods, etc.

Heal (2017: 171) concludes that 'we are leaving our successors less and poorer natural capital – a world with a less stable and hospitable climate, fewer species, less water, and fewer of many other environmental assets. Perhaps this is condemning them to an impoverished lifestyle'. It involves the risk of excluding future generations from the benefits of cost-free waste sinks that the present generation enjoys, and foisting onto the former the dangers and costs of more frequent and intense hazards of nature like hurricanes, floods, and wildfires. The yearly carbon deficit of emissions over withdrawals is huge, hence the accumulated carbon debt is enormous. 'Climate policies that are consistent with the 1.5 °C goal will require upscaling energy system supply-side investments to between US$1.6 trillion and US$3.8 trillion per year globally on average over the 2020–2050 time frame, depending on how rapid energy efficiency and conservation efforts can be ramped up' (UNEP 2019: XII.8; see also IPCC 2018). Stabilizing the atmosphere's carbon content at a level where it does not cause dangerous global warming requires the challenging task of eliminating the environmental deficit and reducing the carbon debt, hopefully paid by companies and individuals that have benefited from it the most. As Hawken (2017) documents, this would result in long-run savings, both environmental and economic, much like paying down the principal on a debt.

Resistance to Full Cost Pricing of Fossil Fuels

Economists argue that no 'single policy would make a greater contribution to solving the climate problem than making coal users pay the full costs of their actions. Full cost accounting doesn't sound like a revolutionary concept – but it is. This is why the coal industry fights it so hard. It's good for the rest of us, good for nature, but bad for them' (Heal 2017: 186). This is true for all fossil fuels. 'Policies to slow climate change are economically simple if politically difficult. They involve raising the price of CO_2 and other greenhouse gases and harmonizing the price across countries' (Nordhaus 2013: 316). Politically difficult to be sure, since that involves including belated environmental and health costs into fossil-fuel prices and all goods using them for their production, called Scope 3 emissions. Solutions to fossil-fuelled global warming that economists find are cheapest to implement and result in most benefit to the economy, like putting a price on carbon pollution, are the most visible to fossil-fuel producers and consumers, hence face the most resistance. Transitioning to low-carbon energy is especially challenging for states that made their economies dependent on the extraction and export of coal and high-emissions crude oil.

Including the full costs of fossil fuels into their price is simple, effective, and inexpensive only if it is universal (Nordhaus 2013). Carbon pricing must be enforced internationally, otherwise companies and countries including costs of environmental damage in prices of their products would be disadvantaged in market competition with those that externalize their pollution costs. A global problem like fossil-fuelled climate change requires a global agreement negotiated, accepted, and enforced by all nations, including fossil fuel extracting and exporting countries. It could be called Paris 2.0. But this will be difficult: policymaking is easy; policy-implementation is hard.

Natural scientists have documented the biophysical dynamics of the fossil-fuelled climate crisis. Economists have documented the cost externalities incentivizing the massive use of fossil fuels, hence the unleashing of global warming, and proposed the disincentive of pricing carbon pollution. This book analyses sociopolitical resistance to solving the problem despite available solutions. There are no easy answers for implementing solutions because the most plausible ones require an increase in the price of fossil fuels that provide the energy for social practices people enjoy or aspire to. Raising prices of fossil fuels to include their full costs confronts powerful fossil-fuel lobby groups and the pre-existing sense of entitlement to cheap gasoline and air fares based on inexpensive jet fuel, low-cost electricity, etc. Feelings of sacrifice concerning practices like driving, flying, cruising, cremating, etc., are likely if and when the full cost of fossil fuels is included in their price. Moreover, it is hard for fossil-fuel users to accept they are underpaying for fossil fuels, hence for gasoline, plane and cruise tickets, goods transported from afar, electricity, cloud servers, etc., and to acknowledge that the remainder will be paid by future generations as costs of environmental damage.

Including the cost of carbon pollution in the price of fossil fuels is resisted by fossil-fuel companies and fossil-fuel consumers especially if the full cost consists of an amount effective in bringing emissions in line with carbon withdrawals. The amount of the full cost is not known precisely, and can only be estimated until the damage caused by fossil fuels occurs. Hence a carbon tax functions as a proxy for costs that will occur later. Usually the more in-your-face the carbon tax is, the more it will be resisted. One attempt to dampen resistance is to place the tax upstream where fossil fuels are extracted, upgraded, and refined. Most of it will, nevertheless, be passed on to consumers. Fossil-fuel companies have power to lobby governments to keep it low, claiming that is needed for them to compete with companies in countries with no carbon tax, or else jobs will be lost.

Environmental economists suggest escalating carbon pollution taxes included in fossil-fuel prices, and in goods based on them. This would replace pollution costs being externalized and paid in terms of environmental damage suffered by others. The sociologist Giddens (2009) suggests indirect measures, as Europe did promoting energy independence and energy security by imposing taxes on imported fossil fuels to finance development of nuclear, wind, and solar energy. This had the side effect of decreasing greenhouse-gas emissions. The political scientist Pielke (2010) suggests an escalating carbon tax dedicated to technological development of clean, renewable energy cheaper than fossil fuels. Valuable as these suggestions are, they all confront the sense of entitlement to cheap fossil fuels, which is especially strong in the USA and Canada. It is extremely difficult to raise fossil-fuel prices everywhere.

Including the costly consequences of fossil fuel use in their price is promoted by economists for its incentive to reduce emissions and innovate, its efficiency, relatively low cost, and low level of government regulation. The Canadian Ecofiscal Commission estimated that, for Canada to meet its 2030 targets, it would have to quadruple its carbon tax to 40 cents a litre for gasoline (with rebates for most low polluters), but the economy would grow by 1.37% annually compared to 0.81% with industry-specific regulations and subsidies (Walsh 2019: A4). You would think this broad-based tax would be the solution of choice for business and conservative politicians with the foresight to take the problem seriously. A few have chosen it in a few countries. But many oppose its implementation: Republicans in the USA and conservative parties in Canada. Their priority is near-term economic and political gain, even at the risk of long-term environmental and economic pain. They claim to have alternatives, but they invariably amount to watered-down, less effective remedies and therefore to discounting future harm. Canadian conservative politicians oppose carbon taxes using the catchy slogan 'punish polluters not commuters' going about their everyday lives. They don't deny anthropogenic climate change but align themselves with voters who avoid responsibility and want someone else to pay to mitigate it. Thus they propose regulating big polluting industries. Valuable as that is, it ignores several deficiencies. First, costs will be passed on by big polluters to consumers, so it won't be cost free after all for commuters. Second, if costs of carbon pollution remediation are paid only by big polluters rather than also by the huge number of small polluters, costs will be enormous for big polluters if a country is to meet its Paris commitments, and they will respond by moving to less environmentally strict jurisdictions, cutting jobs, etc. Third, conservatives typically do not want government regulation of the 'free' market. It is logical to suspect they want mere token regulation, which will fail to reduce emissions. The hidden agenda behind the 'punish polluters not commuters' is: do not make big polluters reduce emissions much, do not make commuters (all other carbon polluters) decrease their emissions, and therefore do renege on the 2015 Paris commitments. Token measures, which give the impression of doing something about climate change, solve the paradox of free-market conservatives preferring government regulations rather than effective market-based remedies.

Pielke's (2010) solution advocating a low, upstream carbon tax to fund innovation is aligned with proposals of conservative politicians who are afraid to tax voters. Alberta is a test case. Leach (2019) demonstrates these conservative proposals could backfire on Alberta because of its huge, high-emissions oil sands industry and its heavy reliance on coal-fired electrical power: 'Large facilities in Alberta produce more emissions than large facilities in all other provinces combined … The rest of Canada might find that a policy focussed on large emitters (polluters) in Alberta is preferable to carbon pricing on themselves (commuters), and that would be a lot worse for Alberta than any carbon tax we've seen proposed'. It likely would result in the oil sands industry, which is already high cost, being no longer profitable nor viable.

Some progressives on the left also oppose pricing carbon pollution, arguing it will hurt the poor and claiming that rebates are lies (Abela 2019). This is part of their opposition to all consumption taxes, such as value-added taxes, sales taxes, taxes on cigarettes and alcohol, etc., as regressive since the poor spend a higher proportion of their income on consumption. The solution of the left is to tax big fossil-fuel polluting companies, which is the same as the solution from the right. That should be done as much as possible but it is dubious that mitigation at the scale and urgency needed can be done without all fossil-fuel beneficiaries, big and small, paying for their share of pollution because of the huge number of small polluters. Moreover, the vulnerable poor will be most affected by climate change. Making exceptions would be a poor solution. It would be better to give some of the revenue raised back to low-income citizens so they won't be hurt by carbon taxes and can benefit financially by finding ways to diminish their fossil-fuel consumption. Implementing the disincentive of carbon taxes and using money raised to offset the cost to the poor is the solution of social democratic countries, which have the best track record for both reducing poverty and combatting climate change. Similarly, cap-and-trade for acid rain was initially opposed by environmentalists as institutionalizing the right to pollute. True, but the cap limits it, pollution shares sold or auctioned by governments make pollution costly, a decreasing cap phases out the pollution, and most importantly, it has been successful. Placing an appropriate price on all carbon pollution is needed so that (i) the price of fossil fuels reflects their full cost, (ii) those who benefit from carbon pollution pay, (iii) research to improve low-carbon energy is financed, and (iv) the carbon debt that future generations have to pay is reduced, and hopefully eliminated

Does Supply or Demand Generate Fossil-Fuel Use and Emissions?

Royal Dutch Shell PLC is the world's second-largest listed energy company, the biggest liquefied natural gas trader, and the largest dividend payer. It has been a target of European climate protests despite its chief executive advocating action on global warming and his company aspiring to halve the intensity of its emissions by 2050. Shell has been targeted because oil and gas remain the backbone of its profits, because its absolute carbon dioxide emissions rose by 2.5% from 2017 to 2018, because its goal seeks merely to reduce its emissions intensity which has typically gone hand in hand with increases of absolute emissions, and because it continues to invest in new fossil-fuel projects that need to operate for decades to be profitable. That chief executive worried that what he called 'the demonization of oil and gas' would hurt investment in Shell and labelled as an unjustified 'red herring' criticism the company's reserves would become economically unviable and stranded if adequate action is taken to reduce emissions. He contended it is 'entirely legitimate to invest in oil and gas because the world demands it. … We have no choice' because demand continues to increase and switching to cleaner energy will take decades (quoted in Bousso and Zhdannikov 2019: B8). That is not good news for the urgent problem of carbon dioxide emissions cumulating and remaining in the atmosphere for a century or more causing more and more global warming.

The demand for fossil fuels is used by extracting companies to legitimate extraction, as if it were a given, but that is a misleading oversimplification. Profit-seeking companies increase demand through advertising products and services requiring fossil fuels, exploiting new deposits to increase supply thereby decreasing price, innovating new fossil-fuelled products and services, etc. For example, companies increased discount and intercontinental jet-fuelled flights, bunker-fuelled cruises and shipping, innovated massive data storage clouds powered by energy-glutton servers, gas-guzzling sports utility vehicles, fossil-fuel powered military equipment with precision destructive capability, higher concrete skyscrapers, and the like. The promotion of these strike a responsive chord in consumers that stimulates demand, but it is an oversimplification to consider only demand without taking into account the effect of increasing supply of all these. In their supply-side analyses, Schnaiberg (1980) and York (2017) correct the imbalance by conceiving of demand as a function of production and supply rather than as a free-floating explanatory factor. Sim (2012) argues that supply of fossil fuels drives their use, and gives the example of the anticipation of taxes and regulations leading oil companies to exploit their reserves quickly before taxes and regulations kick in, which he calls a 'green paradox'. The most significant stimulus for demand of fossil fuels is what is discussed in this chapter, namely selling them at below-cost prices by externalizing to victims carbon pollution costs. In a market economy, consumption is price sensitive. Excluding costly harms of a product from its price drives up demand. That was learned for cigarettes and is true for fossil fuels. Demand for fossil fuels is increased by having users pay only part of their cost upon use, with the rest of the full cost paid belatedly by victims in terms of expensive damage of fossil-fuelled global warming. Contrary to Shell's claims, we do have the choice of reducing demand for fossil fuels, technologically by innovating low-carbon energy and socioeconomically by exercising restraint in fossil-fuelled social practices through pricing carbon pollution, global warming education, moral persuasion, etc.

Stranded Assets or Hazards Stored Safely in the Ground?

The most challenging finding of natural science is that global warming is caused by carbon emissions exceeding withdrawals, and that with current technology, mitigation requires leaving most fossil fuels stored safely in the ground, and certainly not rapidly extracted and combusted as is presently occurring. Even limiting global warming to 2° C would require 'leaving most of the world's fossil fuel reserves plus all the unconventional resources in the ground, at least until carbon capture technology is cheaply available' (Berners-Lee and Clark 2013: 44). Effective, inexpensive carbon capture and storage technology (CCS) is a prerequisite for continuing fossil-fuel extraction and combustion. Timing is a significant issue. To limit global warming to 2 °C, most fossil-fuel reserves must be left in the ground now, and only combusted after carbon capture and storage is deployed. Current extraction and combustion practices do the opposite, namely accelerating them now presuming that cheap, effective CCS and direct air capture technologies (DAC) will be globally implemented in the future.

Unless a miraculous immediate technological breakthrough occurs, the only way to align emissions with carbon withdrawals is by radically reducing global combustion, which implies leaving fossil fuels in the ground, first the most polluting ones, then those that are less harmful. To keep under the 2 °C limit, most present coal reserves and almost half of oil and gas reserves globally must remain stored safely underground until 2050 (McGlade and Ekins 2015: 187–190). The goal is to slow down combusting fossil fuels so that emissions decrease to approach the withdrawal rate, and ideally decrease combustion below the rate of withdrawal so that heat-trapping carbon in the atmosphere declines, as does the carbon debt, and global warming decelerates.

Mark Carney, Governor of the Bank of England and Chair of the Financial Stability Board Task Force on Climate-related Financial Disclosures, stated that a carbon budget consistent with a 2 °C target 'would render the vast majority of reserves 'stranded' — oil, gas and coal that will be literally unburnable without expensive carbon capture technology, which itself alters fossil fuel economics' (quoted in Carbon Tracker

2015). Greater fuel efficiency, a glut of oil because of fracking, and decreasing cost of solar and wind energy have led to predictions that oil must be priced in the future between US$9 and US$20 a barrel to be competitive (Lewis 2019), which will not be possible for high-cost extraction reserves. Nevertheless, oil and gas companies approved $50 billion worth of extraction projects internationally, including big ones in Alberta's oil sands by ExxonMobil that will be profitable only if current emissions policies result in demand leading to a 2.7 °C increase or more, are maintained (Carbon Tracker 2019). They are betting their investments that the world will ignore the Paris commitments. They may win their bet. Low cost of fossil fuels will stimulate demand and combustion, and outcompete wind and solar energy in price. As long as an externalities-free price is paid for fossil fuels by users, with the remainder paid belatedly by victims, oil reserves are more likely to be combusted than stranded. Whether it be by government regulations requiring oil's full cost be paid by users up front or by industry developing innovative clean alternative energy, keeping global warming to 2° C will require a dramatic reduction in its global combustion: by 2050 in some estimates (Jaccard 2018: A13) and 2040 in Rockström et al.'s (2017) estimate. But 2040 and 2050 are only a couple of decades away, less than the lifetime of a pipeline or an oil extraction facility. If a radical reduction in demand does occur, companies extracting oil and building pipelines will have difficulty adapting to this new normal.

Fossil-fuel extracting countries and communities are even less flexible than companies because they are territorially anchored. The temptation is great for countries with fossil-fuel reserves, for communities nearby, and for the industry to extract and use the harmful asset quickly for fear of being forced by government policy or technological alternatives to keep it underground as a permanently stored liability. Concern about fossil-fuelled climate change, especially if there is the innovation of 'lower-cost alternatives to fossil fuels may have the perverse effect of motivating an accelerated extraction of fossil fuels as owners of those resources see no better time than the present to capitalize on their value, as the resources will only be worth less in the future. … What this means is that some form of political commitment to leaving fossil fuels in the ground will likely have to accompany innovation of alternative sources of energy' (Pielke 2010: 224). Full-cost upfront pricing of fossil fuels paid by users would be one form.

Fossil-fuel companies and countries having fossil-fuel reserves are, however, reluctant to leave this fossil-fuelled basis of prosperity in the ground, thereby becoming a stranded asset. The slogan of supporters of the American Republican president is that 'Trump digs coal'. These values go well beyond Trump, as shown by Republican senator Ted Cruz's claim that the USA is the Saudi Arabia of coal and will exploit the resource. The Obama Administration reduced coal for generating electricity by a policy of emissions controls and by promoting hydraulic fracturing to replace coal with less polluting natural gas, but natural gas still emits greenhouse gases, and the controls were subsequently rescinded by the Republican Administration. Governments of fossil fuel rich countries which acknowledge global warming have much difficulty reconciling economic prosperity with the need to phase out fossil fuels, and usually prioritize near-term economic benefits over long-term environmental and economic benefits. Canada's Prime Minister Justin Trudeau stated in January 2017 that the oil sands should be phased out, a statement based on what science demonstrated is needed to act responsibly for safety and sustainability. When he was subsequently criticized by the oil industry and Alberta, however, he quickly claimed he misspoke. He now contends that extracting, combusting, and exporting an increasing amount of Alberta's high emissions, high-cost bitumen can be reconciled with mitigating climate change. Despite good intentions, Trudeau, the Canadian government, and Alberta's previous social democratic government have been captured by the bituminous oil industry. In early March, Trudeau received an award and a standing ovation from oil and gas executives at a conference in Texas for approving pipelines and stating that 'no country would find 173 billion barrels of oil in the ground and just leave them. … The resource will be developed. Our job is to ensure that this is done responsibly, safely, and sustainably' (Trudeau 2017). He omitted the most important prerequisite for doing extraction responsibly, namely do it 'slowly'. How can that job be done when the export by his country of tar sands oil is being increased from 3 million barrels a day to 4 million or more, when its extraction from bituminous sand and upgrading cause more emissions than extracting oil from a well or from the North Sea (see Fig. 2.1 and Table 2.1 in Chapter 2) and when it constitutes the fastest growing source of greenhouse-gas emissions in Canada? This is not just true of Canada. 'To tackle climate change we'll need to focus our attention on slowing the flow of fossil fuels into the global economy. That will require global politics, since no profit-seeking company or self-interested nation is going to abandon its fossil fuel reserves voluntarily' (Berners-Lee and Clark 2013: 81).

However, there are a few exceptions to the refusal to abandon reserves. Norway has an estimated 1.3 to 3 billion barrels of oil and natural gas, worth at least US$82 billion at current prices, near its Arctic Lofoten Islands. It banned drilling in that archipelago because an oil spill would cause disastrous damage in an environmentally sensitive area and because of its Paris climate accord commitments. Surveys show that 44% of Norwegians are willing to leave some oil in the ground to reduce emissions (Holter 2017). Exceptional though it may be, Norway is a country with a vast oil resource there that it is leaving in the ground. At least so far. But leaving it in the ground is a struggle even in Norway, whose oil industry employs 200,000 workers and oil extraction constitutes 12% of its economy. Norway's two largest political parties, backed by the oil industry and its unions, favour oil exploration there, arguing that its North Sea oil resources are being depleted and need to be replaced by new extraction sites for the prosperity of the industry. The issue is whether hydrocarbons left safely in the ground can be made permanent and scaled up globally to meet the carbon emissions/withdrawal balance until technological innovations make fossil-fuels emissions-free. The Obama Administration signed an executive order placing the bulk of American waters in the Arctic off limits for oil and gas leasing, but these regulations were reversed by the Trump Republican Administration.

Will coal, heavy oil, bitumen and eventually light sweet oil and natural gas be viewed as dangerous liabilities safely stored underground or continue to be seen as valuable assets to be exploited quickly and thoroughly rather than stranded in the ground? Keeping a valuable resource, whose use results in harm to the environment or to health, in the ground constitutes conservation but faces strong opposition from consumers and especially from extracting companies and countries for whom it brings major economic benefits. Convincing decision-makers and populations in countries with fossil-fuel deposits to leave them in the ground because they cause too much harm is particularly challenging. Even persuading them to slow down extraction, until technology is developed to make combustion less harmful, is difficult because they would receive the benefits now and the costs of harmful fossil fuels will be paid belatedly by someone else.

The Ineffectiveness of Additionality, Offsetting, and Incrementalism

Three widely used strategies to mitigate anthropogenic global warming are deficient. First, adding low-carbon energy to high-carbon fossil fuels does not mitigate global warming. The media regularly report new innovations in solar and wind energy, electric vehicles, energy storage, etc., that make them more efficient and less expensive. Those innovations, welcome as they are, should not mislead into believing global warming is being solved. Low-carbon energy remains a tiny proportion of the world's supply, but even if a significant amount is added, global warming will continue to worsen if greenhouse-gas emissions are not reduced to carbon withdrawal rates (York 2012). Although the development of low-carbon energy is important for economic growth and to avoid making global warming extremely bad as economies grow, it is the reduction of greenhouse-gas emissions to withdrawal levels that is crucial for mitigating global warming. Replacing fossil fuels with low-carbon energy is key, not adding clean energy to carbon-polluting energy, which means keeping carbon safely in the ground instead of emitting it into the atmosphere. That is, however, technically and socially difficult to achieve in a world with a sense of entitlement to cheap fossil-fuel energy.

A second mitigation strategy involves offsetting. It is usually proposed when the only other way to reduce emissions is restraint concerning fossil-fuel use and when the polluting industry, consumers, and political leaders don't want that. For example, aviation relies totally on fossil fuels. Jet fuel is the only power source capable of lifting a plane as heavy as a Boeing 747 up 10,000 metres high and propel it at almost 1000 kilometres per hour over enormous distances, yet be flexible and cost-effective.

Solar and wind energy stored in batteries aren't up to the job for the foreseeable future. However, it has been estimated that a return flight from New York to San Francisco emits 0.9 metric tonnes of carbon dioxide *per passenger* (Schlossberg 2017), which is about 2000 pounds, so per-person emissions for that one flight are about ten times their body weight. This average varies according to the fuel efficiency of planes, short-haul and long-distance flights because most jet fuel is combusted on lift and descent rather than cruising, seats in spacious business class as opposed to passengers packed in economy (the former accounting for 3 to 9 times more emissions per seat than the latter), etc. There are about 20,000 planes in the world flying continuously, and it is estimated that this number will rise to 50,000 by 2040. Therefore, the carbon dioxide emissions from flying are enormous. And cargo planes and military planes must be taken into account. Offsetting emissions from planes is the only solution if societies refuse to decrease travel by plane. Current offsetting practices involve a small voluntary fee of about $10 used for something of dubious effectiveness such as planting a tree. How could that offset 2000 pounds of carbon dioxide emissions per passenger for the return trip from New York to San Francisco taking into account logging, forest fires, insect infestations, etc., that would move the carbon back from tree to sky. Paying the full cost of flying through offsets of harm caused by jet fuel emissions would be expensive. Critics (Anderson 2012) contend offsetting doesn't work and misleads well-intentioned people to believe the problem is being solved and flying restraint unnecessary. London's Heathrow Airport is investing in restoring peat land near Manchester to offset Heathrow's emissions and make its airport carbon neutral by 2020 (Reality Check team 2019), but Heathrow's initiative was criticized as greenwashing to justify increased flying. Offsetting has been used to avoid emissions reduction through carbon taxes, less nonessential flying, etc. In 2020, both in Trump's speech at Davos and his State of the Union Address, he promised to plant a trillion trees, which is a cute slogan. To mitigate the fossil-fuelled global warming crisis, planting trees to draw down atmospheric carbon has to be in addition to emissions reduction measures, which Trump dismantled.

A third mitigation strategy fails because of the urgency of the problem. The UNEP 2019 Emissions Gap Report concludes that 'we are talking about transformational change now – incremental change simply will not make it. We simply need to transform societies in the next 10 years' (quoted in Farge and Nebehay 2019: A4). Since carbon dioxide remains in the atmosphere a hundred or more years, even completely halting emissions now is like closing the barn door after the horse has escaped. Fossil-fuelled global warming is melting the Arctic ice cover and the permafrost where methane has been safely stored by nature. The permafrost is huge, consisting of about 15 million square kilometres of land mass. Even limiting global warming to 2° C would eventually melt over 40% of this permafrost (Chadburn et al. 2017). Suppose anthropogenic global warming continues until the white Arctic ice cover is replaced by dark water and a huge area of permafrost where carbon had been safely stored had melted. Now suppose after this occurred, by 2050 as an oft-quoted date, emissions from combusting fossil fuels were entirely eliminated, either through technological innovations or social innovations like abandoning fossil fuels, or carbon neutrality and net zero achieved (anthropogenic withdrawals equaling anthropogenic emissions). Would global warming stop? No. It would continue because the dark water of the Arctic Ocean would then absorb more solar energy instead of it being reflected back into space by white ice (albedo effect). It would continue because the long-trapped massive amounts of methane in permafrost would be released into the atmosphere. The human combustion of fossil fuels causing first-order global warming opens the door and unleashes runaway second-order warming by nature's dynamics, even if the first-order warming is belatedly stopped and the door closed. This illustrates the urgency of preventing anthropogenic global warming and demonstrates the appropriateness of the barn door is closed only after the horse escaped analogy. Escalator carbon taxes or progressively tightening carbon caps would have significantly reduced fossil-fuelled climate change if implemented 30 years ago at the initial climate conferences. But now in 2020, after carbonizing the atmosphere for three more decades and continuing until escalator carbon prices push emissions down to withdrawal rates, it will only make a dangerous situation less bad because of accumulated CO_2 in the atmosphere. Of course, less bad is not to be dismissed as worthless. Socially acceptable and omnipresent incremental remedies allow carbon emissions to exceed withdrawals for a long time, with global warming worsening (Latin 2012). Incremental remedies are belated remedies, which come at great long-run cost. Emissions need to be reduced to withdrawal rates promptly so that more carbon does not continue to accumulate in the atmosphere. Moreover, it is necessary to be wary of carbon pollution salespersons who claim they will in the future reduce emissions, such as carbon neutrality by 2050, using carbon capture and storage (CCS) or other technological innovations but who refuse to require efficient CCS now as a precondition for current fossil-fuel extraction and combustion.

References

Abela, P. 2019. Why Progressives Should Reject the Carbon Tax. *Globe and Mail*, 6 May. theglobeandmail.com/opinion/article-why-progressives-should-reject-the-carbon-tax/. Accessed 8 May 2020.

Agyeman, Julian, R.D. Bullard, and Bob Evans. 2003. *Just Sustainabilities: Development in an Unequal World*. Cambridge, MA: Earthscan/MIT Press.

Anderson, Kevin. 2012. The Inconvenient Truth of Carbon Offsets. *Nature* 484 (7392).

Baumol, William J., and Alan S. Blinder. 2010. *Economics: Principles and Policies*, 11th ed. Mason, OH: South-Western Cengage.

Berners-Lee, M., and D. Clark. 2013. *The Burning Question*. London: Profile.

Bousso, Ron, and Dmitry Zhdannikov. 2019. Oil and Gas to Offer Opportunities for Decades to Come, Shell CEO Says. *Globe and Mail*, 16 October: B8.

Bullard, R.D. 2000. *Dumping in Dixie: Race, Class and Environmental Quality*, 3rd ed. Boulder: Westview Press.

Bullard, R.D. 2005. *The Quest for Environmental Justice: Human Rights, and the Politics of Pollution*. San Francisco: Sierra Club Books.

Bullard, R.D., and B. Wright (eds.). 2009. *Race, Place and Environmental Justice After Hurricane Katrina*. Boulder: Westview Press.

Carbon Tracker. 2015. *Mark Carney Warns Investors Face Huge Climate Change Losses*. 30 September. https://carbontracker.org/mark-carney-warns-investors-face-huge-climate-change-losses/. Accessed 15 April 2020.

Carbon Tracker. 2019. *Oil and Gas Companies Approve $50 Billion of Major Projects that Undermine Climate Targets and Risk Shareholder Returns*. 6

September. London. https://carbontracker.org/oil-and-gas-companies-approve-50-billion-of-major-projects-that-undermine-climate-targets-and-risk-shareholder-returns/. Accessed 15 April 2020.

Carolan, Michael. 2018. *The Real Cost of Cheap Food*, 2nd ed. London: Routledge.

Carolan, Michael. 2014. *Cheaponomics: The High Cost of Low Prices*. Abingdon, UK: Earthscan.

Chadburn, S.E., E.J. Burke, P.M. Cox, et al. 2017. An Observation-Based Constraint on Permafrost Loss as a Function of Global Warming. *Nature Climate Change*, 10 April. https://doi.org/10.1038/nclimate3262. Accessed 15 April 2020.

Diamond, Jared. 2019. *Upheaval: Turning Points for Nations in Crisis*. New York: Little, Brown.

Dunlap, Riley E., and Peter J. Jacques. 2013. Climate Change Denial Books and Conservative Think Tanks: Exploring the Connection. *American Behavioral Scientist* 57: 699–731.

Dunlap, R., A. McCright, and J. Yarosh. 2016. The Political Divide on Climate Change: Partisan Polarization Widens in the U.S. *Environment Science and Policy for Sustainable Development* 58 (5): 4–23.

Fairbrother, Malcolm. 2016. Externalities: Why Environmental Sociology Should Bring Them In. *Environmental Sociology* 2 (4): 375–384.

Farge, Emma, and Stephanie Nebehay. 2019. Greenhouse Gas Emissions Hit Record High Last Year, UN Warns. *The Globe and Mail*, 27 November: A4.

Flannery, Tim. 2015. *Atmosphere of Hope: Searching for Solutions to the Climate Crisis*. New York: Atlantic Monthly Press.

Giddens, Anthony. 2009. *The Politics of Climate Change*. Cambridge: Polity Press.

Harvey, Hal, and Robbie Orbis. 2018. *Designing Climate Solutions: A Policy Guide for Low-Carbon Energy*. Washington, DC: Island Press.

Hawken, Paul. 2017. *Drawdown: The Most Comprehensive Plan Ever Proposed to Reverse Global Warming*. New York: Penguin.

Heal, Geoffrey. 2017. *Endangered Economies: How the Neglect of Nature Threatens Our Prosperity*. New York: Columbia University Press.

Holter, Mikael. 2017. Big Oil's Dream of Hidden Crude Off Norwegian Archipelago Fades Away. *Globe and Mail*, 8 August: B2.

IEA International Energy Agency. 2018. Global Energy Demand Grew by 2.1% in 2017, and Carbon Emissions Rose for the First Time Since 2014.

International Energy Agency, 22 March. IEA: Paris. https://www.iea.org/newsroom/news/2018/march/global-energy-demand-grew-by-21-in-2017-and-carbon-emissions-rose-for-the-firs.html. Accessed 23 March 2018.

IPCC. 2018. *Global Warming of 1.5 °C*. http://report.ipcc.ch/sr15/pdf/sr15_spm_final.pdf. Accessed 8 October 2018.

Jaccard, Mark. 2018. PM Logic: Reduce Emissions by Increasing Them. *Globe and Mail*, 20 February: A13.

Jacques, Peter, Riley E. Dunlap, and Mark Freeman. 2008. The Organization of Denial: Conservative Think Tanks and Environmental Scepticism'. *Environmental Politics* 17: 349–385.

Kahn, Mizan R., and J. Timmons Roberts. 2013. Towards a Binding Adaptation Regime: Three Levers and Two Instruments. In *Successful Adaptation to Climate Change: Linking Science and Policy in a Rapidly Changing World*, ed. Susanne C. Moser and Maxwell T. Boykoff, Chapter 8. London: Routledge.

Latin, H. 2012. *Climate Change Policy Failures*. Singapore: World Scientific Publishing.

Leach, Andrew. 2019. How the Commuter vs. Polluter Narrative Could Backfire on Alberta. *Policy Options*, 15 July. https://policyoptions.irpp.org/magaxines/july-2019/how-the-commuter-vs-polluter-narrative-could-backfire-on-alberta/. Accessed 13 November 2019.

Lewis, Mark. 2019. Wells, Wires, and Wheels—EROCI and the Tough road Ahead for Oil. *BNP Paribas Asset Management*, August. Paris: BNP Paribus. https://docfinder.bnpparibas-am.com/api/files/1094E5B9-2FAA-47A3-805D-EF65EAD09A7F. Accessed 5 September 2020.

Mayer, Jane. 2016. *Dark Money: The Hidden History of the Billionaires Behind the Rise of the Radical Right*. New York: Doubleday Penguin Random House.

McGlade, C., and P. Ekins. 2015. The Geographical Distribution of Fossil Fuels Unused When Limiting Global Warming to 2 °C. *Nature* 517: 187–190.

Miranda, M.L., D. Hastings, J. Aldy, and W. Schlesinger. 2011. The Environmental Justice Dimensions of Climate Change. *Environmental Justice* 4 (1): 17–26.

Nordhaus, William. 2013. *The Climate Casino: Risk, Uncertainty, and Economics for a Warming World*. New Haven: Yale University Press.

Pielke, Roger, Jr. 2010. *The Climate Fix*. New York: Basic Books.

Prins, G., et al. 2010. *The Hartwell Paper: A New Direction for Climate Policy After the Crash of 2009*. Oxford: Oxford University Press.

Reality Check team. 2019. Prince Harry and Private Jets: What's the Carbon Footprint? *BBC News*, 20 August. https://www.bbc.com/news/uk-49408915. Accessed 21 August 2019.

Roberts, J. Timmons, and Bradley Parks. 2007. *A Climate of Injustice: Global Inequality, North-South Politics, and Climate Policy*. Cambridge, MA: MIT Press.

Rockström, Johan, Owen Gaffney, Joeri Rogeli, Malte Meinshausen, Nebojsa Nakicenovic, and Hans Schellnhuber. 2017. A Roadmap for Rapid Decarbonisation. *Science* 355 (6331): 1269–1271.

Schlossberg, Titania. 2017. Flying Hurts the Planet: You Can Help It Get Better. *San Francisco Chronicle*, 31 July.

Schnaiberg, A. 1980. *The Environment: From Surplus to Scarcity*. New York: Oxford.

Schulman, Daniel. 2014. *Sons of Wichita: How the Koch Brothers became America's Most Powerful and Private Dynasty*. New York: Grand Central Publishing.

Sim, Hans-Werner. 2012. *The Green Paradox: A Supply-Side Approach to Global Warming*. Cambridge, MA: MIT Press.

Stern, Nicholas. 2009. *A Blueprint for a Safer Planet*. London: Random House.

Stueck, Wendy. 2019. B.C. Wildfires Aren't Stopping Developments in Danger Zones. *Globe and Mail*, 6 July: A7.

Trudeau, Justin. 2017. No Country Would Find 173 Billion Barrels of Oil in the Ground and Just Leave Them There. *McLeans*, 10 March. https://www.macleans.ca/economy/justin-trudeaus-speech-in-houston-read-a-full-transcript/. Accessed 15 April 2020.

UNEP United Nations Environmental Programme. 2010. *Assessing the Environmental Impacts of Production and Consumption*. Nairobi: UNEP.

UNEP United Nations Environmental Programme. 2019. *Emissions Gap Report 2019 Executive Summary*. Nairobi: UNEP. https://wedocs.unep.org/bitstream/handle/20.500.11822/30798/EGR19ESEN.pdf?sequence=13. Accessed 26 November 2019.

Walsh, Marieke. 2019. Canada Should More Than Quadruple Carbon Tax to Meet 2030 Targets, Report from Ecofiscal Commission Says. *The Globe and Mail*, 27 November: A4.

Watt-Cloutier, Shiela. 2019. If We Protect the Arctic, We Save the Planet. *Globe and Mail*, 5 October: O11.

York, R. 2012. Do Alternative Energy Sources Displace Fossil Fuels? *Nature Climate Change* 2 (6): 441–443.

York, R. 2017. Why Petroleum Did Not Save the Whales. *Socius: Sociological Research for a Dynamic World* 3: 1–13.

5

Stuck in Dangerous Carbon-Polluting Practices?

Harvey and Orbis (2018: 12) claim that political will to implement effective climate policies has become strong, and give the example of the Paris climate accord. Such optimism, however, is premature. By 2020, its implementation is in deep trouble. The USA withdrew. Most countries are not on track to meet its goals, including near-term ones and much less more demanding longer-term targets. California is a leader but unfortunately an exception in the USA. Obama's climate and environmental regulations have been undone by Trump. Macron's carbon tax in France have hit a wall of yellow-vested opposition. Australia remains committed to coal as the backbone of its economy, despite suffering severe drought and wildfires. The 2019 COP25 Conference failed. The closing of all coal-fired electricity generation in Ontario by its Liberal government reduced carbon pollution, but it wasn't implemented well. This led to increased electricity rates, which caused a backlash and the election of a conservative government that eliminated most environmental measures. Since Harvey and Orbis (2018) wrote their favourable assessment of the California-Quebec-Ontario cap-and-trade system and

R. Murphy, *The Fossil-Fuelled Climate Crisis*,
https://doi.org/10.1007/978-3-030-53325-0_5

of the Canada's price on carbon, the conservative government in Ontario withdrew its participation and the Canadian government is facing fierce opposition from conservative parties over carbon pricing. Even though wind and solar are developing rapidly, fossil fuels are not decreasing below 80% of global energy supply because of increasing energy demand. Most states are not transitioning their economies off fossil fuels. Rather they are worsening global warming with their emissions exceeding their carbon withdrawals. Global emissions continue to be on an upward trajectory, increasing by 1.5% every year over the past decade (UNEP 2019). Harmful fossil fuels continue to be relatively cheap because their environmental and health costs remain 'externalities' costly to victims rather than included in their price. Risk-taking in fossil-fuel investments continues to bring risk-making for others.

Natural scientists are concerned about runaway greenhouse effects where first-order combustion of fossil fuels unleashes second-order global warming by nature's dynamics. This involves melting of permafrost releasing the powerful greenhouse gas-methane, melting of the reflective ice cover of the Arctic Ocean letting in more sunlight, wildfires emitting greenhouse gases, etc. If humans do not restrain their carbon emissions soon, the dynamics of nature will carry on global warming by themselves even if humans belatedly stop combusting fossil fuels. Unleashing runaway dynamics of nature happens often in the interaction between social practices and nature's dynamics. If carelessness causes a fire, it can be extinguished easily and quickly if caught early. But left unattended, it accelerates on its own and is very difficult and costly to extinguish and causes much damage: think of Notre Dame Cathedral. China censured doctors' warnings of the coronavirus (COVID-19) at the outset to maintain its 'harmonious' society, so now the virus is travelling by plane in human hosts around the world into a pandemic. Italy, Spain, most of Europe, and the USA were slow to contain the viral spread, with the result being disastrous. On the contrary, Taiwan, Hong Kong, and South Korea took decisive measures quickly, so they have many fewer fatalities. Timeliness of response is crucial for the fossil-fuelled climate crisis as well.

The Accelerating Treadmill of Carbon-Polluting Social Practices

There are also runaway effects on the social side. For example, the more fossil fuels are combusted, the hotter the climate becomes, the more air conditioning is used where coal or natural gas is combusted to produce electricity, which results in more carbon emitted into the atmosphere, thereby intensifying the greenhouse effect, making the climate even hotter, prompting people to use more air conditioning, and so on. Communities adapt to the danger of wildfires from fossil-fuelled global warming by cutting down trees, which reduces nature's capacity to absorb carbon from the atmosphere, and setting controlled fires which emit greenhouse gases. The global demand for oil has become so great that its exploitation has gone from wells where extraction results in relatively low emissions, as in Saudi Arabia, to sources that require more fossil-fuel combustion for extraction, upgrading, and refining, such as heavy oil from Venezuela and Alberta oil sands (Davidson and Andrews 2013). Most significantly, there is the treadmill of global economic growth that cancels out the benefits of growth in low-carbon energy. The supposed decarbonization of economies in terms of lower emissions per GDP continues to worsen fossil-fuelled climate change because it lags behind economic growth.

Technological innovation to reduce emissions has not kept up with the accelerating treadmill of fossil-fuel emitting social practices. Urban sprawl and resulting home-to-work commuting, typically one person per car, are intensifying. The affluent today fly to distant countries to see the world in much greater numbers than their grandparents, with most being unaware of how many pounds of carbon dioxide such jet-fuelled journeys emit into the atmosphere. Cruise ships driven by polluting bottom-of-the-barrel bunker fuel became popular over the last two decades. Social media servers that are electricity gluttons, usually based on fossil-fuel energy, enable phones to be smart. Air conditioning, typically powered by fossil fuels, is increasingly cooling buildings and vehicles. These are all good, but they also result in bads, namely the fossil-fuelled climate crisis.

The distinctive contribution of the teenage climate activist Greta Thunberg is that her actions speak louder than her words. When travelling in Europe, she goes in electrically powered trains, which underscores the global warming consequences of combusting jet fuel for short-haul flights. When invited to a United Nations summit on global warming in New York, she experienced how difficult it is to avoid fossil fuels for long-distance travel. To cross the Atlantic, she was offered a racing yacht powered by wind, solar, and underwater turbines built for speed rather than comfort, was warned the journey would still take two weeks and the ride would be choppy, but accepted nevertheless. This was her way of drawing attention to the carbon pollution by long-haul flights combusting jet fuel, which are increasingly numerous.

Fifty years ago, professional sports had few competitors, and events were held in a small number of cities not too distant from one another. Travel was done in economy class if by plane, and in some cases by train or bus. Salaries and prize money were good but relatively modest, and so were profits of owners. Now professional sports are on a treadmill of expansion—of salaries, prize money, and profits, of cities, distances between events, travel by plane, and jet fuel emissions. This resulted in professional sports becoming a major source of carbon pollution, which is obvious for car races (Formula-1, Indy, etc.) and motorboat races and when fans take short haul, discount flights to sporting events. It is also true for teams and athletes themselves. Today the New York Yankees wouldn't think of taking a two-hour, low-carbon train ride to play in Philadelphia or Pittsburgh. They fly. National Football League, National Basketball League, and professional soccer (football) teams fly in lavish charter jets. Profits of team owners, prize money for competitors, and sponsorship payments for marketing have exploded. Players in individual sports, like tennis and golf, are flying continually. Owners and successful competitors now have their own private jets, and do not pay the full cost of their emissions in offsets, even though they could easily afford to. They are members of the 1% in wealth, and invest it to maximize returns, typically without concern for environmental consequences. Athletes often become spokespersons for their favourite causes, but have not tackled fossil-fuelled global warming. Thunberg noted the drought and bush fires caused by fossil-fuelled global warming ravaging Australia in January

2020 during the Australian Open tennis tournament. She found that the world's most successfully branded person, Roger Federer, is sponsored by Credit Suisse, which invests in oil and gas. So she inspired a protest on social media. 'No sector of society is more vulnerable to climate protest than pro sports. … Professional athletes are highly visible, highly remunerated, highly mobile, highly out of touch, and highly sensitive to criticism' (Kelly 2020: B10). But they are also highly supported by fans. It takes courage to criticize the GOAT (Greatest of All Time) Federer? The sports journalist Kelly (2020: B15) concludes that the fossil-fuelled climate 'struggle is no longer about amelioration, it's about abstinence. At some point in the very near future, we are all going to be asked to give up frilly things we enjoy. What's more frilly than sports? … From now on, it's going to be impossible [for teams and athletes] to say nothing about climate change and their role in it'. But will they take action to fully offset their emissions, using their substantial wealth to finance wind farms and solar energy in poor countries?

People in poor countries make use of animate energy (their own physical labour, horses, mules, oxen). But as countries rise out of poverty, they increasingly use fossil fuels to power economic growth. It is usually coal, which is the cheapest, most available, but also the most carbon polluting. The first things wanted by developing countries are efficient transportation, efficient agriculture, smartphones, and air conditioning, but these are powered by fossil fuels and contribute to accelerating the treadmill of carbon-polluting social practices. Population growth also contributes to that acceleration. Although there is much talk about global population growth slowing down, in absolute numbers it is still increasing because of demographic momentum, with more women having children because of past population growth even as the number of children per woman decreases. Implementation of low-carbon energy has not kept up with all these additional emissions, much less driving down emissions to carbon withdrawal rates.

One of the most difficult issues in the climate crisis was raised by South Africa's finance minister. He stated that, because of both economic growth enabling people to escape poverty and population growth, South Africa's economy was two-thirds bigger in 2010 than in 1994 when Nelson Mandela became president. Millions of people previously deprived of electricity are now on the grid, but supply has not kept pace with demand. His government is promoting wind, solar, hydropower, and nuclear to generate electricity. Nevertheless, he concluded that the near-term need for energy trumps long-term threats from greenhouse-gas emissions, and that stunting economic growth in Africa is not the solution to fossil-fuelled climate change. He therefore supports a massive coal-fired electricity plant financed by the World Bank. This is typical of all poor countries with aspirations to enjoy what affluent societies have, but it accelerates the treadmill of carbon pollution practices. Pielke uses this case to illustrate his solution to global warming: the technological innovation of low-carbon energy cheaper and more accessible than fossil fuels, which would outcompete fossil fuels thereby leaving them in the ground. But this requires a leap of faith in technological innovation. The finance minister added that 'the journey [to eliminating greenhouse-gas emissions] will inevitably be costly, requiring massive investments in technology, research and re-engineering the ways in which we live and do business. It will also require a true spirit of consensus and collaboration' (quoted in Pielke 2010: 233). Since poor countries have neither the technology nor financial resources to make massive investments in clean energy, this requires that wealthy societies, which have become rich by combusting fossil fuels and emitting carbon since the industrial revolution, finance the research costs and transfer the technology to poor countries. But as recent climate negotiations demonstrate, resistance remains strong in wealthy, technologically advanced countries to shipping money and free technology to countries that can't pay.

Increased efficiency of aviation and of ground and water transportation based on fossil fuels, with attendant greenhouse-gas emissions, have resulted in a surge of recreational travel. People with money travel to see sights around the world, especially after retirement, which involves a structured escalation of emissions. 'We have channeled our desires, our insecurities, our need to demonstrate our worth and our success, our wanting to fit in and to stand out increasingly into material things – into bigger homes, fancier cars, grander appliances, exotic vacations' (Speth 2009: 161), all of which intensify fossil-fuel combustion. Consumption includes invisible but damaging fossil fuels combusted to 'get away', as Carolan (2014) calls it, by travelling on vacations, often distant ones.

People today drive more than their grandparents did, fly more, and take more cruises. They are even cremated more often when they die. The average cremation consists of combusting about 28 gallons of natural gas, propane, or diesel to incinerate the corpse for about two hours at 1000 °C, reducing it to about 3 lb of ashes, and emitting about 540 lb of carbon dioxide. Cruises are double jeopardy carbon polluters: enormous cruise ships propelled by bunker fuel or diesel pollute, and passengers have to take a polluting plane to arrive at the departure point and fly home. The aviation and cruise industries are growing rapidly, propelled by fossil fuels. The issue is one of scale: if one person chooses to fly, the effect is negligible, but if billions make that decision, the emissions result in significant additions to the accumulation of atmospheric carbon. Societies are on an accelerating treadmill of carbon-polluting social practices, which functions at cross purposes to goals of reducing greenhouse-gas emissions.

Market Innovation as Part of the Problem

Market innovation could in principle be part of the solution to fossil-fuelled climate change. The private sector has been involved in innovating low-carbon energy sources: hydroelectricity, solar and wind energy, nuclear energy, electric energy storage, etc. But most of these have required government feed-in tariffs, tax deferments, subsidies, etc., to take off. Market innovations that reduce emissions are, however, overpowered by those that increase them. The private sector has innovated hydraulic fracking, deepwater drilling, developed ways to extract oil from bituminous (tar) sands, Arctic oil and gas exploration, liquefied natural gas, and pipelines to bring these fossil fuels to market. Profit-seeking companies are great at digging and pumping fossil fuels out from safe storage in the ground, and developing new ways to use them. They are not good at preventing emissions from combustion being dumped in the atmosphere to accumulate, for example, 'clean coal' remains an oxymoron. The reason is because there is no profit to be made, unless government imposes a high price on carbon pollution. 'Although free-market capitalism can help spark innovation in new technologies, our current economic systems demand endless growth, continued consumption, and the most profitable pathways, which often means using the cheapest fuels to extract the most money from energy and production' (Suzuki and Hanington 2017: 251). This has been called the capitalist growth imperative. Most companies steer innovation towards profit-making without consideration of environmental consequences. They have to be pushed to pursue less polluting but more challenging energy resources like solar, wind, hydro, tidal, and geothermic energy. The entrepreneur Hawken (2010: Introduction to the Revised Edition) concludes that 'business is destroying the world, no one does it better'. Whereas the market left to itself speeds up extraction and use of valuable but dangerous resources (coal, oil, asbestos), it is necessary to slow down that treadmill of production (Schnaiberg 1980). As Nordhaus (2013: 257) states, this constitutes market failure, and he concludes that 'free markets will not do the job' of solving the urgent climate problem. Moreover, the relative concentration of decision-making results in profit-seeking companies convincing the population to give priority to near-term economic benefits over long-term environmental and economic benefits. The fossil-fuel industry finances political parties and think tanks (Jacques, Dunlap, and Freeman 2008; McCright and Dunlap 2010; Dunlap and Jacques 2013; Elsasser and Dunlap 2013; Dunlap and Brulle 2015; Dunlap, McCright, and Yarosh 2016) which discount future harm from fossil fuels and oppose solutions to climate change. Many of these right-wing think tanks have misleading names, such as *The Global Warming Policy Foundation* which leads people to believe it constructs policies to combat global warming but is instead a UK lobby group to promote fossil fuels. In the USA, the 'Global Climate Coalition' was created to discredit climate science (Schneider 2009: 120–121). Science is usually seen as being disinterested and apolitical, but when it brings troubling news about the activities of vested interests, it has been described as a 'contact sport' (Schneider 2009) between impact scientists and production scientists. Speth (2009: 116) concludes that 'the planet cannot sustain capitalism as we know it'. Time will tell whether he is right.

Market innovation has given us social media capable of manipulating enormous amounts of data instantaneously, but the powerful computer servers require huge quantities of electricity to operate and be cooled. Since much of its energy comes from combusting fossil fuels, it has worsened global warming. Data storage clouds bring with them century-long greenhouse-gas clouds. Take a recent example of the private sector's innovation of a seemingly pure cultural object. Bitcoins, unlike physical coins, have no tangible existence. They are instead based on algorithms. The complex computations are done by powerful computers, which are only able to function by consuming huge amounts of electricity generating enormous amounts of heat, hence have to be cooled. One of the biggest bitcoin centres is in Dallas Texas because it has cheap electricity provided mainly by combusting coal and natural gas, which emit greenhouse gases causing global warming. Its hot climate requires much electricity for cooling the computers. Even seemingly pure cultural objects like cryptocurrencies with no physical existence are important carbon polluters contributing to global warming. Where on Earth could bitcoin computers be run efficiently using abundant, clean, cheap energy? A few bitcoin companies are now profiting from nature's free services by moving their computers to Iceland to benefit from (i) its abundant geothermal energy for generating electricity and (ii) its cold location by letting the frigid Arctic air cool the computers. This bitcoin industry is not without its critics in Iceland, who argue that the country is 'spending tens or maybe hundreds of megawatts on producing something that has no tangible existence and no real use for humans outside the realm of financial speculation' (Bjarnason 2018).

Apathy Towards the Scientific Forecast of Catastrophe

As scientific evidence accumulated, denial of anthropogenic climate change has in most quarters changed to apathy. In complete apathy, there is indifference to the crisis even though its existence is acknowledged. In partial apathy, something is done but there is a refusal to change fossil-fuelled practices or pay carbon taxes, despite good scientific evidence about the scale and timing needed. In both, danger is discounted, resulting in a failure of foresight, the carbonization of the atmosphere, and the incubation of disaster. Discounting danger is a frequent response when confronting threats where prevention would be costly and disrupt present social practices. Thus disaster researchers have documented many ‘repeat disasters’ (Platt 1999), ‘unnatural disasters’ (Abramovitz 2001), ‘disasters by design’ (Mileti 1999), and the like. In 2009, Schneider (2009: 231) argued that ‘we’ve known pretty well for at least three decades of the climate threat, yet we were paralyzed from much action [to prevent it]’. This still is true a decade later. Paralysis has not resulted from lack of goals and targets, but rather from measures to implement goals into action: ‘aspirational targets are necessary to scale roughly how much to invest, so I am not against targets per se. But without explicit policies and measures to achieve any aspirational targets, it is talk, not action. Remember the problem with the Kyoto Protocol: targets without teeth’ (Schneider 2009: 244). Rand (2014) also demonstrates how paralysis characterizes the response of most societies to climate change.

Lucrative Discounting of Danger

Discounting danger refers to being unresponsive to preventing catastrophes forecast by science, but it can innovative new practices to benefit from global warming and even hasten it. The pursuit of near-term economic benefits trumps predictions of distant catastrophe in inciting action. There is inaction to prevent fossil-fuelled climate change but much activity to profit from it. Deutsche Bank created mutual funds to capitalize on global warming and shift the focus from risk and cost to profiting from its enticing business opportunities (Funk 2014: 2). For his book entitled *Windfall: The Booming Business of Global Warming*, Funk (2014) investigated bonanza profits for companies flying water bombers and manufacturing aerial fire retardant to fight wildfires, businesses profited from water scarcity because of drought and farmland scarcity, others making windfall profits from selling seawalls and reinsurance due to storm surges and deluge, and companies specializing in climate geoengineering and climate genetics. Insurance companies are adapting to fossil-fuelled climate change by refusing coverage and increasing premiums for flooding and wildfires. Funk (2014: 285) concluded as follows. 'Perhaps the most magical assumption of the moment is that our growing belief in climate change will lead to a real effort to stop it. But as I discovered in Canada and Greenland and Sudan and Seattle and all over the globe, that is not automatically true. We are noticing that in this new world, there is new oil to find. There is new cropland to farm. There are new machines to be built. From what I have seen in six years of reporting this book, the climate is changing faster than we are'. His research constitutes a salutary warning of the need to be wary of the enticing side effects of global warming. Global warming is something like a tsunami, which lulls people on the beach into assuming the receding water is safe, before the massive waves crash into them.

At the 2019 Arctic Council, the members, namely Sweden, Norway, Denmark, Finland, Canada, Russia, Iceland, and the Inuit Circumpolar Council signed a declaration stating climate change is a serious threat to the Arctic because its temperature is increasing at twice the rate as the rest of the planet. But the remaining member, the USA led by Secretary of State Mike Pompeo, refused to sign because the melting ice creates the potential for new shipping lanes and exploiting new reserves of oil and natural gas. Where the others saw an environmental threat, Pompeo saw near-term economic opportunity even though he claimed the Trump Republican Administration 'shares your deep commitment to environmental stewardship in the Arctic' (Dickson 2019: A7). Since Inuit already experience adverse consequences of climate change, their leader described Pompeo's refusal as a moral failure.

Freudenburg et al. (2009) documented how New Orleans was given multiple science-based warnings of hurricanes, but they were not heeded because priority was given to economic growth. Canals were built, dredging was done, and other constructions were made for immediate economic gain, which exacerbated vulnerability to hurricanes. 'Predictions of danger that were chillingly similar to what came to pass with Katrina were put forth repeatedly …, however, environmentally damaging projects such as MRGO [which intensified the harmful effects of hurricanes] have continued to be described as necessary "for the good of the economy"' while risks of environmental harm have routinely been dismissed as something 'not to be feared' (Freudenburg et al. 2009: 163–164). Scientific warnings of catastrophe had little effect in promoting safety when they confronted near-term economic priorities. Actions fostered by economic interests made the danger worse. Even emergency management and disaster preparation were woeful.

Freudenburg and Gramling (2011) documented warnings of the likely failure of BP's blowout protector for deepwater drilling in the Gulf of Mexico where pressures are particularly high, but the company and its regulators claimed it was safe. They discounted danger for the profitable operations and went full speed ahead towards a blowout and catastrophic two-month long oil gusher. In other cases as well, pre-disaster reports claiming safety have been called 'fantasy documents' (Clarke 1999). Freudenburg and Gramling (2011: 158) concluded from their study of the Deepwater Horizon blowout that there 'needs to be prevention, not a mistaken belief that we can actually "clean up" such a mess'. But prevention is not being done in North America and in many countries concerning fossil-fuelled global warming. Hence 'literally and figuratively, and both in the Gulf of Mexico and elsewhere, we have been getting into increasingly dangerous waters, doing so without being sufficiently vigilant about the implications of our actions' (Freudenburg and Gramling 2011: xiii).

Backsliding

It would be superficial to assume that the directionality of movement is straightforwardly towards solving global warming. Like any road, there can be backward as well as forward movement. Hoped-for solutions must not blind us to backsliding. One technical type involves the shift from oil wells to extraction of heavy oil and bituminous sands oil which produce more emissions per barrel than oil from a well (Davidson and Andrews 20). Political backsliding is common. From 1992 to 2000, the USA had Vice-president Al Gore whose main priority was mitigating global warming. But mitigation stalled from 2000 to 2008 under the George W. Bush Republican Administration. There followed a great hope for mitigating global warming because 'after 2009, there will be a new U.S. administration that plays a positive role in the process rather than deliberately trying to sabotage it' (Dyer 2008: 174). The Obama Administration did indeed enact policies to diminish carbon emissions from 2008 to 2016, but then came President Trump's belief that 'the concept of global warming was created by and for the Chinese in order to make U.S. manufacturing non-competitive' (Fife 2016: A3). He withdrew the USA from the Paris Agreement to mitigate climate change, named a climate sceptic as head of the American Environmental Protection Agency, dismantled Obama's measures to protect the environment, particularly from fossil-fuel emissions, and called for using more coal to please his voting base in coal-producing states. There is a pattern in American politics concerning the mitigation of global warming: one Democratic step forward followed by one or more Republican steps backward (Dunlap, McCright, and Yarosh 2016). Decisions by the USA whether or not to mitigate global warming then has strong effects on other nations because of American power in the market.

This is an important part of the reason why mitigating fossil-fuelled climate change doesn't get done in democracies. Centre-left governments implement measures to deal with it, then these are undone by right-wing governments in the back-and-forth of electoral politics. In Ontario, Liberals introduced cap-and-trade in cooperation with California and Quebec, promoted wind and solar energy, eliminated coal-fired electricity generation, and did other mitigation measures; then these were reversed by the subsequent conservative government. The Alberta conservative government repealed environmental reforms of the previous social democratic government, such as phasing out coal-burning electricity generation. Backsliding is an integral part of discounting danger when the scientific anticipation of a slow-onset catastrophe confronts near-term economic interests. Even if corrected in a subsequent phase of the electoral cycle, it results in falling further behind the steady accumulation of greenhouse gases in the atmosphere. Thus Lockie and Wong (2018) analyse the conflict between the temporalities of climate change and the temporalities of politics. Schneider (2009: 269) argues that 'we need national mandatory performance standards to be implemented as an urgent priority and thus help to reduce our emissions relative to business as usual *now*, not decades down the road'. In 2009, he was optimistic based on the Waxman-Markey bill, but it was never passed by the American Senate.

Nordhaus (2013: 250–252) suggests an international carbon price regime whereby nations would negotiate a framework agreement with a minimum price for carbon, then each would be free to choose whether to implement a carbon tax, cap and trade, or a hybrid to achieve it. This he argues would be easier than alternatives. However, its implementation is far from politically easy. It is similar to Canada's attempt to achieve its Paris Agreement commitments. It set the national emissions reduction level, and let provinces choose carbon taxes, cap-and-trade, regulations, or a hybrid providing they meet Paris commitments. British Columbia has a successful carbon tax, Quebec and Ontario joined California's cap-and-trade system, the oil province of Alberta began a hybrid of a carbon tax, regulations, phasing out coal, and a cap on emissions from the bituminous sands oil sector. But the harmony fell apart when provinces voted for conservative governments. Alberta and Saskatchewan are fighting the federal initiative in court, Ontario eliminated cap and trade, Alberta repealed the former social democratic government's climate initiatives, and Manitoba pulled out of the agreement. So the Trudeau federal government will have to impose its own carbon tax, and to sell that, it is returning the monies to families in the provinces paying it. On the international level, Nordhaus suggests using trade sanctions to enforce a carbon price on every nation. That might have worked on small nations, but imagine how difficult it would be for the international community to use trade sanctions against Trump's America to force it to place a price on carbon pollution.

The President of the International Institute for Sustainable Development David Runnalls (2008: 3–4) concluded that in 1988–1992, 'Canada was the most advanced country on earth in terms of sustainable development. … we have been going backward, or perhaps more generously, sideways, ever since'. The backsliding of Canada during the period 2006–2015 when the Conservatives were in power is also confirmed by the Yale University Environmental Performance Index (Yale University 2012, 2018) and other indexes (Germanwatch 2012, 2019), by its increasing per capita greenhouse-gas emissions, reneging on its Kyoto

Protocol commitments then withdrawing, promotion of high-emissions oil extraction from tar sands, its strenuous lobbying against European Union labelling of tar sands oil as high emissions, etc. Canada had many enlightened policies to deal with anthropogenic global warming, but they were not implemented, and hence have been aptly described as 'hot air' (Simpson, Jaccard, and Rivers 2007). There is evidence of backsliding in Australia, which rescinded its carbon tax. New Zealand, which was presented by Giddens (2009) as an exemplary case of taking action to prevent anthropogenic climate change, slid back to ineffective measures after a change in government from the Labour Party to the conservative National Party. In the USA, Canada, Australia, New Zealand, and other countries, discounting fossil-fuelled danger has been typical when conservative governments (called Liberal in Australia) come to power.

Fossil fuels tend to be default options when there is resistance to non-carbon-emitting sources of energy. When communities oppose wind farms but don't reduce their energy demand, they rely on fossil fuels. If problems emerge, countries return to fossil fuels. When Japan closed its nuclear reactors after the Fukushima disaster, it turned to liquefied natural gas and other fossil fuels. When Germany lacked energy for electricity generation while phasing out nuclear power, it imported electricity form France's nuclear reactors and used fossil fuels. When the COVID-19 virus struck and practising social distance was counselled, many people abandoned public transit and returned to commuting alone in their private car.

Contemporary conservative parties follow Ronald Reagan's belief that government is the problem, not the solution, for the ills of society (except for the military, police, and jails). American anti-trust laws were eliminated, so monopolization of market power increased. Political parties that advocate redistribution of income and wealth are opposed, so inequality is amplified. Government control of guns, including assault weapons, was weakened by an expansive interpretation of the 2nd amendment of the right to bear arms, so the USA has the most gun violence in the developed world. Libertarianism is interpreted as freedom from government, which results in enfeeblement of the principal institution that can moderate excesses of the market and promote equal opportunity. The lower, working, and middle classes are not free if they are sick and have no health care, and they are not free to develop their talents if they cannot afford quality postsecondary education. Free speech was interpreted as freedom from limits on spending for political parties, so the country became a plutodemocracy where money has an inordinate influence on politics. The government is subjected to powerful fossil-fuel lobbies and prevented from regulating greenhouse-gas emissions, resulting in a climate crisis, and closing off opportunities for future generations who will be deprived of natural capital and nature's services. Heal (2017: 199) argues that 'the external costs that generate them [environmental problems] require government policies and are a logical stake through the heart of belief in unadorned markets. It's hard to believe that we need to solve environmental problems while also believing that the government is the problem and not the solution. … The result is that many conservatives ignore environmental problems, pretending that they don't exist'. They discount danger and oppose government regulations that would ensure the full cost of fossil fuels is paid. Although most pronounced in contemporary USA, this is an outcome of conservative ideology in many countries. Free market ideology and practices, in the sense of free from government regulations, is a significant factor in backsliding on climate action.

Playing Batty Politics

Windmills are often opposed using the claim they kill bats, but it can be greatly diminished by mitigation measures: building wind farms in low-risk areas for bats, shutting off turbines when wind is weak and bat migration occurs, etc. Most important, climate change poses a much greater threat to bats and biodiversity by destroying habitats, as documented by Baerwald (2019), a conservation biologist who specializes in research on bats and wind farms. He concludes that wind farms reduce the climate-change threat to bats by replacing fossil fuels in electricity generation. Nevertheless, the Ontario conservative government cancelled a 29-unit wind farm under construction claiming it would cause irreversible harm to bat populations, despite the opposite conclusion by a scientific committee. The unstated reason for cancellation, at a cost of hundreds of millions of taxpayer dollars, was that some residents feared property values would decrease, which is likely unfounded, and a conservative politician was up for re-election. Baerwald concludes that bats are being used as pawns in batty politics and that this wind farm developer, who is following scientific bat mitigation recommendations and should be emulated by other developers, is being punished.

Every Man for Himself

Funk (2014: 7) argues that the scientific anticipation of global catastrophe is being discounted. 'We hope our collective fear of global warming will push us inevitably toward collective behavior [sustainable green activities]. But what if the [socioeconomic] world as we know it goes on even as the [biophysical] Earth as we know it begins to disappear?' Blumenthal (2014) confirms this is happening in Louisiana: 'the state is also disappearing, overtaken by waters due in part to the drilling, dredging and runoff associated with the state's biggest economic driver [the oil industry]. Not to mention the rising tide from a changing climate'. Funk found evidence supporting a hypothesis opposite to Beck's (2015) cheery postulate that the anticipation of global climate change catastrophe results in a cosmopolitan orientation taking into account needs of distant people and future generations. 'There's another possible response to melting ice caps and rising sea levels, to the reality of climate change – a response that is tribal, primal, profit-driven, short-term, and not at all idealistic. Every man for himself. Every business for itself. Every city for itself. Every country for itself' (Funk 2014: 7–8). It leads to hunkering down in a bunker mentality. He integrates social class and national differences into his analysis. 'Some people – the rich, the northern – will find ways to thrive while others cannot, and many people will wall themselves off from the worst effects of warming while others remain on the wrong side. … The people most responsible for historic greenhouse emissions are also the most likely to succeed in this new reality and the least likely to feel a mortal threat from continued warming. The imbalance between rich and north and poor and south – inherited from history and geography, accelerated by warming – is becoming even more entrenched' (Funk 2014: 288). Even safety is being monopolized.

The potential for exclusion because of global warming is enormous. Already many countries, such as the USA and Eastern Europe, are refusing to accept refugees fleeing war-torn countries, failed states, etc. English resentment against Eastern European immigrants was one important factor motivating Brexit. Fossil-fuelled global warming is predicted to result in sea level rise and inundation of coastal cities, drought and famine (Goodell 2017). Imagine the displacement away from areas made unliveable by climate change. 'Quickly decarbonizing transportation around the globe is essential if humanity is to slow and then reverse the intensifying wildfires, floods, and storms, and the mass migration of people fleeing the worst impacts of climate change' (Rivers and Jaccard 2019: A15). Will prosperous countries that disproportionately caused the problem open their borders and welcome climate refugees, or will they build walls? Moreover, the problem of climate refugees will distract attention from the ongoing carbonization of the atmosphere and from cooperation needed to mitigate it, perhaps even sabotage cooperation.

Top-Down Discounting of Danger

Diversions and apathy have been promoted by the fossil-fuel industry and prominent media commentators, as documented by Freudenburg (2006). Merchants of doubt attempt to confuse the public by constructing disbelief concerning whether fossil fuels cause climate change, like they did about whether cigarettes cause lung cancer (Oreskes and Conway 2010; Dunlap and Brulle 2015). A widely circulated tweet wished an accident would destroy Greta Thunberg's low-carbon boat as it crossed the Atlantic; a meme depicted Trump tipping over the Statue of Liberty to crush her boat as it arrived in New York; the sceptical environmentalist Bjorn Lomborg shared an article which called her a fanatic (Gelin 2019); Russia's petro-state President Putin condemned her. Holman's (2019) large-scale study of the English Canadian mainstream media found that their reports often do not reflect the severity and scope of the fossil-fuelled climate crisis and that 'some sources seemed more interested in convincing Canadians there isn't a climate crisis than actually reporting on it'. Murphy (2015) showed how the scientifically documented environmental problem of global warming becomes a societal non-problem. By focussing on the discursive legitimation of practices that cause anthropogenic climate change, he demonstrated how communication power in the mass media veils the adverse consequences of high-emissions oil extraction and showed how concern about emissions is dampened and quiescence socially constructed. The mediation by media communication power between scientific warnings of danger and fossil-fuelled social practices constitutes an important element explaining why emissions are increasing and why science is having little influence concerning fossil-fuelled social practices.

The media goal of balance, which is a worthy aim for the communication of opinions, is a catalyst for discounting danger concerning complex issues when the preponderance of scientific evidence lies on one side. It gives outlier scientists a megaphone to undermine the scientific consensus and confuse non-scientists. 'If the public is bamboozled into thinking that each "side" in a typical media debate is credible, a typical reaction is to say, "Well, if the experts don't know, how can I know! Let's just wait a while until they figure it out." That is precisely the strategy of the climate deniers. To create public confusion and apathy, which slows policies that help the planet and our children but hurt the special interests who are vested in the status quo' (Schneider 2009: 205). Activities that cause greenhouse-gas emissions are not stopped, so waiting for certainty results in continuing emissions that will remain in the atmosphere for a century and accumulate. And there are double standards at play. The media strive for balance between (i) experts who communicate the preponderance of the evidence demonstrating fossil-fuelled climate change and (ii) people who deny it. But the media wouldn't do the same for the evolution versus creationist debate, or vaccinations causing autism, or now for debate about asbestos or cigarettes causing lung cancer although they did in the past. For complex issues requiring scientific evidence of danger, simple-minded media balance results in discounting, whereas prevention requires an assessment of the credibility of spokespersons founded on the preponderance of the evidence. Concerning global warming, it remains to be seen whether powerful fossil-fuel industries will prevail in discounting danger or whether attempts to discredit scientific findings of danger will fail and foresight emerge as in the cases of CFCs, cigarettes, and asbestos.

Bottom-Up Discounting of Danger

Big oligopolistic fossil-fuel companies are promoting its use and discounting its danger, as are fossil-fuel extracting states that based their economies on it, and large parts of the media. But it is an oversimplification to claim they are the only ones responsible for the fossil-fuelled climate crisis. The population does not consist of mere puppets. People have their own desires. They contribute to emissions by their demand for inexpensive electricity, cheap gasoline, dragging a motor home across the continent, for discount flights, getaway flights of intercontinental tourism, cruises, low-cost data servers, air conditioning, etc. In a democracy, they vote for political parties that oppose taxes on carbon pollution. Even if they alternate their vote in favour of a more environmentally friendly party in the next electoral phase, the one step forward one step back on climate action lags far behind the continual accumulation of carbon emitted into the atmosphere. They demand cheap gasoline, and increase sales of gas-guzzling vehicles over fuel-economical ones.

North Americans have become accustomed, when raising a family, to a single-family house in the suburbs with a yard, and to paying lower prices and less taxes for the same size house compared to the central city. But they then have a long commute from home to work, and most use private cars for reasons of convenience and flexibility, typically one person per car. This results in greater emissions than public transportation in high-density cities like Seoul South Korea. Single-family houses are what their parents had and what they grew up with. Then immigrants see what others have and aspire to the same. When developers construct new inner-city condos, they build them small to make more money by stuffing more in a small city lot. These bachelor-and-widow condos are not suitable for families. American urban sprawl, the long commute, and difficulties in providing flexible, cost-effective public transportation over long distances result in dependence on automobiles, a sense of entitlement to low-cost gasoline and resistance against carbon taxes. This encourages fossil-fuel consumption and greenhouse-gas emissions. Despite densification attempts, sprawl is growing in exurbia. Although most prominent in North America, this pattern is occurring in other countries too. Support by the population for inexpensive fossil-fuelled practices should not be ignored as part of the climate crisis.

Free-Ridership Carbon Pollution as Discounting Danger

Since fossil-fuelled climate change is a global slow-onset problem, the costs paid by any country placing a price on its carbon pollution through carbon taxes, cap and trade, etc., would be immediate but produce few if any near-term benefits for that country, even though it will produce benefits for other countries and future generations. Nordhaus (2013: 318–319) estimates the main climate benefits of placing a price on carbon pollution *now* will occur about *fifty years later*, which he refers to as a 'temporal trade-off'. Hence there is an incentive to reject a price on carbon, keep emitting carbon pollution, hope that other countries reduce their emissions, or have faith that a last-minute, cost-effective technological solution will be found and implemented. Since near-term economic benefits are typically given priority, the optimal financial strategy is to do nothing and let other countries pay the costs of carbon taxes, etc. Mitigating climate change is resisted because it appears to involve a redistribution of wealth from the virtuous country to polluting ones and from this generation to future generations (Keith interviewed in Dyer 2008: 165–167). 'Countries have strong incentives to free ride on the efforts of others because emissions reductions are local and costly while the benefits are diffuse and distant over space and time. … This incentive leads to a noncooperative free-riding equilibrium in which few countries undertake strong climate–change policies – a situation that closely resembles the current international policy environment. … Additionally, there is a tendency for the current generation to ride free by pushing the costs of dealing with climate change onto future generations. Generational free riding occurs because most of the benefits of emissions reductions today would accrue many decades in the future' (Nordhaus 2013: 8). This enticement to discounting danger also applies to companies and individuals.

In a transnational global economy, companies lobby governments against pricing carbon claiming they will be uncompetitive compared to trading partners that are allowed to externalize their costs of carbon pollution. The Canadian government coordinated its carbon pricing with the Obama Administration to meet its 2015 Paris Agreement commitments, but the Trump Administration's withdrawal from that agreement, its elimination of fuel-efficiency requirements, etc., created major competitiveness issues for Canada. Carbon-taxed Canadian companies contend they can not compete against American companies that do not pay a carbon tax. This promotes a race to the bottom, which could be called the economic survival of the dirtiest. Externalizing costs to the environment and therefore to populations distant in space or time is cheaper and better for the bottom line. Even when they agree that fossil-fuelled climate change is a significant threat, companies, governments, and populations have an economic competitive incentive to avoid taking the lead in paying upfront costs required to reduce carbon pollution, and let other countries and companies do it.

However, assuming that polluting countries which place a price on carbon are centrifugally redistributing their wealth to other countries and other generations is a partial view which sees only one side of the issue. The other side is that carbon-polluting societies, companies, and individuals have already partaken of centripetal wealth redistribution by appropriating unduly cheap fossil fuels while shifting the environmental costs of their pollution to all countries and future generations. Polluters deploy the near-term economic strategy of not paying the costs of their degradation of the environment others will need. The issue consists of refusal to pay for costs carbon polluters are causing and shifting costs to others distant in space and time. If a country (or company or individual) combusts fossil fuels but refuses to spend money for carbon capture and storage, carbon taxes, and other measures to eliminate its emissions, then it appropriates the benefits of excessively cheap fossil fuels and foists the costs of its pollution on others. Hence they close off opportunities to others who will be adversely affected by the pollution even in distant places and future generations. This is the way the closure theory perspective analyses fossil-fuelled global warming by taking into account the externalities of harm and costs. This involves a redistribution of costs and benefits, namely (i) letting carbon polluters save money by having victims pay the health and environmental costs of the former's pollution, and (ii) allowing this generation of carbon polluters to indirectly take money from future generations who will have to pay the environmental costs of carbon pollution. It consists of free riding by polluters not paying the cost of their externalities.

The enticements of free riding and refusal to stop it are part of the explanation of why it is exceedingly difficult to implement carbon pricing nationally and internationally, with the result that there remain economic incentives to pollute. Therefore fossil-fuelled climate change continues unabated. Changing freeriding practices by including the cost of pollution in the price of fossil fuels is hard to implement. Polluters who have developed a sense of entitlement to riding freely vigorously combat having to pay the full cost of their pollution. Empathy with victims of fossil-fuelled climate change, cosmopolitanism, and foresight envisioning future harm have been in short supply. This free-rider dilemma can be solved by a rules-based order within and between nations consisting of a binding and enforced international agreement to price carbon pollution, similar to the successful Montreal Protocol phasing out CFCs, and international enforcement practices, such as trade sanctions in the Kigali Agreement. But such international agreements are very difficult to negotiate and even more difficult to implement and enforce, so free riding persists in the case of carbon pollution.

The Dark Side of Adaptation and Resilience Building

Adaptation to fossil-fuelled climate change and building robustness and resilience are necessary, indeed indispensable. But concepts of adaptation and resilience are not as straightforward as they appear. Rather than only postulating overly optmistic concepts, the following negative sides need to be taken into account.

Adaptation and Resilience as Legitimation Devices to Avoid Prevention

Adaptation and resilience construction are popular among those promoting and profiting from the social practices causing climate change and those resistant to including the costs of fossil-fuel externalities in their price. The recent emphasis on adaptation and resilience fits nicely with the growth of the fossil-fuel industry, which claims that instead of using less fossil fuels or innovating expensive technologies to reduce emissions, just rely on end-of-pipe adaptation, robustness, and resilience. For example, building seawalls around New Orleans to defend against hurricanes and buying more powerful pumps to expel water will supposedly enable Louisiana to grow its fossil-fuel economy and have safety too. Adaptation can be promoted to maintain present dangerous social practices by fostering the belief that it will provide the capacity to ride out climate change. By providing a reassuring sense of security, which may be false security, adaptation can inhibit prevention. Thus it can have a consequence similar to the Jevons paradox: the more adaptation is successful, the less people are willing to undertake prevention, and the more persistent are dangerous social practices.

Notwithstanding this dark side, the more prevention is insufficient, the more adaptation and resilience building will be necessary. Moreover, emissions reduction is needed to make both resilience and adaptation successful. 'Enabling climate resilience and sustainable development depends critically on urgent and ambitious emissions reductions coupled with coordinated sustained and increasingly ambitious adaptation actions' (IPCC 2019: SPM-40 C4).

Perverse Adaptation

This consists of adapting to harmful environmental consequences that fossil-fuelled practices cause by intensifying those practices, which will perversely exacerbate global warming. The ice cover over the Arctic Ocean and around Greenland prevented oil and gas extraction and shipping in the past. But fossil-fuelled global warming is melting much of the ice lengthening the season when shipping, oil drilling, and natural gas extraction are possible. Some countries and companies are adapting to this new situation by increasing oil and natural gas extraction there, which itself combusts fossil fuels and will increase the supply of fossil fuels thereby reducing their cost and promoting more combustion, all of which will cause more emissions, exacerbate global warming, and further melt the ice cover (Funk 2014). Planning has begun for shipping between Europe and Asia through the Northwest and Northeast Passages in the Arctic to cut costs of shipping the long way through the Panama Canal. Cruise ships have already begun to visit the Arctic. Russia is particularly active in Arctic extraction and shipping using nuclear-powered icebreakers to break the remaining ice. Shipping in the Arctic combusts fossil fuels and threatens to result in oil spills in a particularly sensitive region where restoration by nature's processes is slower because of the frigid temperature. There are also the perverse effects of last-chance tourism: taking a flight or cruise to see a historic city like Venice or a glacier before they disappear, with the flight or cruise combusting fossil fuels whose emissions contribute to their disappearance. Another pernicious form of adaptation to anthropogenic global warming consists of extracting and combusting fossil fuels quickly to avoid having to leave them in the ground as stranded assets if carbon pricing escalates. Funk (2014) documented many other cases of perverse adaptation to global warming that are intensifying. Perverse adaptation to global warming is arguably the most prevalent form of adaptation at the present time, but nevertheless its study has been largely neglected.

Excuses

One often hears as justification for a country's, an industry's, or a person's greenhouse gas-emissions that their contributions are miniscule compared to that of big emitters. For example, Alberta and its bituminous sands oil industry claim repeatedly that Canada is only responsible for 1.6% of the world's emissions, whereas India and China account for a much higher percentage. This deflects attention away from many significant facts. Canada as a whole only accounts for 0.5% of the world's population, and Alberta for only 0.05%. Canada is among the world's big absolute emitters and one of the biggest in per capita emissions. If Alberta were a country, it would be a big absolute emitter and the world's biggest per capita emitter. The high emissions of India and China are proxies for their high populations despite their low per capita emissions. Moreover, China and India's emissions bring their populations out of extreme poverty, whereas Alberta's and Canada's emissions add to economic growth of already wealthy societies. And India's and China's emissions are recent, whereas Canada and Alberta have been disproportionately placing carbon dioxide in the atmosphere for a century or more. Flannery (2015) shows that this excuse is also used in the high per capita emitting country of Australia because of the importance of coal in its economy. Most importantly, it is incumbent on all emitters, especially high per capita ones, to reduce their emissions, even if they have small populations. Luxembourg is a chronically high per capita emissions country but its total emissions are small because of its small population. Since there are so many small population, high per capita emitters, the cumulative effect is enormous. An enforced international agreement is needed where excuses by the recalcitrant will not be accepted.

There are many problems other than fossil-fuelled climate change in the world: wars, nuclear arms, poverty, racial, gender, religious exclusion, market monopolization, pandemics, etc. If these are used as excuses to avoid dealing with social practices that cause climate change, then danger is discounted, global warming is intensified, and the vulnerable will be threatened even more. Multitasking is possible.

Attachment to Fossil-Fuelled Normality

Beck (1995: 48–49) hypothesized there is a 'death reflex of normality': 'as the hazards increase in extent, and the situation is subjectively perceived as hopeless, there is a growing tendency not merely to accept the hazard, but to deny it by every means at one's disposal'. He argues that bads result in more bads and vicious circles. The attachment to social practices that damage the biophysical environment brings the risk of resulting in irreversible societal collapse, which happened for particular societies (Diamond 2005). There is no reason to conclude this could not occur on a global scale. Beck's normality hypothesis is worryingly close to the present evidence concerning the fossil-fuelled climate crisis. His construction of the reflex-of-normality hypothesis holds up a mirror to let us see what societies are actually doing. Although this is troubling, it is necessary to accurately assess the situation and hopefully take actions needed to allay the crisis. This requires a reality check to determine what is being done. It is indispensable to ask why, when faced with overwhelming scientific evidence about the dangers of fossil-fuelled climate change, there is so little improvement of social practices having harmful side effects. Nordhaus (2013: 325) concludes that 'governments have made little progress in implementing policies [to slow global warming]'. Thus he titles his book *The Climate Casino* implying humanity is rolling the dice and depending on luck. Some obstacles are structural, such as (i) that carbon dioxide is an odorless, invisible gas, (ii) people causing fossil-fuel harm are not the principal victims, (iii) the long lag between both cause and effect and between costly remedial policies and ultimate benefits, and (iv) the need for international cooperation and enforcement. Since populations have developed a sense of entitlement to inexpensive fossil fuels for driving, flying, boating of all sorts, social media servers, electricity, etc., and since charging for harmful carbon pollution will increase their cost, there is strong resistance to a charge even though the money raised could reduce taxes elsewhere, provide more services, finance research to innovate low-carbon technologies, and benefit the economy in the long run.

The danger of fossil-fuelled climate change is that it has the potential to become the mother of all disasters—literally—by spawning multiple catastrophes on a global scale. One possible outcome is that 'the climate-change externality is not corrected. This has been the approach of most nations up to now' (Nordhaus 2013: 245–246). It is the path of fossil-fuel dependency which could be succinctly expressed: societies will continue to enjoy the benefits of fossil fuels; dangers be damned. We are now living in both the golden age of fossil-fuelled practices and the incubation of a global climate change disaster brought about by the massive combustion of fossil fuels. It is likely that latecomers, particularly poor people, poor countries, future generations, and other species, will face a hypercarbonized atmosphere and global warming because of the fossil-fuelled practices of the affluent of the present and previous generations, with all the partly foreseeable, partly unforeseeable adverse consequences that entails. Not only is the habitat of non-human species threatened but also the habitat of humanity. Future forces of nature unleashed by fossil-fuelled climate change could be worse than worse-case scenarios and overwhelm those defenses, as occurred with extreme weather in northeastern North America in 1998 (Murphy 2009), with Hurricane Sandy striking New York City in 2012 (Superstorm Research Lab 2014), and with the 2010 tsunami at Fukushima Japan (Hasegawa 2012). The present period is characterized by a global failure of foresight. Stuck in the fossil-fuelled old normal is the path of least sociopolitical resistance, even when science reveals it brings the greatest threat. This possibility must be confronted rather than discounted in order to take action to get unstuck.

References

Abramovitz, J. 2001. *Unnatural Disasters*. Washington: World Watch Paper.

Baerwald, Erin. 2019. Ontario's Environment Minister Is Playing Batty Politics with Wind-Turbine Decision. *Globe and Mail*, 19 December: A15.

Beck, U. 1995. *Ecological Politics in an Age of Risk*. Cambridge: Polity Press.

Beck, U. 2015. Emancipatory Catastrophism. *Current Sociology* 63 (1): 75–88.

Bjarnason, Egill. 2018. Bitcoin-Mining Boom Gobbles Iceland's Energy Production. *Globe and Mail*, 12 February: B16.

Blumenthal, P. 2014. Mary Landrieu's Fossil Fuel Stance Both a Blessing and a Curse in 2014 Election Bid. *Huffington Post*, 2 October. http://www.huffingtonpost.com/2014/10/02/mary-landrieu-2014-election_n_5921010.html. Accessed 26 August 2016.

Carolan, Michael. 2014. *Cheaponomics: The High Cost of Low Prices*. Abingdon, UK: Earthscan.

Clarke, L. 1999. *Mission Improbable*. Chicago: University of Chicago Press.

Davidson, D., and J. Andrews. 2013. Not All About Consumption. *Science* 339 (6125): 1286–1287.

Diamond, Jared. 2005. *Collapse: How Societies Choose to Fail or Succeed*. New York: Viking.

Dickson, Janice. 2019. Chrystia Freeland Says It's a 'Disappointment' Arctic Council Could Not Issue Joint Communique. *Globe and Mail*, 7 May: A7.

Dunlap, Riley, and Robert Brulle. 2015. *Climate Change and Society: Sociological Perspectives*. New York: Oxford University Press.

Dunlap, Riley E., and Peter J. Jacques. 2013. Climate Change Denial Books and Conservative Think Tanks: Exploring the Connection. *American Behavioral Scientist* 57: 699–731.

Dunlap, R., A. McCright, and J. Yarosh. 2016. The Political Divide on Climate Change: Partisan Polarization Widens in the U.S. *Environment Science and Policy for Sustainable Development* 58 (5): 4–23.

Dyer, Gwynne. 2008. *Climate Wars: The Fight for Survival as the World Overheats*. Toronto: Random House.

Elsasser, Shaun W., and Riley E. Dunlap. 2013. Leading Voices in the Denier Choir: Conservative Columnists' Dismissal of Global Warming and Denigration of Climate Science. *American Behavioral Scientist* 57: 754–776.

Fife, Robert. 2016. Carbon Tax Sets Trudeau at Odds with Trump. *Globe and Mail*, 11 November: A3.

Flannery, Tim. 2015. *Atmosphere of Hope: Searching for Solutions to the Climate Crisis*. New York: Atlantic Monthly Press.

Freudenburg, W. 2006. Environmental Degradation, Disproportionality, and the Double Diversion. *Rural Sociology* 71 (1): 3–32.

Freudenburg, W., and R. Gramling. 2011. *Blowout in the Gulf*. Cambridge, MA: MIT Press.

Freudenburg, William, Robert Gramling, Shirley Laska, and Kai Erikson. 2009. *Catastrophe in the Making*. Washington: Island Press.

Funk, McKenzie. 2014. *Windfall: The Booming Business of Global Warming*. New York: Penguin.

Gelin, Martin. 2019. 'The Misogyny of Climate Deniers'. *The New Republic* August 28. https://newrepublic.com/article/154879/misogyny-climate-deniers. Accessed 6 September 2020.

Germanwatch. 2012. *CCPI The Climate Change Performance Index Results 2012*. http://germanwatch.org/klima/ccpi.pdf. Accessed 5 November 2012.

Germanwatch. 2019. *CCPI The Climate Change Performance Index Results 2019*. https://germanwatch.org/sites/germanwatch.org/files/CCPI-2019-Results-190614-WEB-A4.pdf. Accessed 22 September 2019.

Giddens, Anthony. 2009. *The Politics of Climate Change*. Cambridge: Polity Press.

Goodell, Jeff. 2017. *The Water Will Come: Rising Seas, Sinking Cities, and the Remaking of the Civilized World*. New York: Little, Brown and Company.

Harvey, Hal, and Robbie Orbis. 2018. *Designing Climate Solutions: A Policy Guide for Low-Carbon Energy*. Washington: Island Press.

Hasegawa, Koichi. 2012. Facing Nuclear Risks: Lessons from the Fukushima Nuclear Disaster. *International Journal of Japanese Sociology* 21 (1): 84–91. https://doi.org/10.1111/j.1475-6781.2012.01164.x.

Hawken, Paul. 2010. *The Ecology of Commerce: A Declaration of Sustainability.* Revised Edition. New York: Harper Business.

Heal, Geoffrey. 2017. *Endangered Economies: How the Neglect of Nature Threatens Our Prosperity*. New York: Columbia University Press.

Holman, Sean. 2019. Dear Journalists of Canada: Start Reporting Climate Change as an Emergency. *The Tyee*, 28 May. https://thetyee.ca/Mediacheck/2019/05/28/Start-Reporting-Climate-Change-Emergency/. Accessed 7 July 2019.

IPCC. 2019. *The Ocean and Cryosphere in a Changing Planet*. https://report.ipcc.ch/srocc/pdf/SROCC_SPM_Approved.pdf. Accessed 25 September 2019.

Jacques, Peter, Riley E. Dunlap, and Mark Freeman. 2008. The Organization of Denial: Conservative Think Tanks and Environmental Scepticism. *Environmental Politics* 17: 349–385.

Kelly, Kathal. 2020. Federer Fails to Return Thunberg's Serve on Climate Issues. *The Globe and Mail*, 13 January 2020: B10, B15.

Lockie, Stewart, and Catherine Wong. 2018. Conflicting Temporalities of Social and Environmental Change. In *Environment and Society: Concepts and Challenges*, ed. Magnus Boström and Debra Davidson, 327–350. London: Palgrave Macmillan.

McCright, A., and R. Dunlap. 2010. Anti-Reflexivity. *Theory, Culture, and Society* 27 (2–3): 100–133.

Mileti, D. 1999. *Disasters by Design.* Washington: James Henry.

Murphy, R. 2009. *Leadership in Disaster: Learning for a Future with Global Climate Change*. Montreal: McGill-Queens University Press.

Murphy, R. 2015. The Media Construction of Climate Change Quiescence: Veiling the Visibility of a Super Emitter. *Canadian Journal of Sociology* 40 (3): 331–354.

Nordhaus, William. 2013. *The Climate Casino: Risk, Uncertainty, and Economics for a Warming World*. New Haven: Yale University Press.

Oreskes, Naomi, and Erik Conway. 2010. *Merchants of Doubt: How a Handful of Scientists Obscured the Truth on Issues from Tobacco Smoke to Global Warming*. New York: Bloomsbury Press.

Pielke, Roger, Jr. 2010. *The Climate Fix*. New York: Basic Books.

Platt, R. 1999. *Disasters and Democracy*. Washington: Island Press.

Rand, Tom. 2014. *Waking the Frog: Solutions for Our Climate Change Paralysis*. Toronto: ECW Press.

Rivers, Nic, and Mark Jaccard. 2019. Greener Fuels Not as Expensive as Some Claim. *Vancouver Sun*, 7 June: A15.

Runnalls, David. 2008. Why Aren't We There Yet?: Twenty Years of Sustainable Development. *International Institute for Sustainable Development Commentary*, April 2008: 3–4, 9.

Schnaiberg, A. 1980. *The Environment: From Surplus to Scarcity*. New York: Oxford.

Schneider, S. 2009. *Science as a Contact Sport*. Washington: National Geographic.

Simpson, Jeffrey, Mark Jaccard, and Nic Rivers. 2007. *Hot Air: Meeting Canada's Climate Change Challenge*. Toronto: Emblem McClelland & Stewart.

Speth, James Gustave. 2009. *The Bridge at the Edge of the World: Capitalism, the Environment, and Crossing from Crisis to Sustainability*. New Haven: Yale University Press.

Superstorm Research Lab. 2014. A Tale of Two Sandys. *White Paper*, December. New York. https://superstormresearchlab.files.wordpress.com/2013/10/srl-a-tale-of-two-sandys.pdf. Accessed 13 February 2015.

Suzuki, David, and Ian Hanington. 2017. *Just Cool It: The Climate Crisis and What We Can Do*. Vancouver and Berkeley: Greystone Books.

UNEP United Nations Environmental Programme. 2019. *Emissions Gap Report 2019 Executive Summary*. Nairobi: UNEP. https://wedocs.unep.org/bitstream/handle/20.500.11822/30798/EGR19ESEN.pdf?sequence=13. Accessed 26 November 2019.

Yale University. 2012. *EPI Environmental Performance Index 2012*. http://epi.yale.edu. Accessed 8 November 2012.

Yale University. 2018. *EPI Environmental Performance Index 2018*. https://epi.envirocenter.yale.edu/epi-country-report/CAN. Accessed 22 September 2019.

6

A Pattern When Exploiting Valuable but Dangerous Resources

Natural science provided an understanding of the causes and consequences of anthropogenic climate change. It also provided the solution. 'We now know that long before 2050 we must have eliminated the burning of all fossil fuels, if we are to stave off a climate crisis' (Flannery 2015: 204). Steffen (2015) argues that eighty per cent of fossil fuels reserves already listed on stock exchanges must not be combusted to stay within 2 °C of global warming and must be left safely in the ground. Natural science even presented a roadmap to keep global warming to 2 °C, which would prevent the most serious calamities. To meet the 2015 Paris Agreement commitments, the least carbon-polluting fossil fuels need to be used, as slowly as possible, and most coal reserves and much heavy oil reserves need to remain safely underground. Some believe this is happening. A study by the investment firm PNB Paribas (Lewis 2019: 3) concludes that 'the oil industry has never before in its history faced the kind of threat that renewable electricity in tandem with EVs [electrical vehicles] poses to its business model … and could replace up to 40% of global oil demand if it had the necessary scale'. Lewis argues this constitutes 'the death toll for petrol' and that economics are transforming

R. Murphy, *The Fossil-Fuelled Climate Crisis*,
https://doi.org/10.1007/978-3-030-53325-0_6

that industry into relentless and irreversible decline over the next twenty-five years. Energy storage and EVs are becoming cleaner and cheaper than fossil fuels and combustion engines, rendering long-term investments in the latter risky. 'If all of this sounds far-fetched, then the speed with which the competitive landscape of the European utility industry has been reshaped over the last decade by the rollout of wind and solar power – and the billions of euros of fossil-fuel generation assets that this has stranded –should be a flashing red light on the oil industry's dashboard' (Lewis 2019: 3). Butts (2019: O8) concludes that 'just like the whale oil that came before it, there was a time and place for fossil fuels, which had their day. But the writing is on the wall'. If Butts and Lewis prove to be correct, fossil-fuel reserves and companies holding them will become devalued, with shareholders facing loses, and countries exporting crude oil will have to find a different driver for their economies.

But is this correct? Are societies on the road specified by the natural science roadmap to a solution? It is unlikely that the fossil-fuel industry will give up without a struggle. The present chapter documents the road actually being travelled concerning one type of unconventional oil. It specifies a pattern of response to scientifically documented danger and of the tension between local near-term economic growth and long-run global environmental safety.

The case to be examined involves Canada's and Alberta's exploitation of its bituminous sands (also called oil sands and tar sands). They contain the world's third largest known oil reserves. Alberta's tar sands oil and Venezuelan heavy oil (the biggest reserve) emit more greenhouse gases per barrel to extract and upgrade than oil from wells in the other large reserve (Saudi Arabia), and the trend is to extract more oil from these unconventional sources (Davidson and Andrews 2013). Alberta has visions of becoming an oil superpower—Saudi Arabia of the north. In Canada, provinces rather than the federal government have control over natural resources; only they and not the federal government receive royalties from their exploitation.

Of 220 countries in the world, Canada was the eighth highest absolute emitter of carbon dioxide equivalent greenhouse gases (Harvey and Orbis 2018: 29–30) and ranks even higher in terms of per capita emissions (CDIAC 2018). Since 1990, Canada has been emitting between

600 and 750 megatonnes of carbon dioxide equivalent every year into the atmosphere, where it will remain and accumulate for a century. Canada ranked 25th out of 180 countries on the 2018 Yale University Environmental Performance Index of countries, including 6th on environmental health, 4th on air quality, 1st on drinking water, but 137th on climate and energy (Yale University 2018). In the latter, it ranked 132nd in total emissions intensity, and poorly in methane, black carbon, and N_2O emissions intensity. On the Energy Transition Index for 2020 (World Economic Forum 2020), Canada ranked 28th, slightly ahead of the USA (32nd) and Australia (36th) but well behind the leaders in Nordic countries, Switzerland, Austria, Britain, and France.

Canada has one of the world's highest proportions of electricity produced from zero-emissions primary energy, principally hydro and nuclear. Its most populous province of Ontario closed all its coal-fired electricity generation. However, emissions remain high because its cities are characterized by urban sprawl—'75 per cent of new housing in Canada over the past decade has been built as sprawl' (Keesmaat, McKenzie, and Florida 2020: O5)—with the automobile used for the home-work commute, and especially because of oil extraction from Alberta's bituminous sands, which is the country's fastest growing source of emissions. Extraction and upgrading requires energy hitherto supplied by fossil fuels. Recent research (Liggio et al. 2019) even strongly indicates that oil sands emissions have been underestimated. Although Alberta extracts and exports almost 4 million barrels of crude oil a day (most being diluted bitumen from the tar sands), it wants to export more. Land-locked Alberta blames inability to do so on lack of pipelines through neighbouring provinces due to opposition from them, from environmentalists, and from indigenous groups. A growing obstacle is the high cost, high emissions required to extract oil from sand in a remote location. 'Contrary to pundits eager to vilify climate policy and a lack of pipelines, Alberta's oil patch woes are macroeconomic [declining world oil price] and its heavy oil is destined to be uncompetitive' (Rand 2018: B4). For its investments, Shell states that it now 'targets new projects that can be profitable at oil prices of US$20 to US$30 a barrel and which emit relatively low greenhouse emissions' (Bousso and Zhdannikov 2019: B8). Alberta's oil industry is already struggling for profitability at US$60 a barrel, and extracting oil from bituminous sand and upgrading it results in more emissions than pumping it from a well. McKibben (2019: O8) argues that 'keeping Alberta's oil underground is going to become as crucial a global priority as keeping the Amazon rain forest standing tall'. Alberta has extensive solar energy potential in its south, much wind and hydroelectric potential coming off the Rocky Mountains onto its prairies, massive natural gas reserves in its north, and some geothermal energy potential. But Alberta insists on extracting and exporting crude oil from its tar sands.

In 1997, Canada ratified the Kyoto Protocol and committed to reduce its emissions by 6% compared to its 1990 level by 2012, then proceeded to increase its emissions and officially reneged on its commitment in 2011. Canada's policies regarding global warming were characterized as 'hot air' (Simpson, Jaccard, and Rivers 2007). Canada signed the 2015 Paris agreement on climate change, committed to reduce its greenhouse-gas emissions 20% by 2020, 30% by 2030, and 80% by 2050, but in 2017 its government endorsed construction of pipelines carrying emissions-intensive bitumen from Alberta to the USA and Asia. In a study of the progress of G7 countries towards ending government support for fossil fuels, Canada placed 1st for ending support for coal mining and replacing fossil-fuel based electrical power, but placed dead last for ending support for oil and gas production (Whitley et al. 2018).[1] This reflects government support for high-emissions extraction of oil from the bituminous sands, which is the Achilles heel of Canadian climate change mitigation.

Alberta has had conservative governments inspired by anti-government ideology relying on private companies to find, extract, and sell oil. It made itself dependent mainly on foreign and particularly American oil companies, which did not develop a local value-added petrochemical sector to maximize benefits for the Alberta population, instead exporting as much crude oil as possible to heavy-oil refineries in Texas and Washington State. Like most crude oil exporting states, Alberta is vulnerable to oil price swings and boom-and-bust economic cycles, unlike Texas where low prices for crude oil make its refineries more profitable and compensate losses in its extracting sector. In 2015, Alberta changed from a conservative government to a social democratic one, which implemented measures to reduce emissions intensity in order to have a social license to build pipelines to export more bituminous oil. But in 2019, it reverted to a conservative government that refuses a price on carbon, seeks federal tax reductions for fossil fuels because they are non-renewable, and continues combusting coal for electricity generation.

With borders on three oceans, Canada is particularly vulnerable to sea level rise, and is experiencing warming about twice the global rate and northern Canada about three times the global rate, already reaching 2.3 °C (Bush and Lemmen 2019). Much like Australia and California, Western Canada has experienced massive forest fires. It has also experienced floods and invasions of non-indigenous insects no longer killed by cold winters, causing health risks like Lyme disease and asthma. Nevertheless, as a cold country many Canadians question what is bad about warming. Stoddart and Smith (2016: 266) give a valuable analysis of the 'leaders and laggards' and 'localized cosmopolitanism' in Canada's diverse relationships to fossil-fuelled climate change. Canada is a particularly interesting case to investigate tension between the pursuit of near-term economic prosperity and long-term environmental and economic sustainability, between foresight and discounting danger. It is studied to examine the depth of the climate crisis, but should not be interpreted as claiming that Alberta and Canada are uniquely bad carbon-polluting states. Russia is forging ahead with fossil-fuel exploitation of the Arctic opened up by global warming. The USA is proceeding at full speed with hydraulic fracturing and has withdrawn from the 2015 Paris Agreement. On the Energy Transition Index, 'fuel importing countries continue to outperform fuel exporting countries, as the gap between their average scores increased' (World Economic Forum 2020: Executive Summary). Resistance to decreasing carbon pollution and to reducing the rate of fossil-fuel extraction is likely greater in countries whose economies are totally dependent on the extraction of fossil fuels, such as Russia and Saudi Arabia, than in Canada which has a more diverse economy.

Near-Term Economic Benefits Bringing Long-Term Harm: A Pattern

Are Canada and Alberta taking the road specified by natural science in its roadmap to mitigate global warming, namely having the foresight to transition to low-carbon energy sources of economic prosperity? Or are they discounting scientific warnings and taking the road of accelerating the extraction of bitumen to achieve near-term economic growth. The road actually being travelled fits the following pattern of steps.

Step 1[2]—The Geological Pre-human Epoch

It is important to start from nature's services in producing resources and avoid the oversimplification that extractive industries produce them. 'Fossil fuels contained hundreds of millions of years' worth of sunlight. The energy was stored in carbon-based molecules formed when ancient plants decomposed anaerobically under layers of sediment to form incredibly energy-dense solids, liquids and gases: coal, crude oil and natural gas' (Berners-Lee and Clark 2013: 8). This carbon accumulated safely underground. One form was bitumen containing oil. When Europeans first came to Alberta and saw the black gooey viscous sand, they called it 'tar sands'.

Step 2—Trial and Error

Applied science, driven by visions of profit, sought technological innovations for decades during the early and mid-twentieth century to make extraction of oil from bituminous sands profitable, but failed. Hence it remained safely in the ground.

Step 3—Euphoria

In the 1990s, the Canadian government granted tax incentives to the industry. Those tax breaks and the rising price of crude oil made extraction profitable. Technology was developed using combustion of natural gas to boil water to flush bitumen from the sand. It was then diluted with light crude oil or natural gas condensates to make it flow and enable it to be pumped in pipelines. The result was an oil rush by transnational oil companies to extract the oil, which by 2017 resulted in the export of 3 million barrels a day with projections increasing to 6 million barrels.

Step 4—Eye Opener to Problems

The combustion of large amounts of natural gas results in huge quantities of greenhouse-gas emissions. The oil industry clear cuts large tracts of Boreal forest, labelled 'overburden', to get at the bitumen below, devastating the habitat of the northern Alberta caribou herd. Water-assisted and steam-assisted extraction leaves polluted tailings ponds. Only a small fraction of that land has been renatured (Leahy 2019: 20). Oil extraction from tar sands, now euphemistically labelled oil sands, is a high emissions, high-cost process. Profits are sensitive to highs and lows of crude oil prices and many bankruptcies of small extracting companies occurred. This resulted in many 'orphan wells', which need to be cleaned but is costly and constitute a huge, growing unpaid cost of oil extraction.

Step 5—Denial of Problems

Deforestation and tailings ponds are visible and impossible to deny, but can be ignored since they are located in remote areas. Carbon dioxide, though, is invisible to the senses. Claims were made that global warming was not occurring, only normal changes from cold to warm weather. Whatever warming occurred was not due to fossil fuels but to changes in the Earth's orbit, the sun's radiation, etc. The Canadian Association of Petroleum Producers (McMillan 2019b) highlighted improved intensity-based measures indicating somewhat lowering of emissions per barrel of oil extracted, although still higher than from wells. This falsely implied warming is being mitigated even while it is worsened by increasing the number of barrels extracted. Behind-the-scenes lobbying of government decision-makers and overt lobbying of the population by the industry has been massive (McKay, Pourbaix, and Evens 2019). It brands itself as the 'energy sector', thereby masking differences between low-carbon energy of wind, solar, and hydro compared to high-carbon tar sands oil.

Step 6—The Accumulation of Both Danger and Scientific Evidence

During the slow-onset incubation of harm period, scientific evidence builds up of cumulative atmospheric carbon and higher atmospheric temperature. Science increasingly confirms that fossil-fuel emissions cause global warming, that deforestation reduces nature's capacity to withdraw carbon from the atmosphere, and that extracting oil from bituminous sands powered by combusting natural gas emits more greenhouse gases than extracting oil from wells. Evidence also occurs globally in the form of more frequent and severe hurricanes, floods, and drought-generated wildfires. People see melting of glaciers and of Arctic ice cover, first-hand for some and through the media for others.

Step 7—Displacement of Blame

The scientific evidence led to the acceptance by most of the population that combustion of fossil fuels is causing global warming. Debate then shifts to who is responsible. Everyone who uses fossil fuels is responsible for man-made global warming to an extent, but some countries, companies, and individuals are disproportionately more responsible than others (Freudenburg 2006).

Proponents of Canada's bituminous sands claim it only produces part of Canada's emissions, that Canada produces only 1.6% of the world's emissions, hence keeping their oil in the ground 'can have only one effect: countries with much lower environmental standards will fill the void left by Canada, only too happy to meet global energy demand with their products' (McMillan 2019b; see also Wente 2018: A15). If Canada doesn't extract fossil fuels, another country will. This is the master argument used to legitimate carbon pollution that every country can use: my country shouldn't do the mitigation scientists recommend because other countries won't. Blame other countries, not mine. Hence oil extraction increases, decreasing its price, increasing its consumption, combustion, and carbon pollution. This strategy of displacing blame could be called the Hummer fallacy. There are so few Hummers that collectively they produce a small percentage of emissions, but they each emit more than lighter, smaller, fuel-efficient vehicles. Similarly for the tar sands, the significant issue is whether extracting oil from bituminous sand causes more emissions per barrel than extracting it from wells, and it does, or more emissions than using low-carbon energy. In a multicausal world, any one source only produces a fraction of global emissions. Claims that Alberta has higher environmental standards than countries like Saudi Arabia and Norway are belied by the fact that its emissions per barrel are higher. Because of its exploitation of tar sands and use of coal, Alberta emits far more greenhouse gases than its share of the world's population would warrant, so its claims about environmental standards ring hollow. Flannery (2015: 70) documents that the same fallacy is deployed in the coal country of Australia to justify its disproportionately high-carbon emissions, adding 'that 180 individual nations out of the world's 193 are each responsible for less than 2% of global emissions, which is precisely why it's important for all of those "rounding errors" to work together' to reduce emissions because they add up.

The rebuttal to these someone-else-should-solve-it claims, which prioritize near-term economic growth at any cost, involves enforcing binding international agreements, with achieving the 2015 Paris Agreement commitments as a start. This is what occurred with CFCs in the Montreal Protocol and was successful. Canada did not try to justify its use of a small fraction of the world's CFCs; instead it led in phasing them out and negotiating the Montreal Protocol, which should be a template for mitigating carbon pollution.

The Canadian finger is usually wagged at China for its high emissions (Murphy 2015). But this ignores that (i) China's emissions are a proxy of its large population and that Canada's per capita emissions are four times greater, (ii) that China's emissions are growing to enable people in extreme poverty to become less poor whereas Canada's emissions enable the affluent to become more prosperous, and (iii) China's increased emissions are recent, whereas Canada has been emitting carbon into the atmosphere for a long time where it remains for a century and accumulates. The claim is also made that Canadian bituminous oil is no worse than Venezuelan heavy oil, which is true but implies a race to the bottom because they are both higher emissions sources than from wells.

A distraction strategy was next used by drawing on Canada's overall good international reputation. Since Canada does not persecute gays and lesbians and is a leader in women's rights, unlike Saudi Arabia, Iran, etc., proponents of bituminous sands exploitation claim that Canada extracts and exports 'ethical oil' (Levant 2010; McMillan 2019b) despite more emissions extracting from bituminous sands than from Saudi wells (see Chapter 2 Table 2.1 and Murphy 2015). Agreed that the exclusionary practices of Middle Eastern countries are unethical, but they are not intrinsically about oil and should not be used as an excuse to pollute. Moreover, the tar sands oil industry is pushing to build pipelines to sell its oil to China (Willis 2019: B1, B5). If it is unethical to buy oil from a country with an oppressive regime, why is it not unethical to sell oil to a country governed by the dictatorship of the Communist Party thereby providing it with the energy to persecute Ugers and repress Hong Kong? This strategy of misplaced ethics to legitimate the tar sands could backfire on Alberta.

Step 8—The Empire Strikes Back

Most Canadian support for mitigating fossil-fuelled climate change and pricing carbon pollution comes from Eastern Canada. In response, the book entitled *Let the Eastern Bastards Freeze in the Dark* (Janigan 2012) was a big success in Alberta. The principal Canadian science educator, Dr. David Suzuki, has become the whipping boy of tar sands proponents, just like Al Gore in the USA. The *modus operandi* of proponents is: if you criticize my industry, I will attack you personally. Conservative politicians attempted to diminish the Canadian environment minister, an attractive blond woman who negotiated the Paris Agreement, by labelling her 'climate Barbie'. Their followers spray painted the slur 'cunt' on her election office and yelled obscenities when she walked with her children (Blanchfield 2019). She requires security because of threats. Another Conservative politician who supports the tar sands industry depicted teenage climate activist Greta Thunberg, who has Asperger syndrome, as mentally unstable. He claimed she is an alarmist, a threat to Canada's prosperity and civilization, and should be attacked (Mason 2019b: A13).

When Europe's largest bank HSBC Holdings PLC decided to curtail lending to new projects in high-emissions Alberta oil sands in its commitment to mitigate global warming, it was attacked by oil sands proponents (Milke 2019). The head of the Canadian Association of Petroleum Producers wrote articles entitled 'HSBC's stunning energy and human-rights hypocrisy: It ditched oilpatch only to embrace Saudi Arabia' and 'HSBC would rather deal with oligarchs than responsible Canadian energy companies' (McMillan 2019a and b). This conveniently ignores (i) that oilpatch oil extracted from tar sands has a significantly higher carbon footprint than oil from Saudi wells, and (ii) that Alberta oil companies are pressuring governments to construct pipelines to export their oil to Chinese oligarchs. Alberta's conservative premier ended all government business with HSBC. France's BNP Paribas SA and its giant AXA Equitable Financial Service LLC, Switzerland's Zurich Insurance Group Ltd., and several American state pension funds also divested from the oil sands. The premier dismissed this preoccupation with climate risk as 'the flavour of the day' (McCarthy 2019a: B3). His slogan attacking carbon taxes claimed it punished people for heating their homes and driving to work. That electoral sound bite doesn't make sense. The oil extracting country of Norway has a carbon tax and its people heat their well-insulated homes, travel to work in electrified public transit, and are more prosperous than those in Alberta. Norwegians recognize that a carbon tax finances home insulation and public transit and thereby mitigates punishing their grandchildren by bequeathing them a degraded atmosphere. The Alberta premier attacked British Columbia and Quebec for preventing additional pipelines across their territories, threatening neighbouring British Columbia with stopping supplies of oil and natural gas. A significant segment of Alberta's population is threatening to separate from Canada if pipelines to tidewater are not built promptly, but separation would only exacerbate its land-locked status. Alberta's government is launching a legal battle to prevent a national price on carbon pollution. It paid consultants to battle Greenpeace in court. The Alberta fossil-fuel industry incited the population to attack the previous social democratic premier for her environmental measures, and Canadian Prime Minister Trudeau for not ramming through more pipelines against the will of other provinces and indigenous groups. Trudeau was their scapegoat despite approving two pipelines and paying billions of taxpayers' dollars to purchase and enlarge a third, which amounted to a massive subsidy for the fossil-fuel industry.

Step 9—Discounting Danger

More pipelines, namely the Keystone XL (extra large) opposed by Obama but approved by Trump and Line 3, will increase export of bitumen beyond its present 3 million barrels daily. Oil sands proponents assume higher prices in Asia. So a private company planned to triple the capacity of an existing pipeline, but British Columbia and coastal indigenous groups oppose it because it will lead to three times more oil tankers in environmentally sensitive waters and they fear a spill of difficult-to-clean diluted bitumen. Global warming is rarely mentioned but underlies opposition to exporting more high-emissions oil. To get the pipeline built, the federal government bought it and will pay to construct the bigger adjacent one for $12 billion. For that it is being criticized by both sides. Tar sands promotors claimed that 'Trudeau can't solve climate change and have his pipeline too' (Morton 2018: A13), arguing that, since the Paris Accord, Trudeau has assisted the anti-pipeline movement by stigmatizing the bituminous sands, repeating nonsense about social license, imposing carbon taxes, closing down coal-fired electricity, and subsidizing renewables. Trudeau could only reconcile extraction and environment by speaking out of both sides of his mouth. Proponents of reducing emissions come to similar conclusions for different reasons.

Homer-Dixon and Strauch (2018) contend that Canada's gamble of increasing extraction from oil sands and building pipelines, yet acting responsibly by fulfilling its Paris commitment, involves double-speak and wilful self-delusion. If the rest of the world succeeds in cutting emissions, oil demand, and price will be reduced, hence the first oil left in the ground will be high emissions, high-cost oil from bituminous sands. Canada's policy of increasing extraction and building long-lasting pipelines will succeed economically only if the Paris Accord fails. It amounts to approval now for a pipeline resulting in more emissions in exchange for uncertain cuts of future emissions. 'We can't have our lucrative oil sands profits and a safe climate too' (Homer-Dixon and Strauch (2018: O3). Instead of using $12 billion to subsidize fossil fuels, much solar, wind, hydroelectric, and geothermal energy could have been built in Alberta promoting economic growth and jobs in the clean energy sector.

Reconciling growth of the bituminous sands oil sector and mitigating climate change is problematic because the first is easy but the second is sociopolitically difficult. It is uncertain (i) whether escalating carbon taxes sufficient to reduce emissions will be accepted by industry and the population, (ii) whether coal phase out and reduction of methane venting and leakage will be implemented, and (iii) whether subsequent governments will implement commitments. The worse-case emissions scenario would be building pipelines promoting growth of bitumen extraction but rejection of commitments to reduce emissions. Even with well-intentioned governments in Alberta and Canada, promotion of bitumen extraction is strong but overall emissions reduction is puny. Moreover, more pipelines facilitate emissions after export in other countries.

In a state whose economy is based on oil extraction, even social democratic governments are pushed into being nationalist salesperson for its oil, even if that involves disproportionate carbon pollution. Alberta's previous social democratic premier argued that 'the choice is whether that oil will come from a responsible, progressive and forward-looking energy producer like Alberta, or from somewhere with no standards to speak of, like Russia' (McCarthy and Tait 2017: A14). How then does Alberta have one of the world's most carbon-intensive types of oil? As shown in Table 1 in Chapter 2, extracting oil from Alberta's tar sands results in more emissions than extracting oil in Russia because more natural gas has to be combusted to extract oil from sand. Only by mistaking policy discourse for actual emissions could she argue that Alberta is a responsible, progressive, and forward-looking energy producer. Lower emissions would only occur if regulations were sufficiently strict to offset differences in physical properties between Albertan and Russian oil.

Big greenhouse-gas emitters have demonstrated a capacity to capture their regulators. In 2019, a joint federal-Alberta review panel recommended approval of an oil sands extraction project despite finding it would destroy wetlands and old growth forests, threaten vulnerable species, and emit four million tons of carbon dioxide yearly during its 41-year operation. The panel predicted it would extract 260,000 barrels of bitumen daily creating jobs and contributing royalties and taxes to governments, concluding that adverse impacts were outweighed by economic benefits, and that environmental damage is justified (McCarthy 2019b).

The media have also been captured. Drought resulted in devastating wildfires in the tar sands oil region at Ft. McMurray in 2016, a year earlier at neighbouring La Ronge, in 2011 at nearby Slave Lake, in 2017 in neighbouring British Columbia, and in 2019 wildfires caused severe air quality problems in Edmonton. The Canadian media rarely attributed the wildfires to global warming caused by fossil-fuel emissions, of which Canada's fastest growing source is the tar sands oil industry.

Alberta's present conservative government is lobbying the Canadian government to use Article 6 of the Paris Accord, called 'internationally transferred mitigation outcomes', to receive emissions credits for exporting liquefied natural gas (LNG) to Asia to displace coal there (Jang 2019: B2; Cryderman 2019). If permitted, then credits could be used to offset Alberta's emissions, perhaps tar sands emissions would disappear in creative carbon accounting statements. But natural gas produces significant carbon pollution if methane leaks, emissions from its liquefaction and shipping long distances, etc., are counted. And Article 6 is intended to reduce global carbon pollution, for example if Alberta used part of its oil wealth to finance tree planting or wind and solar energy in poor countries, like Norway does. Alberta's attempt raises also issues of double counting emissions reductions: countries where reductions occurred hitherto received credits; now Alberta wants states supplying natural gas to receive credits; but it must not be both. If Canada is permitted to use this loophole, then all LNG exporting countries would be permitted. Canada could also be credited with reducing American coal emissions because it sells massive amounts of hydroelectricity to the USA, France could be credited with reducing German coal emissions when selling nuclear energy to Germany. This manoeuvre shows how far oil extracting states will go to justify their carbon pollution instead of reducing domestic emissions.

Step 10—An Oil Identity

Material interests lead to ideal interests (Stoddart 2012). Alberta's prosperity based on oil exports has resulted in an oil identity, which emphasizes economic benefits and discounts danger. Companies, communities, and workers in the bitumen industry claim its good name is slandered when science presents evidence that fossil fuels cause harmful climate change. They react angrily when shown to be polluters. Milke (2019: B4) claims that HSBC's policy of divesting from oil sands 'blackens Alberta's reputation'. Suggestions for transition to low-carbon renewable energy, of which there is much in Alberta, are perceived as insulting. Climate science criticism of fossil fuels is interpreted as criticizing Alberta's identity, implying its prosperity is bringing harm to others including future generations of Albertans. This is hard to accept. Feelings are hurt. Defending against criticism of fossil fuels involves defending both Alberta's economic interests and its identity.

Alberta's conservative government created a $2.5 million public inquiry into foreign-funded environmental critics of the oil sands for having defamed the industry (Giovannetti 2019b: A6). This despite the fact that less than 10% of their funding comes from foreign sources, whereas 43% of Alberta's oil industry is foreign owned (Kopecky 2019: O9). Audits found no wrongdoing by environmental groups. Alberta's government allotted $30 million for a rapid-response war room to combat green left critics of the oil sands, specifically *National Geographic* for an article (Leahy 2019) arguing that Alberta's oil sands is the world's most destructive oil operation. That government threatens lawsuits to silence environmental organizations and strip away their charitable status. It claims Alberta is a world leader in decreasing its environmental footprint, which is patently false. Alberta's oil sands are the country's fastest growing source of emissions, thereby enlarging not only Alberta's but also Canada's harmful environmental footprint. Oil executives and conservative politicians created the *Modern Miracle Network* to propagate the narrative of natural gas and oil as modern miracles and deflect attention away from their dangers. They equate a cap on emissions as a cap on prosperity, thereby having no insight that a cap can induce innovation which brings low-carbon, long-term prosperity.

Another strategy of rhetorical purification involves branding. Their original name 'tar sands' was changed to 'oil sands' to sound more appealing. Fossil fuels are not mentioned, rather the word 'energy' is used, thereby masking the difference between clean, low-carbon energy, and carbon-polluting energy. The Conservative Party is creating a 'Canadian clean' brand claiming that 'the world needs more "clean" Canadian oil and gas' (Clarke-Whistler 2019: B4). But Alberta's regulations fail to reduce emissions to Saudi levels because extraction from bituminous sand produces more emissions than extraction from wells. Calling it 'Canadian clean' is a public relations attempt to misrepresent the difficulty of cleaning the resource the industry starts with.

Economic interests are important, but cultural factors and identity politics are involved as well. Renewable energy can be innovated to replace fossil fuels, but it is harder to innovate a new identity to replace Alberta's proud history of producing seemingly safe fossil fuels. A similar identification with coal in Appalachia was fostered by that industry (Bell and York 2010).

Step 11—Social Injustice

Alberta is by far the wealthiest province in Canada, as measured by GDP per capita (McGugan 2019), because it extracts and exports more than three million barrels of crude oil per day mainly from its tar sands. It had high GDP even in 2018, when low international oil prices resulted in Alberta having higher unemployment than the national average. Oil wealth is monopolized in Alberta because of its low corporate and personal income taxes, with large segments of the population excluded from prosperity and hurting. GDP per capita is a misleading indicator of overall well-being when wealth is not distributed equitably.

Furthermore, fossil-fuelled climate change consequences will be suffered mainly by the world's vulnerable, especially low-lying poor countries and future generations. Not only oil is being exported, but also harmful consequences. Nevertheless, proponents of fossil fuels turn the social justice argument upside down, claiming fossil fuels are cheap and help the poor. The CEOs of three fossil-fuel companies claimed in an ad in thirty newspapers that exporting tar sands oil contributes to 'a growing global economy that is expected to lift three billion people out of poverty in the decades ahead' (McKay, Pourbaix, and Evans 2019: A6). Opponents of carbon taxes claim they punish the poor. This can be avoided by recycling money raised from carbon taxes to the poor in ways enabling them to make money by reducing their emissions. And global poverty could be reduced by sending some oil profits to poor countries for offsetting, as Norway does.

To gain indigenous acceptance of fossil-fuel extraction and pipelines, poor indigenous communities in the bituminous sands area and along pipeline routes are being solicited by big extraction corporations to own shares in pipelines and form companies doing contract work (Alexis and Poscente 2019). Some indigenous communities have accepted for economic reasons, but others have not for fear of polluting their traditional environment. This has divided indigenous communities.

Step 12—Dependence on Technological Solutionism

Technological innovations resulting in emissions-free exploitation of bituminous sands are hoped for (Giovannetti 2019a). Cold solvent catalysts producing no carbon emissions were suggested to replace combusting natural gas. Black (2018) claims we can have it all with no pain using that technology: 'Time to do away with carbon taxes, to save the planet, and spur growth in the oil sands'. So far it hasn't succeeded, and emissions keep rising. Canadian conservative parties oppose carbon taxes, and their solution would require big carbon polluters to invest in technology, but with no assurance that technological innovation will reduce emissions.

Existing technologies could entirely eliminate upstream emissions, making bitumen-extracting emissions lower than extraction from wells. Natural gas could be replaced by the abundant hydroelectricity from neighbouring provinces of Manitoba or British Columbia to supply energy for extraction. A nuclear reactor could supply emissions-free energy, which was planned but dropped because of upfront costs (Flannery 2015: 115), even though the massive bitumen reserves would render it economical over the long run. Wind or solar farms could also be used to replace the combustion of natural gas.

Step 13—Use It or Lose It

Alberta is already exporting over 3 million barrels of oil daily and wants to increase that to 6 million. Pipelines are being built south (Line 3 and Keystone extra large) to heavy oil refineries in the USA, and west (Trans Mountain) to a port then shipped to heavy oil refineries in Asia. A planned Energy East pipeline to ship crude oil to European refineries was withdrawn because of low oil prices and opposition along the pipeline route. With all these pipelines, it is as if Alberta and Canada seek to drain the bituminous sands of oil quickly before renewables supplant oil or binding international limits are placed on oil. 'As a high-cost, high-emissions resource, oil sands will not expand as oil prices fall in a declining oil market' (Jaccard 2018: A13). Bitumen could become undesirable oil because of cost or reducing emissions (van Lierop 2018). The International Energy Agency forecasts global oil demand will peak between 2023 and 2030, plateau for several years, then decline because of the development of renewable energy, storage, and electric vehicles (Mason 2019a). Prices will decrease, especially if Venezuela, Iran, Iraq, and Nigeria return as major suppliers. Exporting oil quickly before it is replaced by low-carbon energy is Alberta's version of what Sim (2012) calls the 'green paradox'. Increasing extraction, is likely a prelude to Canada's failure to achieve its Paris commitments. Since Alberta's tar sands constitute the world's third largest oil reserves, emitting all that carbon pollution would result in a huge addition to global warming. If on the contrary, Canada's Paris goal of an 80% reduction in emissions from its 2005 level is implemented, 'every other industry and province would need to go to near-zero emissions to allow room for increased – or even existing – bitumen production in Alberta' (Rand 2018: B4). This would put enormous pressure on 95% of the economy to allow 5% to pollute. It would also likely wreck the Canadian confederation by forcing a sky-high-carbon tax on other provinces to meet Canada's Paris commitments. Two outcomes are to either drain the bituminous sands of its oil before implementing the Paris Agreement or renege on that agreement, like Canada did with the Kyoto Accord. A third possibility would be for Alberta to exploit its ample wind, hydro, and solar energy. But developing those resources requires innovation and economic risk-taking, whereas extracting fossil fuels involves continuing on the same path, which is more convenient in the near term but more environmentally risky in the long run.

Step 14—To Be Determined

No descriptive label is yet possible for this step because there are opposing dynamics at work: pursuing near-term fossil-fuelled economic goals versus long-term low-carbon ones. Albertans and their government are pushing hard to extract and export more bitumen than the present 3 million barrels a day. But external pressure is forcing them to implement a small carbon tax on big carbon polluters. The Canadian government is trying to implement a broad-based carbon tax. Internationally there is pressure to divest from high-emissions oil. The outcome of these opposing dynamics is uncertain. The first thirteen steps demonstrate there is a long, winding, arduous road in the transition from the entrenched path of a fossil-fuel economy to a low-carbon economy, especially for countries, companies, communities, and workers with valuable but dangerous fossil-fuel reserves. Thus Stoddart, Mattoni, and McLevey

(2020) ask '*Can Oil Extraction and Nature Conservation Co-Exist?*' as the subtitle of their book.

A Comparison

Are there precedents for phasing out otherwise profitable, useful resources when they are scientifically demonstrated to be harmful, precedents that could serve as guides to the difficulties and possibilities involved? To better understand the challenges, the socioeconomic dynamics that occurred when another valuable resource was found by science to be harmful will be investigated. Wynn (2016: xvi) astutely asks: 'should not this story of asbestos in Quebec lead us to ask whether there are parallels in the ongoing development of the oil and gas industry in Western Canada even as scientific evidence of the global perils of greenhouse gas emissions mounts?'. Bituminous oil and asbestos are similar in that they are highly valued for their properties, but are threatening because those properties cause slow-onset harm to the environment and human health, respectively. Both bring near-term benefits and long-term harm. Is the political, economic, and cultural pattern documented for bituminous oil similar to the one that emerged in learning that asbestos is harmful? This is particularly important because the end point of the sequence for asbestos is known, but for fossil fuels it is as yet unforeseeable. Since the world's largest deposits of both asbestos and bituminous oil are in Canada, this societal variable is controlled and a direct comparison is possible.

Obviously there are differences. Societies never became as dependent on asbestos as they are on oil. Nevertheless, in the early and mid-twentieth century asbestos was becoming an almost indispensable resource for modern safety, however hard that is to imagine today after its abandonment. Extraction and use of asbestos result in mainly health threats with environmental threats being secondary, whereas fossil fuels result in mainly environmental threats with health threats being secondary. Furthermore, the harm of asbestos to health was more visible than adverse consequences of fossil-fuelled global warming.

Near-Term Economic Benefits Bringing Long-Term Harm: A Similar Pattern

Step 1—The Geological Pre-human Epoch

Nature's dynamics create raw materials (asbestos, fossil fuels, lead, uranium, etc.) over a geologically long period and they accumulate safely underground. Asbestos is fibrous rock, which can be broken to resemble raw cotton, withstand temperatures in excess of 3000 °F, and be added to other substances to augment resistance to burning, rust, and decay.

Step 2—Trial and Error

Many of asbestos' properties were known in the nineteenth century. Profit-driven applied science sought technological innovations to render it useful and profitable, but failed. In this early period, it largely remained stored safely in the ground.

Step 3—Euphoria

By the 1870s, manufacturers were using asbestos-based building products to satisfy increasing demand in the industrializing USA. New applications were continually developed: adding asbestos to lead paint to fireproof walls, applying it to roofing shingles to contain fires, etc. A US patent was granted for asbestos wall plaster, which revolutionized the building industry, with homes, schools and hospitals coated with durable, soundproof, fireproof asbestos. The increasing deployment of electricity prompted the growing use of asbestos insulation around wires and in walls. Asbestos use spread from America and Britain to other countries. In the 1870s, a major deposit was discovered in Quebec. By 1905, asbestos mining became Quebec's most profitable industry and one of its regions contributed eighty per cent of the world's supply. The industry leader, New York's Johns-Manville Company, purchased a profitable mine in Quebec which became the world's largest.

Extraction and use were vastly accelerated by World War I, when asbestos was inserted into uniforms of soldiers and firefighting equipment. International demand grew further after the war for use in fireproof clothing, fire-retardant housing materials, durable cement structures, and automobile parts particularly brake linings, and was seen as essential to rebuilding Europe's cities. 'Asbestos promised safety of those who used it and profits for those who sold it. Chrysotile asbestos from Quebec once made up 95 per cent of the global trade in the mineral' (van Horssen 2016: 6). Euphoria is shown by the title of the 1920 book *Asbestos and the Asbestos Industry: The World's Most Wonderful Mineral and Other Fireproof Materials* (Summers 1920). Opencast mining came first, but it visibly damaged the land, which became hidden when mining went underground. By 1939, asbestos was portrayed as indispensable to modern life, which was described as the 'Asbestos Age'. It 'had gained a global reputation that was synonymous with safety, and manufacturers were constantly adding it to new products such as oxygen bottles in airplanes and hospital ceiling tiles' (van Horssen 2016: 55). During World War II 'the US Army and Navy Munitions Board had asbestos on its "critical minerals" list and was prepared to protect its Canadian suppliers by invasion if enemy powers took control of the mines' (van Horssen 2016: 83). In 1952, Canadian miners were extracting seventy per cent of the global supply of asbestos. 'By 1953, the global market price for asbestos was higher than for gold' (van Horssen 2016: 120) and Johns-Manville Corporation had record profits throughout the nineteen-fifties, and the Canadian Mining Journal claimed that asbestos mining was ensured as one of Quebec's major industries for at least another century. Between 1955 and 1965, the global demand for asbestos doubled. New technology was developed to extract asbestos and applications were innovated to include it into ever more valuable commodities.

Step 4—Eye Opener to Problems

However, asbestos fibres were ubiquitous in the air of mines and workplaces where asbestos was used, and employees were coughing after working short periods. 'The first documented case of an American Johns-Manville employee becoming ill because of the mineral was in the 1920s' (van Horssen 2016: 182). But serious health problems had a long incubation period and took decades to manifest themselves. Only later was it confirmed that asbestos causes asbestosis, mesothelioma, and lung cancer, as well as skin, breast, ovary, and colon cancer. British medical researchers were leaders in uncovering the health problems. But even they agreed that asbestos 'was too important to the industrial and economic development of the Western world, especially during the Great Depression, to prohibit its trade' (van Horssen 2016: 57).

Step 5—Denial of Problems

Serious harm was slow onset, hence it took a long time for science to confirm that asbestos was the cause. Vested interest groups claimed that the risk was unproven. The asbestos industry refused to accept that asbestos causes cancer and other respiratory afflictions, and accelerated its production.

Step 6—The Accumulation of Both Danger and Scientific Evidence

Slow-onset respiratory illness eventually resulted in more people dying from asbestos. Companies introduced filters and supplied workers with individual respirators, but especially in asbestos mines there were so many fibres that respirators became clogged and workers refused to wear them because they could not breathe. Companies complained they could not reduce fibres further without great expense. The air was filled with fibres even in towns next to mines. 'Dust' was the euphemism used for asbestos fibres to disguise their toxicity by associating them with an innocuous everyday substance around the home. Cumulative scientific evidence slowly but overwhelmingly confirmed the causal connection between asbestos and cancer. By the 1970s, the slow-onset harms of asbestos were emerging everywhere, and the Western world was rejecting asbestos. The American Environmental Protection Agency began a campaign against asbestos.

Step 7—Displacement of Blame

Asbestos companies countered scientific evidence of danger with public relations campaigns claiming Canadian chrysotile asbestos was inert and safe, and only African crocidolite asbestos caused health problems. This claim quieted concern for decades. There was also growing evidence that cigarette smoking causes lung cancer, so asbestos companies launched public relations campaigns claiming their blue-collar workers are smokers, and it is not asbestos but rather cigarette smoking that caused their lung cancer. This blaming of victims diverted some attention away from the dangers of asbestos, but scientists increasingly documented that both cigarettes and asbestos cause cancer.

Step 8—The Empire Strikes Back

As lawsuits by workers seeking compensation increased, asbestos companies lashed out at what they labelled 'quack doctors' and 'shyster lawyers' complicit in maligning their product. They hired their own doctors to examine workers, suppressed knowledge of ill effects, and funded researchers to conduct studies showing asbestos was safe. One was a researcher at the renowned McGill University. When the New York Times wrote that he was linked to the asbestos industry, he sued, and his connections to the mining association had to be pried out of him. His lawsuit backfired because the subsequent publicity increased awareness of the health risks of asbestos. Doctors and engineers employed by asbestos companies attacked doctors documenting risk and attempted to undermine their conclusions. Uncertainty of future harm and of risk was played up. Like for fossil fuels, science became 'a contact sport' (Schneider 2009) between teams of company doctors versus teams of impact doctors.

Step 9—Discounting Danger

Risk was partially acknowledged by asbestos companies because of the scientific evidence and visible afflictions, but near-term economic benefits were given priority over long-term danger, which was discounted. Companies formed the Asbestos Information Committee, an international association with members especially from the USA, United Kingdom, and Canada, to promote asbestos. Johns-Manville Corporation dismissed health warnings as being based on media sensationalism and outdated information before safety improvements were implemented (Markowitz and Rosner 1994, 2002). Products containing asbestos continued to be relatively cheap because their environmental and health costs remained externalized to be suffered by others, rather than included in their price. The Quebec government eventually nationalized the industry so that it could continue. The Canadian government criticized European nations that banned asbestos, and promoted exports of asbestos products to developing nations (Attaran, Boyd, and Stanbrook 2008). In both 2004 and 2011 at the UN Rotterdam Convention, Canada prevented chrysotile asbestos from being placed on the Prior Informed Consent list of harmful minerals, which could have impeded trade in it. 'Canada exploited its generally positive international image to cast shadows over medical reports proving the dangers of asbestos, to avoid strict regulation of chrysotile in global markets, and to sell the mineral to developing countries, where workers and other citizens were neither adequately informed about the risks nor protected from them' (van Horssen 2016: 13). The asbestos industry, its workers, the community of Asbestos, the Quebec, and Canadian governments all acted in ways that confirmed Beck's (1995: 48–49) hypothesis of the 'death reflex of normality', literally. It states that 'as the hazards increase in extent, and the situation is subjectively perceived as hopeless, there is a growing tendency not merely to accept the hazard, but to deny it by every means at one's disposal'. Mitigation of danger that requires costly, lifestyle, and identity altering remedies prompt a response of clinging to normality. There was a misfit between these hazards and their acknowledgement, even when the risks were devastating. The dominance of economic goals over environmental and health concerns occurred despite discourse claiming that the three were being balanced responsibly.

Step 10—An Asbestos Identity

The local community supported its asbestos industry and opposed keeping asbestos safely in the ground, which was certainly based on jobs and the economy, but also on their desire to be proud of the industry upon which their identity depended. Asbestos workers and asbestos communities had supplied the world with a product declared indispensable to modern activities and made the world safer from fire. Pride was the reason why inhabitants of the community with the world's biggest asbestos mine had named their town 'Asbestos'. In 2006, after evidence had accumulated that 90,000 people were dying globally each year from asbestos-related health problems, the town debated whether to change its name, but refused. When the Canadian Cancer Society urged the government to stop supporting the asbestos industry, the town cancelled its annual fund-raising drive for cancer research. The region and the workers lobbied the Quebec and Canadian governments to support the industry, and both gave unwavering support until 2012 despite all the medical evidence of harm being done by asbestos, particularly to the workers themselves. The prosperity of this region based on the extraction of asbestos resulted in an asbestos identity, hence second-order identity politics grow out of first-order economic interests. Criticism of asbestos for causing health problems was perceived as insulting. Feelings were hurt. Not only was the region's basis of prosperity being challenged, but also its identity with the implication that its prosperity has brought harm to others. Rejecting criticism of asbestos implied not only defending the region's economic interests, but also its identity. The asbestos case showed that workers, companies, and states which extract a harmful resource will fight to keep extracting it, and even increase extraction, because it is central to their current economy and their identity. Products were eventually innovated to replace asbestos, but it was more difficult to innovate a new identity to replace the community's proud history of producing a product that appeared to increase safety.

Step 11—Social Injustice

Use of asbestos was eventually banned in wealthy countries that enacted strict health and environmental regulations. Asbestos was exported from Canada to poor, developing countries even though strict rules were implemented against its use in Canada. Canada's otherwise good international reputation was used to give legitimacy to the extraction and sale of asbestos. Asbestos products were claimed to be essential for developing countries because they were cheaper than alternatives (McCulloch and Tweedale 2008). Since the alternatives are more expensive, by 2011 the countries that used the most asbestos had become the poorer countries of China, India, Russia, Brazil, and Kazakhstan in that order (Rice 2014: Table 1). The principal producers by that date had become Russia, and then China, Brazil, Kazakhstan, and Canada. Harm had been displaced from wealthy countries to developing ones.

Step 12—Use It or Lose It

One lesson from the asbestos case is to be wary of the temptation, when evidence of danger is mounting and regulations are foreseen, to extract and sell the harmful resource quickly for fear of it becoming a 'stranded asset' left permanently stored safely underground, what others call a dangerous liability. Foreseeing the demise of their industry, asbestos companies, extraction communities, and their governments 'wanted to extract as much asbestos as they could from the mine before the industry collapsed' (van Horssen 2016: 129). This 'green paradox' (Sim 2012) results in a surge of harm before regulations are implemented.

Step 13—Dependence on Technological Solutionism

Despite harms becoming evident through both scientific assessment and asbestos workers eventually developing cancer, production continued based only on faith that future technological innovations would make asbestos safe for human health and the environment, even when there was no basis for such a belief. This case demonstrated that the biophysical properties of an otherwise valuable substance can be so harmful that no technical way can be found to make it safe, and that the only way to ensure safety is to leave it in the ground and pay the financial and social costs of developing alternatives. Forcing companies to pay full health costs of their products through lawsuits then prompted technological innovations that successfully devised safer alternatives.

Step 14—Harmful Yet Valuable Resource Kept Safely in the Ground

Anticipating 52,000 new lawsuits in the USA alone for compensation, the Johns-Manville Corporation filed for bankruptcy protection in the USA in 1982, and its mine was sold to former managers. It was not until 2012, however, that the Canadian and Quebec governments gave up the fight to save the industry because it had become clearly visible that almost a hundred thousand people globally were dying from asbestos each year and because of the resulting international reaction. The local workers, companies, and governments tried to prolong use of asbestos despite the danger, and only were dissuaded when consequences became too visible, lawsuits too punitive, and the international reaction too strong. Better to leave it safely underground and promote innovation of replacements even if they are more costly and require social change. It took a century of accumulating scientific evidence and increasing visibility of harm to implement prevention because of the reaction of powerful companies, their workers, and governments placing near-term economic priorities above health considerations.

Even though asbestos was unsafe for workers who handled it, asbestos-based products still fostered safety by reducing fire hazards. Cost/benefit calculations were transformed only when mounting externalities became visible in terms of (i) impacts on human health, (ii) financially in terms of litigation, and (iii) increasing government regulations in purchaser nations. Lawsuits forced inclusion of previously externalized health costs into the expenditures of asbestos companies. Asbestos companies went bankrupt and asbestos jobs were lost. No technical, cost-effective solution to make asbestos harmless could be found. The only realistic solution was prevention and prohibition thereby leaving asbestos underground. This maximized the possibility of innovating safer, more sustainable alternatives. Companies innovated alternatives, thereby stimulating economic growth and new jobs were created, even in the town still called 'Asbestos'. This creative destruction eventually led to a reconciliation of economy, health, and environment, and therefore responsible action. Foresight finally prevailed over discounting future harm. The asbestos case shows that, however problematic and costly the transition is to leaving a valuable but harmful substance underground, it can be done successfully and sustainable prosperity enhanced. It wasn't market competition that drove changes to safety, but instead findings by scientists (medical doctors), action by civil society, and regulations and international agreements by consumer governments that obliged the abandonment of asbestos by extractor states. Canada, Quebec, and the town still have huge reserves of valuable but dangerous asbestos safely stored underground.

The Uncertain Fourteenth Step for Fossil Fuels

These steps describe the response pattern that occurs when dealing with valuable but harmful resources of nature. The sequence proceeded from naïve use of a seemingly safe and valuable raw material, to irresponsible use where risk and harm are known but discounted and externalized to be suffered by others, and finally to foresight of either (i) responsible use where future health and environmental costs are included in its price, or (ii) to abandonment if responsible use is not technologically and economically possible.

The pattern of steps on the road actually travelled for the valuable but dangerous resources of bituminous oil and asbestos are remarkably similar, except that the 14th step of leaving the valuable but harmful resource safely in the ground has been achieved for asbestos but not for bituminous oil. The Canadian and Quebec governments—where asbestos reserves are—promoted its extraction and export long after it was proven dangerous by science; similarly the Canadian and Alberta governments—where tar sands are—promote the extraction and export of bitumen long after it has been proven that fossil-fuel combustion causes climate change and that oil extraction from bituminous sands is more emissions-intensive than pumping oil from wells. Those governments talk about balancing economy and environment, but give priority to near term, path-dependent economic growth to the detriment of environmental and long-term economic goals. Canada's natural resources minister claimed that more pipelines to export additional bituminous oil are needed to finance the transition to a low-carbon economy (McCarthy 2017: B3). This is a variant of a more general claim of some economists: wealthy countries have more resources to deal with environmental problems, so countries should become wealthy even by dangerous means and then market-based technological innovation and wealthy governments will solve the problems (Simon 1981, 1996; Lomborg 2001, 2010, 2019; Desrochers and Szurmak 2018). This is like saying that profits from asbestos are needed to finance medical solutions to lung cancer. So far very little if any of the fossil-fuelled wealth generated has been used to solve fossil-fuelled global warming. No evidence supports assumptions that the benefits of exporting more bitumen will be used to pay for transitioning to a low-carbon economy. Most will likely become profits for companies and general revenue for governments. What is certain is that the pipelines will pump diluted bitumen for a half-century, and that carbon emissions upstream in extraction and downstream when used will remain in the atmosphere for a century (Jaccard 2018). Scientific knowledge of the climate crisis is discounted when there is dependence on the oil industry for profits, jobs, and government royalties. The asbestos case confirmed that social pressure to enhance safety comes from groups, consumers, and states different from immediate producers. York (2017: 6) similarly documented that the whaling industry only relented hunting whales to extinction after a growing international outcry resulted in an enforced moratorium on commercial whaling.

I would hypothesize that this pattern of steps for resources as different as bituminous oil and asbestos was also similar in its broad outlines for the transition away from other valuable but harmful substances, such as CFCs, DDT, lead additives, etc. For these substances and asbestos, decisions in favour of safety and abandonment have been made. But for coal, and especially oil and natural gas, environmental and health concerns remain subordinate to near-term socioeconomic desires, current identities, lifestyle practices, and predispositions. If the transition to safety is this onerous for asbestos, then how much more difficult will it be for crude oil, which remains the inanimate energy source hitherto judged indispensable for modern life. Nevertheless, lessons can be learned from the asbestos case to neither underestimate nor exaggerate difficulties involved in leaving seemingly indispensable commodities underground to prevent them causing slow-onset harm. It is hopeful for learning that the problematic transition to safety can eventually succeed, albeit after much opposition and the valuable but harmful resource causing much harm.

From the Depth of the Problem to an Assessment of Solutions

This documentation of the depth of the fossil-fuelled climate crisis, and that societies have been largely unresponsive to warnings of science, does not mean it will necessarily continue. Rather its analysis should be seen as a self-denying prophecy of the incubation of disaster to be avoided. It is a necessary counterweight to wishful thinking. Presumptions that scientific warnings will necessarily lead to resolving the fossil-fuelled climate crisis can inadvertently promote complacency and magical thinking. They can constitute obstacles to solutions. Documentation of failure to heed scientific forewarnings, discount future harm, refusal to change fossil-fuelled practices, and take preventive action, may be more effective in prompting prevention by showing the dangerous consequences of present practices.

The analysis of (i) what is not being done to mitigate fossil-fuelled climate change and (ii) how those problems are worsening, valid as it is, admittedly runs the risk of resulting in fatalism and the culture of get-it-while-you-can. It could lead to believing the problem is too big to solve, that everything has to be replaced (the market, consumption, size of population) to solve anything. Scientific warnings of catastrophe incite remedial action in some groups but apathy in others. The overall resultant of these opposing responses, characterized by either foresight or short-sightedness, remains to be determined. Analyses need to conceive of the future as open-ended rather than having a predestined happy ending or cataclysm. Although it is important to analyse the fossil-fuelled climate crisis in all its depth, it is also important to present and evaluate remedies. The book now turns to assessing proposed solutions.

Notes

1. The United States placed second last for ending government support for oil and gas production.
2. Note that the steps are empirically grounded ideal types (Weber 1978), which are reference points for understanding the complexities of the real world. There is empirical overlap between steps and no theory of necessary stages.

References

Alexis, Tony, and Paul Poscente. 2019. Indigenous Ownership of TMX: Too Important to Get Wrong. *The Globe and Mail*, 26 July: A11.

Attaran, Amir, David R. Boyd, and Matthew B. Stanbrook. 2008. Asbestos Mortality: A Canadian Export. *Canadian Medical Association Journal* 179 (9): 871–872.

Beck, U. 1995. *Ecological Politics in an Age of Risk*. Cambridge: Polity Press.

Bell, Shannon Elizabeth, and Richard York. 2010. Community Economic Identity: The Coal Industry and Ideological Construction in West Virginia. *Rural Sociology* 75 (1): 111–143.

Berners-Lee, M., and D. Clark. 2013. *The Burning Question*. London: Profile.

Black, David. 2018. Time to Do Away with Carbon Taxes, to Save the Planet, and Spur Growth in the Oil Sands. *The Globe and Mail*, 15 May: B4.

Blanchfield, Mike. 2019. Police Called After Vulgar Slur Spray-Painted on Front of McKenna's Office. *The Globe and Mail*, 25 October: A6.

Bousso, Ron, and Dmitry Zhdannikov. 2019. Oil and Gas to Offer Opportunities for Decades to Come, Shell CEO Says. *The Globe and Mail*, 16 October: B8.

Bush, E., and D.S. Lemmen. 2019. *Canada's Changing Climate Report—Executive Summary*. Ottawa: Environment and Climate Change Canada. https://www.nrcan.gc.ca/sites/www.nrcan.gc.ca/files/energy/Climate-change/pdf/CCCR_ExecSumm-EN-033119-FINAL.pdf. Accessed 2 April 2019.

Butts, Ed. 2019. The Cautionary Tale of Whale Oil. *The Globe and Mail*, 5 October: O8.

CDIAC. 2018. *Global Carbon Dioxide Budget*. Berkeley: Carbon Dioxide Information Analysis Center.

Clarke-Whistler, Karen. 2019. The Climate Plan We've Been Waiting For. *The Globe and Mail*, 2 July: B4.

Cryderman, Kelly. 2019. Can Alberta's Gas Cut China's Coal Demand? *The Globe and Mail*, 30 November 2019. A7.

Davidson, D., and J. Andrews. 2013. Not All About Consumption. *Science* 339 (6125): 1286–1287.

Desrochers, Pierre, and Joanna Szurmak. 2018. *Population Bombed: Exploding the Link Between Population and Climate Change*. Toronto: The Global Warming Policy Foundation.

Flannery, Tim. 2015. *Atmosphere of Hope: Searching for Solutions to the Climate Crisis*. New York: Atlantic Monthly Press.

Freudenburg, W. 2006. Environmental Degradation, Disproportionality, and the Double Diversion. *Rural Sociology* 71 (1): 3–32.

Giovannetti, Justin. 2019a. UCP's Climate Change Plan Puts Focus on Technology, Industry, Minister Says. *The Globe and Mail*, 22 May: A4.

Giovannetti, Justin. 2019b. Alberta to Probe Environmental Groups. *The Globe and Mail*, 5 July: A6.

Harvey, Hal, and Robbie Orbis. 2018. *Designing Climate Solutions: A Policy Guide for Low-Carbon Energy*. Washington: Island Press.

Homer-Dixon, Thomas, and Yonatan Strauch. 2018. The Great Canadian Climate Delusion. *The Globe and Mail*, 2 June: O3.

Jaccard, Mark. 2018. PM Logic: Reduce Emissions by Increasing Them. *The Globe and Mail*, 20 February: A13.

Jang, Brent. 2019. Shell, Petronas Back Ottawa's Push for Paris Climate Accord Emissions Credits. *The Globe and Mail*, 11 September: B2.

Janigan, Mary. 2012. *Let the Eastern Bastards Freeze in the Dark: The West Versus the Rest Since Confederation*. Toronto: Random House Canada.

Keesmaat, Jennifer, Kwame McKenzie, and Richard Florida. 2020. No Going Back: Canada's New Normal Starts in Cities. *The Globe and Mail*, 23 May 2020: O5.

Kopecky, Arno. 2019. Stuck in the Middle. *The Globe and Mail*, 13 July: O9.

Leahy, Stephen. 2019. This Is the World's Most Destructive Oil Operation and It's Growing. *National Geographic*, 11 April. https://www.nationalgeographic.com/environment/2019/04/alberta-canadas-tar-sands-is-growing-but-indigenous-people-fight-back/. Accessed 26 January 2020.

Levant, E. 2010. *Ethical Oil*. Toronto: McClelland & Stewart.

Lewis, Mark. 2019. Wells, Wires, and Wheels—EROCI and the Tough Road Ahead for Oil. *BNP Paribas Asset Management*, August. Paris: BNP Paribus. https://docfinder.bnpparibas-am.com/api/files/1094E5B9-2FAA-47A3-805D-EF65EAD09A7F. Accessed 6 September 2020.

Liggio, John, Shao-Meng Li, Ralf M. Staebler, Katherine Hayden, Andrea Darlington, Richard L. Mittermeier, Jason O'Brien, Robert McLaren, Mengistu Wolde, Doug Worthy, and Felix Vogel. 2019. Measured Canadian Oil Sands CO_2 Emissions Are Higher Than Estimates Made Using Internationally Recommended Methods. *Nature Communications* 10: 1863. https://doi.org/10.1038/s41467-019-09714-9; www.nature.com/naturecommunications. Accessed 23 April 2019.

Lomborg, B. 2001. *The Skeptical Environmentalist*. Cambridge: Cambridge University Press.

Lomborg, Bjorn. 2010. *Smart Solutions to Climate Change: Comparing Costs and Benefits*. Cambridge: Cambridge University Press.

Lomborg, Bjorn. 2019. Empty Gestures Trivialize the Very Serious Challenge of Climate Change. *The Globe and Mail*, 28 December: O11.

Markowitz, G., and D. Rosner. 1994. *Deadly Dust: Silicosis and the Politics of Industrial Disease in Twentieth-Century America*. Princeton: Princeton University Press.

Markowitz, G., and D. Rosner. 2002. *Deceit and Denial: The Deadly Politics of Environmental Pollution*. Berkeley: University of California Press.

Mason, Gary. 2019a. How Many Pipelines Do We Really Need. *The Globe and Mail*, 15 November: A13.

Mason, Gary. 2019b. The Ugly Campaign to Discredit Thunberg. *The Globe and Mail*, 4 September: A13.

McCarthy, Shawn. 2017. Indigenous Voices Rise to the Fore Amid National Energy Debate. *The Globe and Mail*, 17 October: B3.

McCarthy, Shawn. 2019a. Zurich Insurance to Pivot Investment Away from Oil Sands. *The Globe and Mail*, 26 June: B3.

McCarthy, Shawn. 2019b. Review Panel Backs Teck Oil Sands Mine, Saying Jobs, Economic Benefit Outweigh Environmental Impact. *The Globe and Mail*, 26 July: A1.

McCarthy, Shawn, and Carrie Tait. 2017. Notley, Carr to Make Pipeline Case in B.C. *The Globe and Mail*, 25 November: A14.

McCulloch, Jock, and Geoffrey Tweedale. 2008. *Defending the Indefensible: The Global Asbestos Industry and Its Fight for Survival*. New York, NY: Oxford University Press.

McGugan, Ian. 2019. On Productivity, Canadian Provinces Trail Far Behind Southern Neighbours. *The Globe and Mail*, 12 November: B8.

McKay, Tim, Alex Pourbaix, and Derek Evens. 2019. We Have Big Decisions to Make as a Country, and There Is an Opportunity for Each of You to Influence the Outcome. *The Globe and Mail*, 1 August: A6.

McKibben, Bill. 2019. Our Climate Change Bill Is Coming Due. *The Globe and Mail*, 28 December: O8.

McMillan, Tim. 2019a. HSBC's Stunning Energy and Human-Rights Hypocrisy: It Ditched Oilpatch Only to Embrace Saudi Arabia. *Financial Post*, 27 November. https://business.financialpost.com/opinion/hsbcs-stunning-energy-and-human-rights-hypocrisy-as-it-ditches-oilpatch-only-to-embrace-saudi-arabia. Accessed 29 April 2020.

McMillan, Tim. 2019b. HSBC Would Rather Deal with Oligarchs Than Responsible Canadian Energy Companies. *Financial Post*. https://business.financialpost.com/opinion/hsbc-would-rather-deal-with-oligarchs-than-responsible-canadian-energy-companies. Accessed 29 April 2020.

Milke, Mark. 2019. Given the Social Contradictions, HSBC's Anti-Oil Sands Position Is Irresponsible. *The Globe and Mail*, 29 May: B4.

Morton, Ted. 2018. Trudeau Can't Solve Climate Change and Have His Pipeline, Too. *The Globe and Mail*, 1 June: A13.

Murphy, R. 2015. The Media Construction of Climate Change Quiescence: Veiling the Visibility of a Super Emitter. *Canadian Journal of Sociology* 40 (3): 331–354.

Rand, Tom. 2018. A Sound Debate About Canada's Emissions, Brought on by Fury over Trans Mountain. *The Globe and Mail*, 3 May: B4.

Rice, James. 2014. Asbestos and the Globalization of an Occupational and Environmental Hazard, 1960–2011. *Environmental Justice* 7 (1): 1–8.

Schneider, S. 2009. *Science as a Contact Sport*. Washington: National Geographic.

Sim, Hans-Werner. 2012. *The Green Paradox: A Supply-Side Approach to Global Warming*. Cambridge, MA: MIT Press.

Simon, Julian. 1981. *The Ultimate Resource*. Princeton: Princeton University Press.

Simon, Julian. 1996. *The Ultimate Resource 2*. Princeton: Princeton University Press.

Simpson, Jeffrey, Mark Jaccard, and Nic Rivers. 2007. *Hot Air: Meeting Canada's Climate Change Challenge*. Toronto: Emblem McClelland & Stewart.

Steffen, W. 2015. Unburnable Carbon: Why We Need to Leave Fossil Fuels in the Ground. *Climate Council,* 23 April. Climatecouncil.org.au/unburnable-carbon-why-weneed-to-leave-fossil-fuels-in-the-ground.

Stoddart, Mark C.J. 2012. *Making Meaning Out of Mountains: The Political Ecology of Skiing*. Vancouver: University of British Columbia Press.

Stoddart, Mark C.J., and Jillian Smith. 2016. The Endangered Arctic, the Arctic as Resource Frontier: News Media Narratives of Climate Change and the North. *Canadian Review of Sociology* 53 (3): 316–336.

Stoddart, Mark C.J., Alice Mattoni, and John McLevey. 2020. *Industrial Development and Eco-Tourism: Can Oil Extraction and Nature Conservation Co-Exist?* Palgrave Macmillan.

Summers, A. Leonard. 1920. *Asbestos and the Asbestos Industry: The World's Most Wonderful Mineral and Other Fireproof Materials*. London: Sir Isaac Pitman and Sons.

van Horssen, Jessica. 2016. *A Town Called Asbestos: Environmental Contamination, Health, and Resilience in a Resource Community*. Vancouver: University of British Columbia Press.

van Lierop, Wal. 2018. Energy Is Evolving, and Albertans Need Facts. *The Globe and Mail*, 21 May: B4.

Weber, Max. (1922) 1978. *Economy and Society*, ed. Guenther Roth and Claus Wittich. Berkeley: University of California Press.

Wente, Margaret. 2018. The Trans Mountain Pipeline Will Never Be Built. *The Globe and Mail*, February: A15.

Whitley, Shelagh, Han Chen, Alex Doukas, Ipek Gençsü, Ivetta Gerasimchuk, Yanick Touchette, and Leah Worrall. 2018. G7 Fossil Fuel Subsidy Scorecard: Tracking the Phase-Out of Fiscal Support and Public Finance for Oil, Gas and Coal. *Overseas Development Institute of the UK*, June. https://www.odi.org/sites/odi.org.uk/files/resource-documents/12222.pdf. Accessed 4 June 2018.

Willis, Andrew. 2019. The Future Is Plastics: Murray Edwards, Li Ka-Shing Add to Oil Patch Holdings as Others Flee. *The Globe and Mail*, 7 July: B1, B5.

World Economic Forum. 2020. *Fostering Effective Energy Transition*. Geneva: World Economic Forum. http://www3.weforum.org/docs/WEF_Fostering_Effective_Energy_Transition_2020_Edition.pdf. Accessed 15 May 2020.

Wynn, Graeme. 2016. Foreward: The Long Dying. In *A Town Called Asbestos: Environmental Contamination, Health, and Resilience in a Resource Community*, ed. Jessica van Horssen. Vancouver: University of British Columbia Press.

Yale University. 2018. *EPI Environmental Performance Index 2018*. https://epi.envirocenter.yale.edu/epi-country-report/CAN. Accessed 22 September 2019.

York, R. 2017. Why Petroleum Did Not Save the Whales. *Socius: Sociological Research for a Dynamic World* 3: 1–13.

Part II

Assessing Solutions

7

Risk and Safety; Real and Staged

The physical properties of carbon dioxide are that it is invisible, odorless, and remains in the atmosphere for a century or more, hence carbon pollution accumulates there causing global warming. Production science has made fossil fuels emitting carbon dioxide more available through its innovations of hydraulic fracturing, deepwater drilling, oil extraction from tar sands and the Arctic, etc. Global warming is being locked-in with each new well drilled, oil reserve discovered, each bit of shale hydraulically fractured, and pipeline built because mitigating it would require not only a fossil-fuel write-off but also an infrastructure write-off. Fossil-fuelled social practices, for example flying and cruising, are increasing rapidly. Impact science has demonstrated the risk that will result from these practices and exploiting these valuable but dangerous resources. Berners-Lee and Clark (2013: 2) conclude that 'the choice we face is between taking unimaginable risks with the planet and leaving vastly valuable fossil fuels in the ground'. Even in the transition period away from fossil fuels towards low-carbon energy, greenhouse-gas emissions would continue to exceed carbon withdrawal and accumulate in the atmosphere. This worsens the fossil-fuelled climate change risk.

R. Murphy, *The Fossil-Fuelled Climate Crisis*,
https://doi.org/10.1007/978-3-030-53325-0_7

Most social action is founded on extrapolating from what succeeded and was safe in the past, but fossil-fuelled climate change is bringing discontinuities. The most general lesson to be learned is that extrapolations from the past can no longer be taken for granted as valid. As greenhouse gases accumulate in the atmosphere, delayed-action solutions rather than prompt ones constitute dangerous brinkmanship on a global scale based on discounting danger. Brinkmanship and blind faith in last-minute technological solutions amount to a failure of foresight and a maximization of risk. The fossil-fuelled climate crisis brings risk and uncertainty, whereas solutions would bring safety and sustainability. Part II of the book now presents an assessment of proposed solutions, hence it begins with an analysis of risk, uncertainty, safety, and assessments (see also Lockie and Wong 2017; Wong and Lockie 2018).

Risk Assessment and Material Risk

Ulrich Beck (1992, 1995, 2007) was the pre-eminent social scientist of risk in the late twentieth and early twenty-first centuries and an influential public intellectual whose writings were disseminated to the general public and served on advisory committees in his native Germany. He (Beck 1992) portrayed modern society as the 'risk society' where the very successes of science, technology, the market, and organization result in the creation of new risks. The increasing capacity to produce 'goods' results in the side effect of also producing 'bads'. In his many publications, Beck was not always consistent in his use of the concept 'risk'. In some instances, he used it the way it is used in the wider society, as a synonym for danger. In other instances, he used it in a narrower sense referring to the subset of dangers that are calculable and are calculated. A simple version of the latter is as follows: risk = impact x probability. In the scientific sense, this conception constitutes assessed risk, which consists of a fallible estimate of a dangerous material referent. Extrapolations are made from past events (floods, wildfires, hurricanes, ice storms, etc.) and from assumptions and modelling scenarios in order to assess the probability and likely impact of future harmful events, hence the risk is calculated. To use an analogy, if a coin is tossed many times, the evidence indicates that heads will occur 50% of the time and tails 50%, so even though the outcome is unknown, future action can be premised on the probability of 1 out of 2. This logic is used to calculate and manage complex risks in society. If the historical record documents that a high-impact flood occurs in a particular area every three years, then it will be called a flood plain and only parks or farming fields will be sited there. But if the historical record shows that such floods only occur every century, then risk-takers may construct a housing development there together with robust storm sewers, levees, etc., whereas the risk averse will avoid the area. Risk in this sense involves known unknowns: the outcome of tossing the coin is unknown, but the probability is known; similarly, whether a storm, wildfire, drought, etc., will occur in any specific year is unknown, but the chances are known over a large number of observed cases. A hundred-year flood could occur in year 99 or year 2, but at least the historical evidence demonstrates that on average it will occur once every hundred years. Governments, insurance companies, investors, etc., can usually manage risk in this sense, for example, investors can diversify their portfolios.

Modern societies talk more about risk, and their technological successes have brought new risks into being. Nuclear physics gave the world nuclear bombs and reactor meltdowns, which didn't occur before. Biochemistry gave antibiotics, which resulted in antibiotic-resistant bacteria. They also made old risks more threatening, for example cyclones, drought, and wildfires are intensified by climate change caused by combustion of fossil fuels, land use changes, etc. The modern risk society consists of a hybrid entanglement on the empirical level of risk discourse (including risk calculations) and new material risks. Hence it is necessary to analyse both the cultural assessment of risk and material risk, and especially relationships between them. Beck's (2015) concept of 'catastrophism' refers to the anticipation of catastrophe, which is an extreme form of risk assessment, but it cannot be assumed that risk assessment based on scientific evidence is straightforwardly accepted by the public and decision-makers. And even if risk is acknowledged, it cannot be assumed that decision-makers and the population will act on this risk assessment in a context of competing priorities, particularly near-term economic ones, interests, and cultural predispositions.

It should not be presumed that enlightened discourse results necessarily and straightforwardly in enlightened practices. Such discourse can be what Simpson, Jaccard, and Rivers (2007) called global warming 'hot air' and what the teenage climate activist Thunberg called 'empty words'. Discourse and material practices can be relatively autonomous. It is social practices that are consequential for the material environment. Words only have material outcomes if they affect social practices. Blurring the distinction between material risk and cultural assessments of risk obscures the paradoxical present situation of improved discourse, including policy discourse, coexisting with worsening fossil-fuelled climate change.

Does the assessment of risk correspond to the material reality of risk as potential harm, or is there a mismatch between material risk and risk assessment? Disaster research documented that when a mismatch occurs, so does a failure of foresight and the incubation of man-made disaster (Turner and Pigeon 1978; Vaughan 1996). Assessments of risk or safety are just as real as material risk or material safety, but one should not be reduced to the other. It is necessary to probe more deeply what is meant by 'risk', which is a widely used concept not only in the social sciences but also in society. Although the increase in risk calculations and risk discourse constitutes significant change in society, the underlying transformation augmenting biophysical danger is more fundamental: technological innovations unleashing more frequent, intense, or novel dynamics of nature; population growth and movement placing more people in harm's way of nature's dangerous forces; etc.

Risk and Risk Assessment

The Oxford English Dictionary defines 'risk' as the (exposure to) the possibility of loss, injury, or other adverse or unwelcome circumstance; a chance or situation involving such a possibility. It then distinguishes 'risk' from 'risk perception', which consists of the judgement people make about the severity and probability of a risk, and may vary from person to person and group to group. Thus in society at large a distinction is made between (a) the exposure to the possibility of adverse consequences and

(b) judgements people make about that exposure. In this meaning, risk is the objective, material exposure to a dangerous possibility, whether it is known or unknown, acknowledged, ignored, denied, or exaggerated. Unknown risk refers to threats unknown at the time, and may well be known later, for example after a disaster has occurred. 'Risk perception' is used to refer to beliefs in the exposure to danger or on the contrary to disbelief, or some intermediary state such as it being discounted, or even to imagined inexistent threats. The absence of putative risk is only known later, as when the twenty-first century began without the supposedly catastrophic millennium bug causing serious adverse consequences. The Oxford Dictionary captured the commonly understood conceptions of risk and risk perception. Nevertheless, a qualification is needed. Although the adverse consequences of falling off a cliff are perceived as we walk along its edge, modern risks such as the depletion of the ozone layer and fossil-fuelled climate change are not directly perceived. Awareness of these threats depends on scientific knowledge. Hence 'risk perception' should be replaced by 'risk assessment' in order to include modern global threats like these. Such risk assessments could be scientific, as in formal calculations of probability multiplied by impact. They could also be by non-scientists, as the Inuit observing the shrinking ice cover of the Arctic Ocean or mountain inhabitants noting the receding glaciers and inferring risk of more loss to come. These are all socially constructed fallible assessments of material risk, which is their referent.

Take the following example. The innovation of deepwater oil drilling under enormous ocean pressures creates new risks in the material objective sense of exposure to the possibility of adverse consequences. British Petroleum claimed in its assessments to the regulator that the probability of those harmful consequences was so miniscule because of its blowout protector, which it described as failsafe, that their occurrence was effectively impossible. But discounting risk was literally and catastrophically blown out of the water by deepwater ocean pressures that made its blowout protector fail and become unsafe (Freudenburg and Gramling 2011). Similarly NASA calculated that its safety devices protected the Challenger Space Shuttle from the enormous physical forces acting on it, so it launched the shuttle. But it underestimated risk with calamitous results (Vaughan 1996).

These cases illustrate that reticence to anticipate catastrophe where danger lurks and resulting lack of foresight leads to vulnerability. They demonstrate that the assessment of risk must not be straightforwardly equated with risk, nor the assessment of safety equated with safety. Risk and its opposite, safety, are not necessarily what we think they are. It is necessary to distinguish the cultural assessment of risk or safety from material risk or safety. Risk is material, risk assessment is cultural, and the interaction of the two occurs in the hybrid cultural–material world. Sometimes the cultural assessment is of safety, which steers social practices to continue doing what has been done, even where danger exists, and the consequence is catastrophe (Chernobyl, Challenger Space Shuttle, BP Deepwater Horizon, etc.). Or the cultural assessment is of risk and this determines social practices, even where there is no material risk (millennium bug, claims that vaccines cause autism). The assessment of risk is fallible, with science being the best available way to assess risk in the material world, and it recognizes its fallibility. The other way is by experience, either that of safety or disaster. Sayer (2001: 969) argues that 'the most compelling reason for accepting the basic realist premise of the independence or otherness of the world is the experience of making mistakes, of having one's expectations confounded and of crashing into things unexpectedly - in other words, the experience of falsification'. This is particularly true but unfortunate if the mistakes and crash result in catastrophe, and in the case of fossil-fuelled climate change, perhaps an irreversible global one.

Some risks are sociotechnical constructions with unintended consequences, for example innovation of CFCs depleting the ozone layer, but some are constructions of nature, such as the earthquake destroying Messina, Sicily in 1908, the Mount Pelée volcano destroying St. Pierre in 1902, and the storm surge drowning 6000 residents of Galveston Texas in 2000 (Zebrowski 1997; Murphy 2010). The assessment of risk or safety is always a social construction, typically one based on extrapolation from past experiences. There is also risk that is unforeseeable even with the most advanced science (Murphy 2009), which raises the question of how to confront the unforeseeable (CST 1999). A distinction must be made between the underlying material danger and the sociocultural estimation of it, whether by scientific or lay means. In order to facilitate communication with the lay population, this book will follow the Oxford dictionary by retaining the distinction between risk (and its synonym danger) and risk assessment (often called risk perception) to study their relationships and gradations.

Beck's (2007: 13) states that 'the same risk looks like a dragon to some, but like a worm to others … [and] acceptable risks are those which are accepted'. These statements are underpinned by the difference between 'risk' and 'looks like'. Similarly Renn (2008: 2) states that the 'link between risk as a mental concept and reality is forged through the experience of actual harm (the consequence of risk). … What counts as a risk to someone may be an act of God to someone else or even an opportunity for a third party'. Risk as a mental construct—and its interpretation as undesirable or as an opportunity—is different from risk as material reality. The first constitutes risk assessment, the second consists of risk or danger itself. In order to analyse relationships between the two, material risk needs to be clearly distinguished from its cultural assessment. Conflating the two leads to confusion. Risk-takers rue the day they accepted risk and suffer buyers' remorse if catastrophe strikes. Defining a threatening dragon as an innocuous worm does not make it so.

Beck makes the astute assertion concerning risk assessment that the 'risks which we believe we recognize and which fill us with fear are mirror images of our selves, of our cultural perceptions' (Beck 2007: 13). But it needs to be complemented by a similar assertion about safety assessments, namely that the safety which we believe we recognize and which fill us with complacency and apathy is a mirror image of ourselves, of our cultural perceptions. These are fallible cultural assessments of risk or safety in a real material world with its own dynamics. Whether the assessment is accurate determines vulnerability or robustness and hence the material consequences of social practices.

Beck (1995: 50–51) gave a devastating critique of mistaking verbal designations for their biophysical referents. If material hazards are not dealt with, 'there remains only the social construction of non-toxicity. It does not, admittedly, inhibit the effect, but only its designation. … That might be a momentary consolation but it is no help against poisoning'. Disaster research documented the superficiality of conflating risk and safety assessments with their material referents. Nature's dynamics presumed safe can shock humans into awareness of their erroneous presumptions, as when the extremely improbable tsunami flowed over the supposedly failsafe protective barrier at Fukushima Daiichi drowning thousands of inhabitants and striking the nuclear reactor resulting in a meltdown (Hasegawa 2012, 2015). Sayer (1997: 482) reminded social scientists not to confuse 'social constructs or interpretations with their material products or referents'.

The Actualization of Risk Potential into Disaster

Beck (2007: 9–10) stated that 'the moment risks become real, when a nuclear power station explodes or a terrorist attack occurs, they become catastrophes. Risks are always *future* events that *may* occur, that *threaten* us'. 'Real' is a poorly chosen word, and should be replaced by 'actualized'. Before the explosion, the risk was real and threatening, more with lax safety measures than stringent ones, and more than if there were no nuclear station. The risk only became actualized from a potential into a material catastrophe when the explosion occurred. Risks have a reality similar to potential energy. They may become actualized into material catastrophes, much like potential energy may become actualized into forces and kinetic energy. Or action can be taken to prevent risks from becoming actualized into catastrophes: a bomb can be dismantled before it detonates; CFCs with the potential to deplete the ozone layer can be phased out by the Montreal Protocol; fossil fuels have the potential to result in a greenhouse effect pushing global warming to more than 2 °C but can be prevented by replacing them with low-carbon energy, etc. Decision-makers assessed New Orleans as a safe city, hence defensive and preventive measures were not implemented and its risk potential, clearly indicated by prior scientific predictions (Freudenburg et al. 2009), was actualized into a catastrophe when hurricane Katrina struck. Culturally varying assessments of risk interact with the material potential of nature's dynamics in different ways: by acknowledging risk or denying it; by taking action to prevent disaster or discounting it as too far in the future and choosing to run the risk for economic reasons; by preparing for it by reducing vulnerability; or even by imagining a threat where none exists.

Risk Calculability and Cultural Perceptions

Beck (2007: 12) postulates that the 'global anticipation of catastrophe for the most part resists the methods of scientific calculation. The less calculable risk becomes, however, the more weight culturally shifting perceptions of risk acquire, with the result that the distinction between *risk* and *cultural perception of risk* becomes blurred'. The first assertion needs to be unpacked, especially concerning fossil-fuelled climate change. The anticipation of climate catastrophe has been rigorously calculated by science. What resists calculation are the specific harms, timing, location, tipping points, etc. Concerning the second affirmation, it is true that incalculable risks can be a prolific breeding ground for risk perceptions. The predicted catastrophe of the millennium bug was not calculable, yet that risk was terrifyingly anticipated. Social scares immediately prior to the change of millennium were abundant: computers were predicted to go haywire resulting in planes falling from the sky at midnight 2000, and missile sites and machines in hospital operating rooms running amok. This cultural assessment of risk was wiped out when clocks struck midnight by the evident absence of adverse consequences. However, the blurring of the difference between risk and its cultural perception usually has little to do with the calculability of risk. The risk of catastrophe for New Orleans situated below sea level on the coast of the Gulf of Mexico between an enormous lake and the Mississippi River in an area prone to hurricanes was extensively calculated, but that did not prevent the calculations and risk from being ignored and social practices continued as if the cultural perception was of safety. Near-term economic interests had priority, and the result was catastrophic hurricane Katrina striking unprepared New Orleans (Freudenburg et al. 2009). Fossil-fuelled climate change also demonstrates that its risk or safety assessment varies between countries and cultures even though its scientific risk assessment has been robustly calculated by the IPCC and disseminated internationally (Likskog et al. 2010). The cultural assessment of safety or risk has more to do with interests and social predispositions than with the calculability of risk. This is important because decision-makers often legitimate their complacent actions by claiming that risks were not calculable before a disaster, as

George W. Bush did after Hurricane Katrina, but they were rigorously calculated yet ignored. Will this be the case for scientific assessments of the fossil-fuelled climate crisis?

Uncertainty

Calculating impacts and probabilities, hence risk, depends on having evidence about what occurred previously and on continuity between occurrences in the past and in the future. This is what distinguishes risk from uncertainty. To use the previous analogy of the coin, suppose someone weights it and there is no evidence concerning tosses of the weighted coin. Now the evidence from past tosses of the unweighted coin would be misleading for predicting outcomes of tosses in the future. Instead of continuity, there would be discontinuity. Both the outcome (heads or tails) and the probability of occurrence would be unknown. 'In an uncertain situation, there's not enough information to establish probabilities. Not only is the outcome unknown, but there's no reliable way to weigh possible outcomes' (Agrba 2019: 9). If both probability and impact are unknown, then the situation is a complex one of multiple unknowns. Fossil-fuelled global warming is changing the frequency, intensity, and timing of storms, hurricanes, floods, wildfires, sea level rise, melting Arctic ice cover and permafrost, etc. Hence, extrapolations from the past are misleading, risk calculations based on them are invalid, and fossil-fuelled societies have placed themselves in a situation of uncertainty. What was previously a hundred-year flood in Houston now could occur every ten years. Fifty-year wildfires in California and Australia might occur every five years. Worse still, there is no reliable evidence on how the frequency of occurrences and their impacts are transforming. Because climate change is global, strategies of managing outcomes by diversification don't work either because the whole world is being affected. Discontinuities of fossil-fuelled climate change are altering the dynamics of nature such that the concept 'risk society' needs to be replaced by 'uncertainty society'. Science informs societies of the fossil-fuelled danger, but does not currently have the capacity to provide precise measures of its future impact or probability, which remain uncertain.

More complex still, fossil-fuelled climate change engenders different dimensions of uncertainty. One consists of the inability to calculate the probabilities, impacts, location, and timing of *acute*, *sudden* disasters like hurricanes and wildfires. Another dimension of uncertainty involves assessing the timing (Lockie and Wong 2017) and impact of *chronic*, *slow-onset* catastrophes, such as melting of Antarctica's and Greenland's glaciers and subsequent ocean level rise, displacement of the jet stream, gulf stream, and polar vortex, which will have widely varying impacts and probabilities in different geographical and topographical locations. The biggest uncertainty is whether fossil-fuelled global warming will irreversibly tip the planet into a less hospitable habitat for humanity.

There is one caveat to this discussion of unknowns and uncertainty. The preponderance of the scientific evidence indicates that there is little uncertainty about the overall trend line of increasing global warming. Uncertainty concerning specific harms should never be used as an excuse for not changing socioeconomic practices and technologies to mitigate the trend line of fossil-fuelled global warming. George W. Bush's excuse for lack of mitigation and preparation for hurricanes in New Orleans (Freudenburg et al. 2009) must not be emulated for fossil-fuelled climate change.

The Staging of Risk: 'When You Ride ALONE, You Ride with Hitler'

These clarifications of the concepts 'risk', 'safety', 'uncertainty' and 'assessment' are necessary to assess solutions, which are underpinned by all these concepts. The climate crisis has been caused by fossil-fuelled practices, including voting practices concerning carbon taxes, etc., and solutions will require changes in those practices. 'We are aware of the existential stakes and the urgency, but even when we know that a war for our survival is waging, we don't feel immersed in it. The distance between awareness and feeling can make it very difficult for even thoughtful and politically engaged people – people who *want* to act - to act. … If we don't act until we feel the crisis that we rather curiously call "environmental" – as if the destruction of our planet were merely a context – everyone will be committed to solving a problem that can no longer be solved' (Safran Foer 2019: 13). Presenting research papers and scientific modelling to non-scientists is unlikely to change the social practices of the latter. Staging a problem is important for transforming what science has presented intellectually into one that people perceive, feel, and act upon. The scientific finding of ozone layer depletion was staged as a 'hole' in the ozone layer. The danger of smoking was staged with warnings on cigarette packages, and in some places, gory images. Staging is particularly necessary where there is a slow-onset threat that is invisible, has catastrophic consequences, is caused by embedded social practices, and where the risk disproportionally affects those who did not cause it, particularly the poor and future generations.

During the Second World War, 'the [US] government enacted – and Americans accepted – price controls. … Gasoline was severely regulated, and a speed limit of 35 miles per hour was imposed nationally to reduce gas and rubber consumption. U.S. government posters advocating carpooling declared, "When you ride ALONE, you ride with Hitler"' (Safran Foer 2019: 8). Taxes were increased dramatically to support the war effort. Material risk has the *potential* to influence expectations, actions, politics, and transform the world, but whether this potential is *actualized* depends on many factors, including the staging of risk in competition with the staging of safety. Safran Foer (2019) gave examples of staging that helped alleviate problems, such as the social construction of the iconic status of Rosa Parks on the bus as both a true story and a fable to combat racial discrimination in the USA.

What is staging? Following (Beck 2007: 10), '"staging" here is not intended in the colloquial sense of the deliberate falsification of reality by exaggerating "unreal" risks. … For only by imagining and staging world risk does the future catastrophe become present – often with the goal of averting it by influencing present decisions. Then the diagnosis of risk would be "a self-refuting prophecy" – a prime example being the debate on climate change which is supposed to prevent climate change'. Staging consists of drawing attention to an environmental problem and making it relevant. Beck (2007: 98) argues that now the 'political site of the world risk society is not the street but *television*, the Internet – in short, the old and the new media. … Its function is assumed by the *staging of cultural symbols in the mass media*'. His examples of staging are the 'Greenpeace people [who] are multinational media professionals who know how self-contradictions between pronouncements and violations of safety and surveillance norms can be presented …. [and he sees] Greenpeace using the instruments of the media age to stage worldwide mass civil disobedience' (Beck (2007: 99).

The Socioculturally Staged and the Materially Real

Is risk a material reality of potential harm or a staged, socially constructed belief? They are entangled empirically, but analytically blurring the difference between them and conflating the material and the cultural, or worse still reducing the material to the cultural, obstructs understanding of relations between them. Material risk and the staged assessment of risk should not be assumed to be identical. Staged perceptions of risk can promote action that reduces vulnerability, increases prevention and therefore safety when there is material risk, but staged claims of safety and discounting future harm in the presence of material risk fosters apathy and complacency, thereby aggravating risk. The population's assessment of global risk is influenced by the staging of risk versus the opposite staging of safety and discounting danger. Material risk can be diminished by its accurate assessment followed by corrective action. Or it can be exacerbated when safety is persuasively staged and future harm discounted where danger lurks. Risk assessment and action depend on both material risk and on the staging of risk in competition with the staging of discounting danger. The contact sport Schneider (2009) described between impact scientists and production scientists is also played between the stagers of risk and discounters of danger.

Beck's previous hypothesis that danger shapes expectations and becomes a political force coexists awkwardly with his hypothesis here that risk needs to be staged and that it is staging which shapes expectations and becomes a political force. The two can be reconciled by clearly stating that staging is a crucial intervening variable between (i) material danger or its absence, and (ii) expectations and political action. Beck shows that the staging of risk intervenes between scientific knowledge of material risk and cultural assessments of risk. 'As a result, their "reality" can be dramatized or minimized, transformed or simply denied according to the norms which decide what is known and what is not. They are products of struggles and conflicts over definitions within the context of specific relations of definitional power, hence the (in varying degrees successful) results of stagings' (Beck 2007: 30). Staging leads to fallible assessments of risk or safety, whether scientific or lay, which should not be straightforwardly equated with material risk or safety. Sometimes dynamics of nature, which are autonomous from socially constructed definitions of risk or safety, upstage the staged definitions and, contrary to socially constructed expectations, undermine the definitional power of assessments by producing disasters in cases staged as safe, or safety in cases staged as disastrous, as the illustrations given above show.

Suitability for Staging?

Beck (2007: 72) argued that 'there is a striking discrepancy between the material destruction being wrought by climate change, which is *irreversibly* transforming conditions of life on the planet, and its *suitability for staging in the mass media*'. Whereas visible evidence of terrorists blowing people up lends itself to media staging and action against terrorism, global warming relies on top-down documentation by scientists, dissemination of research results by the media, and staging by social movements to distinguish fossil-fuelled causes of this creeping global catastrophe from nature's local disasters which have previously occurred. Planes crashing into the twin towers of New York's World Trade Center provided an iconic image of terrorism understood everywhere. People aren't moved by the fate of famished polar bears as much as the fate of people on fire jumping from the 100th floor of the burning towers. 'The planetary crisis – abstract and eclectic as it is, slow as it is, and lacking in iconic figures and moments – seems impossible to describe in a way that is both truthful and enthralling' (Safran Foer 2019: 14). Carbon dioxide is invisible to the senses, as is the excess of emissions over carbon withdrawals, which is cumulative and threatens to unleash further autonomous dynamics of nature that will lead to runaway climate change if social or technological corrective measures are not taken or taken belatedly. This can only be made visible by impact science. Global warming involves an urgency unfelt by the population and decision-makers, which occurs behind their backs if they are unwilling to turn around and see it through scientific conclusions. The problem worsens even as remedial measures are being taken when they are too little, too late.

Erroneous staging of safety is especially likely where danger is slow onset, as in fossil-fuelled climate change. Safety claims based on extrapolation from a safe past are more easily staged than tipping into dangerous discontinuities a century from now, no matter how well documented the science upon which the latter is based. Assumptions of safety drawing on extrapolation from past experience are subtle and taken for granted, much like breathing clean air is taken for granted until it is seriously polluted. Expert evidence resulting in forecasts of harm to peoples distant in time and space can be easily discounted. It is not only the cleverness at staging risk or safety that is important, but also the underlying material conditions that facilitate discounting danger over staging risk, and make the staging of safety easy and the staging of danger for future generations complicated. Near-term economic priorities, cultural predispositions, and normal conditions favour the staging of safety and discounting of future harm, even when they result in practices that produce environmental danger.

Stagings to Mitigate Fossil-Fuelled Danger

Scare 'em to Death

The above shows the difficulty of convincing consumers to change their fossil-fuelled consumption practices and voters to vote for policies that mitigate climate change. One solution is to scare them into climate-friendly practices. The choice of book titles and presentations are done to frighten readers into changing their fossil-fuelled practices. Take a brief sampling of titles: *The World Without Us* (Weisman 2007); *Seven Years to Save the Planet* (McGuire 2008); *Climate Wars: The Fight for Survival as the World Overheats* (Dyer 2008); *Now or Never* (Flannery 2009); *Requiem for a Species* (Hamilton 2010); *Challenging Legitimacy at the Precipice of Energy Calamity* (Davidson and Gismondi 2011); *The Sixth Extinction* (Kolbert 2014); and *The Uninhabital Earth: Life After Warming* (Wallace-Wells 2019). Beck's (2015) theory of emancipatory catastrophism also assumes that people can be scared into emancipating themselves from fossil fuels by discourse about a coming global catastrophe. The content of these books is often based on the best available natural scientific understanding of fossil-fuelled climate change, which is admittedly terrifying, and has much to offer. Their frightening titles may prove right in the long run, but they have not succeeded in convincing leaders and populations to reduce greenhouse-gas emissions, which continue to rise. Setting expiry dates is particularly dubious. The 'seven years' after the McGuire book was published expired in 2015, so readers might conclude it is now too late to save the planet. Since fossil-fuel practices were not changed 'now' in 2009 as Flannery insisted, readers could infer that it will be 'never'. If societies were 'at the precipice of energy calamity' in 2011 as Davidson and Gismondi contended, they must have fallen into the abyss since then because fossil-fuel combustion intensified.

Failed predictions of catastrophe have a long history. The 1972 staging of *Limits to Growth* (Meadows et al. 1972) predicted a planetary collapse, and yet humanity is more prosperous than ever almost a half-century later in 2020. The Malthusian catastrophe (Malthus 1798) has not yet occurred despite more than two centuries of exponential population growth. However well intentioned, these scare tactics didn't work. Only a small proportion of populations have viscerally experienced calamities from fossil-fuelled climate change until now. All the rest persisted in their greenhouse-gas emitting practices, and so did even the victims of floods, wildfires, Arctic ice melting, etc. Failed stagings undermine the credibility of predictions of harmful consequences. The important issue is the relationship between staged assessments and the underlying material risk. The present book is based on the premise that the evidence should be presented as accurately and dispassionately as possible. If that scares people into action, then all the better, but scaring is not used here as a tactic. After six more years of increasing emissions and worsening global warming since his 2009 book, Flannery (2015) concluded that hope has more potential than fear, so he published a book entitled *Atmosphere of Hope*.

Finding a Dramatic Iconic Image

Some social scientists claim that staging ozone layer depletion as a 'hole' inspired action to construct measures to phase out CFCs (Grundmann 2001). That hypothesis strikes me as at best a minor contributing factor. Whether it be the ozone 'hole' or 'riding with Hitler', an evaluation of these stagings would require comparison with action without them. That has not been done, so it is difficult to determine their influence compared to the influence of, respectively, the scientific documentation of ozone layer depletion (and innovation of technological alternatives) and World War II itself. Moreover, in the three decades since a scientific consensus emerged concerning dangerous global warming, no staging of it has incited safer practices. Images of bleached coral in the Great Barrier Reef haven't persuaded Australians to go easy on coal nor convinced other populations to restrain their fossil-fuelled practices such as cruising, flying, etc. However, a COVID-19 virus that threatens them personally and immediately has had more effect on flying and cruising, despite having nothing to do with fossil-fuelled global warming.

Presenting the Climate Problem as Opportunities

It is possible that people will not be motivated to change deeply ingrained fossil-fuelled social practices by dangers distant in space or in time or by solutions calling for replacing the market or radically restricting consumption. A different solution is to deal with the problem by positive thinking. This assumes that the population and decision-makers will be more motivated by focussing on the near-term, local benefits of the transition to renewable energy rather than the dangers of fossil fuels. Hence, immediate benefits need to be emphasized. Much like the transition from horses to fossil fuels resulted in new, better jobs, the next energy transition from fossil fuels to low-carbon energy can be portrayed as leading to cleaner, better-paying jobs (Rand 2014; Hawken 2017; Carolan 2018, 2020). In a competitive market economy, the transition can be staged as a race to the top of the green economy, with the countries, companies, and workers that win the race reaping most of the rewards.

A related strategy is to stage mitigation as an indirect consequence rather than tackling global warming head-on. Giddens (2009) makes much of the possibility that policies implemented to attain other goals can better solve the climate change problem as a side effect. He gives the example of Europe's goal of achieving energy independence and energy security at the time of the 1970s OPEC oil crisis. This prompted Europe to implement high petrol prices and develop efficient public transportation, mostly electrified, and resulted in France developing nuclear energy. The increased petrol prices were not sold as carbon taxes to combat fossil-fuelled climate change, but had that consequence. Giddens generalizes this as a methodology to solve fossil-fuelled climate change: emphasize near-term economic benefits of clean growth rather than long-term environmental dangers of fossil fuels. Harvey and Orbis (2018: 63) argue that most 'policies that aim to reduce greenhouse gas emissions have co-benefits: positive effects for society other than mitigating climate change. The most important co-benefit is usually an improvement in public health'. Pielke (2010: 232) quotes John Kerry's logic in 2010 when he was an American Democratic senator arguing in favour of climate legislation: '"It's primarily a jobs bill, and an energy independence bill and a pollution reduction-health-clean air bill. Climate sort of follows. It's on for the ride"'.

Drawing attention to near-term benefits such as jobs from the development of renewable energy is important, and has been frequently tried by decision-makers attempting to mitigate global warming. But although helpful, unfortunately it has not succeeded in decreasing emissions to withdrawal rates. Kerry's ride proved to be bumpy on a two-way street. The Trump Republican Administration undid all the Democratic climate legislation and promoted jobs and energy independence by deregulating and expanding the fossil-fuel industry. Near-term economic benefits can be achieved with carbon-polluting fossil fuels. Energy security and independence were attained with hydraulic fracturing and deepwater drilling in the United States, by extracting oil from bituminous (tar) sands in Canada, by prioritizing coal and liquefied natural gas in Australia, etc. Texas has one of the world's largest wind farms, as well as one of the world's biggest petrochemical industries. China seized the opportunity of leading the development of solar energy, but remains the world's biggest carbon polluter by combusting coal. These states added low-carbon energy, but did not reduce their carbon-polluting energy. The greenhouse effect will only be mitigated by decreasing emissions to carbon withdrawal rates, unless there is a miraculous technological breakthrough rendering the combustion of fossil fuels emissions-free. Solutions emphasizing opportunities and indirect solutions have to be assessed according to what is needed. It is necessary to purposefully steer innovations to eliminate dangerous fossil-fuel emissions. Moreover, any strategy of soft-pedaling scientific assessments of danger is disingenuous because it can lead people to claim that they didn't restrain fossil-fuelled practices because they didn't know the seriousness of the problem.

Staged assessments of risk are often contested. The debate can then be about the staging itself rather than the underlying material risk, whether the staging represents real threats or are exaggerated or even imaginary unreal allegations. Even for real material danger, staging can have perverse sociocultural consequences that backfire and decrease its recognition. Examples are the scary book titles with expiry dates that undermine credibility when the date has passed with little consequence.

Staging of Safety and Discounting Danger

Beck (2007) analysed the staging of risk, but ignored the staging of safety and discounting of danger. He thereby sees the mouse and is blind to the elephant. Hence it is necessary to turn Beck's approach right side up by examining the staging of safety. In Canada, where Greenpeace originated, it has been ineffective in staging the risk of extracting four million barrels of diluted bitumen from the tar sands daily and exporting it by rail, pipeline, and ship. On the other hand, the Canadian Association of Petroleum Producers (CAPP) have the resources to employ professionals in media staging to divert attention away from the atmospheric carbon pollution, freshwater contamination, deforestation, etc., of tar sands exploitation, more euphemistically renaming it 'oil sands', conflating carbon-polluting fossil fuels with low-carbon wind and solar by using the word 'energy, labelling the thousands of square kilometres of bitumen an "oil patch" to create the impression of smallness' (Murphy 2011), and shifting attention away from its long-term danger to its near-term economic benefits, thereby legitimating it (Davidson and Gismondi 2011). Who is winning the media staging battle of risk versus discounting danger? Since anthropogenic climate change results from carbon emissions, the rapid rise of Canadian emissions from tar sands exploitation, notwithstanding the formation of Greenpeace in Canada in 1971, indicates the winner. Despite best efforts by environmentalists at staging the risk of global climate change, the other side discounting danger claiming near-term economic benefits of becoming an oil superpower and staging safety is winning (Murphy 2015). This is true concerning fossil fuels in most countries, so they continue to be combusted at high rates.

The full complexity of staging needs to be analysed. Staging (i) can consist of presenting the best available evidence, both scientific and lay, in ways understandable to non-scientists, or it (ii) can involve misrepresentation, either deliberate or unintentional because of ignorance. Staging of safety and discounting scientifically documented predictions of danger can block the formation of risk expectations and corrective action where there is catastrophic risk, as when authorities in New Orleans claimed it was safe to build canals for economic growth, which amplified vulnerability to Hurricane Katrina (Freudenburg et al. 2009). Research (Jacques, Dunlap, and Freeman 2008; Dunlap, McCright, and Yarosh 2016) has documented how think tanks have been formed and financed by the fossil-fuel industry, tobacco industry, CFC chemical industry, etc., to downplay risks inherent in their products. Freudenburg (2006) showed how scientifically documented environmental problems are staged as non-problems using a 'double diversion'. If risk makers succeed in convincing the public to discount danger, then preventive action is not taken, and vulnerability to harm is increased. This constitutes the staging of safety where danger lies in wait. The staging of risk confronts its opposite performative construction: the staging of safety and discounting danger.

Beck argues that diagnosing, imagining, and staging the world risk of catastrophe, such as anthropogenic climate change, creates a self-denying prophecy which helps avert it. But the converse is also true. Staging, imagining, performing, and acting as if safety will prevail, and as if scientific predictions of future harm should be discounted, are leading to the opposite self-denying prophecy that brings on catastrophe. Claims of safety-as-usual and disregarding danger result in failure to enact precautionary measures in the context of dangerous forces of nature, hence susceptibility to harm persists, and safety threatens to be replaced by catastrophe. The prophecy of safety incites practices that result in its denial in fact. This occurred for local disasters. In communications with regulators, BP claimed the blowout protector on its Deepwater Horizon oil rig was failsafe. However, the material reality of risk under deep ocean water pressures was at odds with this staged discourse, the blowout protector failed to protect safety, and the result was a catastrophic explosion, workers killed, and a two-month-long oil gusher that contaminated the Gulf of Mexico (Freudenburg and Gramling 2011). Concerning fossil-fuelled climate change, if one judges by social practices resulting in a net increase in the carbon content and temperature of the atmosphere despite scientific warnings, globally the staging of safety is winning out over the staging of risk and danger is being discounted. The following are examples of strategies that are being used to stage safety.

Staging of Faith in Market Miracles

Simon (1981, 1996) and Lomborg (2001, 2007, 2010) present an economist's formulation of staging safety widely believed in society. They dismissed risks of depleting resources by claiming that as long as technological innovations and market dynamics of supply and demand are fostered, there will never be scarcity because reason will always invent new products, create substitutes when resources become scarce, and devise solutions to problems as they arise. If technological innovation and the market are allowed to work their magic, they will bring abundance and affluence to everyone forever. The main risk is that irrational social movements will shackle the invisible hand of the market and impede technological ingenuity. These assertions are, however, refuted by the evidence because profit-driven technological innovations have hitherto failed to reduce carbon emissions and have worsened such pollution by innovating new ways to extract carbon from the ground, where it has been stored safely by nature's processes, and emitting it into the atmosphere where it becomes dangerous. There are no substitutes for the atmosphere and oceans as carbon sinks. Market-based technological innovations have filled them with carbon having deleterious consequences. Wishing for last-minute, end-of-pipe, profit-driven technological solutions, so as to avoid restraining fossil-fuelled practices, constitutes dangerous brinkmanship.

Staging Mitigation as a Job Killer

Fossil-fuel promoters often depict policies to solve the climate crisis as job killers. Thus carbon taxes, cap and trade, fuel-efficiency standards, and the like are demonized as destroying jobs and prosperity. Such staging is contradicted by documentation that the transition from fossil fuels to low-carbon energy results not in jobs lost but rather jobs shifted (i) from high carbon-emitting fossil fuels to low-carbon sources of energy like solar, wind, hydroelectric, and geothermal, (ii) from the pursuit of extraction to the pursuit of efficiency, and (iii) from the manufacturing sector to the service sector. Far from being killed, jobs are re-energized and could become more abundant and more lucrative, as occurred with other energy transformations. The best climate performing states with high taxes on fossil fuels, such as Sweden and Switzerland, are also the most prosperous (Yale University 2018).

Staging Fossil Fuels as Poverty Reduction

Another claim is that fossil fuels reduce poverty (Lomborg 2019: A15). Whereas those concerned about fossil-fuelled climate change want carbon taxes, which will impact the poor, fossil-fuel extractors keep fuel prices low by increasing supply through the growth of extraction. In his attack on Greta Thunberg, Lomborg contends that it is fossil-fuel tycoons, not activists like Thunberg, who are alleviating poverty. The degradation of the natural environment and of nature's services by fossil-fuelled climate change threatens, however, to increase poverty in the long run. And carbon taxes can be recycled back to the poor in ways that incite them to use public transit and fuel-efficient vehicles.

Staging Blamelessness

For the fossil-fuelled climate crisis, blamelessness takes many forms. A significant one is that fossil-fuelled companies are irreproachable because consumers demand fossil fuels. This demand–supply issue was examined in Chapter 4. Another is the assertion that we are all fossil-fuel sinners. Everyone who drives a car, takes a plane, turns on a furnace or air conditioner, or uses a social media server is at fault for fossil-fuelled climate change. Disproportionality (Freudenburg 2006) of blame is ignored. The difference between big and little blameworthiness is intentionally blurred. Coal miners toiling underground are as much to blame as billionaire coal mine owners who make fortunes by not paying the health and environmental costs caused by their carbon pollution, and then use those fortunes to lobby politicians to oppose regulating carbon pollution, e.g. the Koch brothers in the USA. If everyone is equally to blame, then no one is to blame. The grain of truth in such stagings masks how misleading they are.

One reason for the predominance of blamelessness is politeness. For example, all the participating countries at the 2019 Arctic Council except the USA agreed with a declaration that recognized climate change as a serious threat to the Arctic. The Americans led by Secretary of State Mike Pompeo did not want any hindrance to the exploitation of oil and gas in the Arctic. The Inuit representative called the American position 'a moral failure'. However, the Finnish Foreign Minister, who chaired the meeting, recognized the urgent need to mitigate, adapt, and strengthen resilience but said 'he did not want to name and blame anyone' (Dickson 2019: A7). Those who promote, disproportionately use, or profit from fossil fuels are mainly responsible for fossil-fuelled climate change, but people are reluctant to lay blame in polite conversation.

Staging Fossil-Fuel Critics as Hypocrites

A particularly aggressive staging is to label as a hypocrite anyone who points out the adverse consequences of fossil fuels yet drives a car, flies in a plane, uses air conditioning and social media servers. The American Al Gore, Canadian David Suzuki, climate scientist Stephen Schneider, and other fossil-fuel critics have been pilloried by fossil-fuel promoters for taking planes to give speeches on the dangers of global warming. At present, there is no alternative to flying for necessary, long-distance travel. These detractors of climate scientists and activists are using the rhetoric of hypocrisy to shut them up. The implication is that the way to avoid hypocrisy is to deny that fossil-fuel combustion causes climate change, then you can fly as much as you want without hypocrisy. But biophysical dynamics don't cooperate with such discourse, the fossil-fuelled greenhouse effect marches ahead unimpeded, and future generations suffer the consequences.

Staging of safety when conditions are dangerous leads to failure to acknowledge the underlying material risk and retards improvements of social practices and of technologies to avoid calamities. Other strategies to frame fossil fuels and climate change as non-problems, which involve various forms of staging, are examined in Murphy (2015).

Learning from Disaster Research

It is important to learn from disasters. Researchers (Turner and Pidgeon 1978; Vaughan 1996) have long found that disasters are typically preceded by a 'failure of foresight' when evidence of the impending disaster was available but discounted, and by an 'incubation period' when the disaster could have been averted or at least greatly diminished if corrective action had been taken. For example, promoters of economic growth in Louisiana claimed there was little risk even though evidence suggested otherwise, and the result was the Hurricane Katrina and BP Deepwater Horizon calamities (Freudenberg et al. 2009; Freudenberg and Gramling 2011). These were therefore 'man-made disasters' (Turner and Pidgeon 1978) based on a failure of foresight even though they involved powerful forces of nature because the available scientific evidence of danger was discounted during the incubation of avoidable disasters. This is well-expressed by the title of Freudenburg et al.'s study of the Hurricane Katrina disaster: *Catastrophe in the Making: The Engineering of Katrina and the Disasters of Tomorrow*. This shows the importance of foresight which takes into account rather than discounting the available evidence, and acts according to it to prevent 'man-made' complacency and reduce vulnerability in order to promote safety and sustainability.

Studies of disasters typically end with 'lessons learned'. The traumatic experience usually results in better technical protection against the kind of catastrophe that occurred in the recent past. After Hurricane Katrina, levees were repaired and reinforced and more efficient pumps were installed (Freudenburg et al. 2009). After the BP disaster, better blowout protectors were mounted on oil rigs (Freudenburg and Gramling 2011). Earthquakes motivated charting of geological fault lines and earthquake zones and building codes were then implemented. Floods promoted mapping of flood plains, building dykes, constructing river bypasses to protect cities (e.g. Winnipeg), and restricting development on flood plains. Wildfires prompted the documentation of areas prone to them and preparations to reduce fatalities and property damage. The enormous cost in lives and money of the Indian Ocean 2004 Boxing Day tsunami inspired countries to pay for tsunami early warning systems. Disasters are important prompts to make infrastructures more robust to withstand hazards similar to those which have already occurred, called 'yesterday's disasters', in order to reduce vulnerabilities, and to make societies more resilient so they can bounce back (Murphy 2009).

But the prompt is not always acted upon except in a minor technical way. After the experience of a disaster, the question is: will adequate mitigation to avoid recurrence be done or will the prompt be minimized when such mitigation involves cost or disruption of social practices? In many cases the danger is discounted and the risk is run. It is more convenient to believe that the hazard will not appear with destructive force for another century, or at least not during the mandate of political and economic leaders nor during the lifetime of citizens who must pay the cost, so why bother. The costs of prevention and disruption of social practices are immediate whereas the dangers are distant in time and can be contested. Hence 'repeat disasters' happen (Platt 1999).

With fossil-fuelled climate change, the threat is not a repeat of a disaster that has already happened but rather the scientific prediction of a catastrophe that is likely to occur. If there is a reluctance to change and pay the costs of mitigation after the experience of disaster, then imagine the reticence concerning the mere prediction of disaster, scientific though it may be. Nevertheless, the most general lesson learned from disaster studies is a sociocultural one: pay attention to warning signs and modify social practices and technologies that threaten to lead to calamity. Applying this broad lesson to fossil-fuelled climate change means taking seriously the scientific evidence as well as visible indications of danger (receding glaciers, melting Arctic ice cover, more frequent and/or intense extreme weather such as wildfires, floods, and hurricanes, etc.), and letting them incite less risky social practices and safer technologies.

The Uncertainty of Whether There Will Be Foresight or Danger Will Be Discounted

Flannery (2015: 194) documents how much best practices using existing technologies can do to limit fossil-fuelled climate change and concludes that 'we clearly have the tools needed to avoid more than 2 °C of warming. But will we use them?' Writing about the fossil-fuelled climate crisis, Rand (2010: 212) argues that 'the ten clean technologies in this book are a roadmap of where to go. Getting there is up to all of us'. Is the road being taken and are we getting there? Harvey and Orbis

(2018: 299, 303) state that 'although the technology exists today and is falling in cost, significantly reducing global greenhouse gas emissions is a Herculean task, one that will not simply happen on its own even with cheap clean technology. … So the challenge is not technical, nor even economic, but rather is a matter of enacting the right policies and ensuring they are properly designed and enforced'. To accomplish such a Herculean task, everyone, especially those with power, have to do their part. No free-riders allowed. If our generation does not practice fossil-fuel restraint through enforced measures like carbon taxes, then our grandchildren could be forced to practice fossil-fuel abstinence (little if any cruising, flying, even driving). 'The habit of focusing on the present and discounting the future leads away from a thoughtful appraisal of long-term consequences and the world we are making' (Speth 2012: 6–7). Short-sightedness is related to the priority given to near-term economic benefits and cultural predispositions, leading to danger being discounted.

Note the difference between this conception of discounting future harm and the theoretical conception of discounting the future by economists. In the latter, a hundred dollars today is worth more than a hundred dollars in ten years because the money could be used immediately, in particular to accumulate more money, hence the fixed amount of future money is discounted. However, this logic does not apply to people being killed or species being wiped out by extreme weather. A flood today is worse than the same flood in ten or even a hundred years only if the threat of a future flood is not discounted, that is, only if measures are taken in the interval to mitigate it by modifying the cause of the flooding, restoring wetlands, building flood walls, etc. Discounting danger of future harm is invalid if present practices lead the danger to cumulate, like fossil-fuelled global warming, and if preventive measures are not implemented. The increase in carbon emissions which have been documented and accumulated demonstrates that the kind of discounting danger which is currently being practised consists of refusing adequate mitigation in order to enjoy the near-term benefits of inexpensive fossil-fuel combustion. In such cases, discounting scientific predictions of future harm is not only invalid but also dangerous. It is this form of discounting danger that is referred to by the subtitle of the book.

Societies are currently conducting an unplanned, trial-and-error alteration of the global climate, with the possibility of tragic error looming large. Nevertheless the solutions to fossil-fuelled climate change are known and not very difficult technically. They consist of reducing emissions to withdrawal rates by regulations and pricing carbon pollution, increasing carbon withdrawal, and doing this globally by implementing a binding, enforced international agreement to avoid free-riders, much like what the Montreal Protocol accomplished for CFCs depleting the ozone layer. Making the polluter pay for externalities would act as a powerful stimulus to innovate and implement emissions-free, clean renewable technologies. This would result in decarbonizing transportation, industry, and the energy supply. Increasing withdrawal of atmospheric carbon could be done by replanting many of the forests cut over the past two centuries and using environmentally friendly agriculture. Technological barriers and knowledge insufficiencies will only become major problems if humanity continues fossil-fuelled social practices at all levels that open further the Pandora's Box of nature's safe storage underground of valuable but dangerous hydrocarbons and unleash runaway climate change. Solutions to fossil-fuelled climate change are technically available. The incubation of disaster results from a failure of foresight and social resistance to changing fossil-fuelled social practices and to paying for needed improvements.

References

Agrba, Liza. 2019. The Uncertainty of Climate Change. *Report on Business* (November): 9–10.

Beck, U. 1992. *Risk Society*. London: Sage.

Beck, U. 1995. *Ecological Politics in an Age of Risk*. Cambridge: Polity.

Beck, U. 2007. *World at Risk*. Cambridge: Polity Press.

Beck, U. 2015. Emancipatory Catastrophism. *Current Sociology* 63 (1): 75–88.

Berners-Lee, M., and D. Clark. 2013. *The Burning Question*. London: Profile.

Carolan, M. 2018. *The Food Sharing Revolution: How Start-Ups, Pop-Ups, and Co-Ops Are Changing the Way We Eat*. Washington, DC: Island Press.

Carolan, M. 2020. *Society and the Environment: Pragmatic Solutions to Ecological Issues*, 3rd ed. Boulder: Westview Press.

CST Commission scientific et technique. 1999. *Pour Affronter L'imprévisible*. Québec: Gouvernement du Québec.

Davidson, D., and M. Gismondi. 2011. *Challenging Legitimacy at the Precipice of Energy Calamity*. New York: Springer.

Dickson, Janice. 2019. Chrystia Freeland Says It's a "Disappointment" Arctic Council Could Not Issue Joint Communique. *Globe and Mail*, 7 May: A7.

Dunlap, R., A. McCright, and J. Yarosh. 2016. The Political Divide on Climate Change: Partisan Polarization Widens in the U.S. *Environment Science and Policy for Sustainable Development* 58 (5): 4–23.

Dyer, Gwynne. 2008. *Climate Wars: The Fight for Survival as the World Overheats*. Toronto: Random House.

Flannery, Tim. 2009. *Now or Never*. Toronto: HarperCollins.

Flannery, Tim. 2015. *Atmosphere of Hope: Searching for Solutions to the Climate Crisis*. New York: Atlantic Monthly Press.

Freudenburg, W. 2006. Environmental Degradation, Disproportionality, and the Double Diversion. *Rural Sociology* 71 (1): 3–32.

Freudenburg, William, Robert Gramling, Shirley Laska, and Kai Erikson. 2009. *Catastrophe in the Making*. Washington: Island Press.

Freudenburg, W., and R. Gramling. 2011. *Blowout in the Gulf*. Cambridge, Mass.: MIT Press.

Giddens, Anthony. 2009. *The Politics of Climate Change*. Cambridge: Polity Press.

Grundmann, Reiner. 2001. *Transnational Environmental Policy*. London: Routledge.

Hamilton, Clive. 2010. *Requiem for a Species*. London: Earthscan.

Harvey, Hal, and Robbie Orbis. 2018. *Designing Climate Solutions: A Policy Guide for Low-Carbon Energy*. Washington: Island Press.

Hasegawa, Koichi. 2012. Facing Nuclear Risks: Lessons from the Fukushima Nuclear Disaster. *International Journal of Japanese Sociology* 21 (1): 84–91. https://doi.org/10.1111/j.1475-6781.2012.01164.x.

Hasegawa, Koichi. 2015. *Beyond Fukushima*. Melbourne: Trans Pacific Press.

Hawken, Paul. 2017. *Drawdown: The Most Comprehensive Plan Ever Proposed to Reverse Global Warming*. New York: Penguin.

Jacques, Peter, Riley E. Dunlap, and Mark Freeman. 2008. The Organization of Denial: Conservative Think Tanks and Environmental Scepticism. *Environmental Politics* 17: 349–385.

Kolbert, Elizabeth. 2014. *The Sixth Extinction*. New York: Henry Holt and Company.

Lidskog, Rolf, et al. 2010. *Transboundary Risk Governance*. London: Earthscan.

Lockie, Stewart, and Catherine Wong. 2017. Risk, Sustainability, and Time: Sociological Perspectives. In *Social Science and Sustainability*, ed. Heinz Schandl and Iain Walker, 187–198. Melbourne: CSIRO Publishing.

Lomborg, B. 2001. *The Skeptical Environmentalist*. Cambridge: Cambridge University Press.

Lomborg, B. 2007. *Cool It*. New York: Alfred A. Knopf.

Lomborg, Bjorn. 2010. *Smart Solutions to Climate Change: Comparing Costs and Benefits*. Cambridge: Cambridge University Press.

Lomborg, Bjorn. 2019. On Climate Change, Humanity Is Not "Evil". *The Globe and Mail*, 27 September: A15.

Malthus, Thomas. 1798. *An Essay on the Principle of Population*. London: Pantianos Classics.

McGuire, Bill. 2008. *Seven Years to Save the Planet*. London: Weidenfeld & Nicolson.

Meadows, Donella, Dennis Meadows, Jorgen Randers, and William Behrens. 1972. *The Limits to Growth*. New York: Universe Books.

Murphy, Raymond. 2009. *Leadership in Disaster: Learning for a Future with Global Climate Change*. Montreal: McGill-Queens University Press.

Murphy, Raymond. 2010. Environmental Hazards and Human Disasters. In *The International Handbook of Environmental Sociology*, 2nd ed, ed. Michael Redclift and Graham Woodgate, 276–291. Cheltenham, UK: Edward Elgar.

Murphy, R. 2011. The Challenge of Anthropogenic Climate Change for the Social Sciences. *International Review of Social Research* (1): 167–181.

Murphy, R. 2015. The Media Construction of Climate Change Quiescence: Veiling the Visibility of a Super Emitter. *Canadian Journal of Sociology* 40 (3): 331–354.

Pielke, Roger Jr. 2010. *The Climate Fix*. New York: Basic Books.

Platt, R. 1999. *Disasters and Democracy*. Washington: Island Press.

Rand, Tom. 2010. *Kick the Fossil Fuel Habit: Ten Clean Technologies to Save our World*. Toronto: Eco Ten Publishing.

Rand, Tom. 2014. *Waking the Frog: Solutions for our Climate Change Paralysis*. Toronto: ECW Press.

Renn, Ortwin. 2008. *Risk Governance: Coping with Uncertainty in a Complex World*. London: Earthscan.

Safran Foer, Jonathan. 2019. *We Are the Weather: Saving the Planet Begins at Breakfast*. New York: Farrar, Strauss, and Giroux.

Sayer, A. 1997. Essentialism, Social Constructionism, and Beyond. *The Sociological Review* 45 (3): 453–487.

Sayer, A. 2001. Reply to Holmwood. *Sociology* 35: 967–984.

Schneider, S. 2009. *Science as a Contact Sport*. Washington: National Geographic.

Simon, Julian. 1981. *The Ultimate Resource*. Princeton: Princeton University Press.

Simon, Julian. 1996. *The Ultimate Resource 2*. Princeton: Princeton University Press.

Simpson, Jeffrey, Mark Jaccard, and Nic Rivers. 2007. *Hot Air: Meeting Canada's Climate Change Challenge*. Toronto: Emblem McClelland & Stewart.

Speth, James Gustave. 2012. *America the Possible: Manifest for a New Economy*. New Haven: Yale University Press.

Turner, B., and N. Pidgeon. 1978. *Man-Made Disasters*. London: Wykeham.

Vaughan, D. 1996. *The Challenger Launch Decision*. Chicago: University of Chicago Press.

Wallace-Wells, David. 2019. *The Uninhabitable Earth: Life After Warming*. New York: Penguin Random House.

Weisman, Alan. 2007. *The World Without Us*. Toronto: HarperCollins.

Wong, C., and S. Lockie. 2018. Sociology, Risk and the Environment: A Material-Semiotic Approach. *Journal of Risk Research* 21 (9): 1077–1092.

Yale University. 2018. *EPI Environmental Performance Index 2018*. https://epi.envirocenter.yale.edu/epi-country-report/CAN. Accessed 22 September 2019.

Zebrowski, Ernest, Jr. 1997. *Perils of a Restless Planet: Scientific Perspectives on Natural Disasters*. Cambridge: Cambridge University Press.

8

Are Safe Social Practices on the Horizon?

Dealing with acute problems requires accepting chronic problems of monitoring, expense, and action to make practices safe (Tenner 1997). The air and water in most cities today are cleaner than they were near the beginning of industrialization because the adverse effects of local pollution were acknowledged and action taken to remedy them, sometimes at great cost. However, the problem of mobilizing action to deal with environmental and health problems is much greater when the pollution can not be perceived by the senses, like carbon dioxide, and when adverse consequences do not occur where pollution is caused but instead are distant in space or time. Today's global risks are different in kind from early industrial risks that had only local effects where pollution affected the polluter. Now victims inhabit distant countries, or will live in the future. In many cases technical remedies cannot be found, so the only solution is prevention: abandoning the innovation (DDT, CFCs) or leaving the useful but harmful substance (asbestos) safely underground. Will this be the fate of fossil fuels, starting with coal, heavy unconventional oil, etc.? That is, however, seen as an economic "opportunity cost" and a "stranded asset", so there is much resistance.

R. Murphy, *The Fossil-Fuelled Climate Crisis*,
https://doi.org/10.1007/978-3-030-53325-0_8

Grounds for Hope

Even the most deadly forces of nature can be dealt with and their consequences mitigated if scientific warnings are accepted, if future harm is not discounted, and if there is sufficient foresight to make changes in technologies, which is sometimes expensive, and/or in social practices. For example, in the past the contagious variola virus that causes smallpox killed 3 of every 10 people who contracted it and left survivors with severe scars (CDC 2016). It devastated populations since the third century BC, transmission followed trade routes, and made cities the deadliest places to live. In 1796, Dr. Edward Jenner noticed that milkmaids who contracted cowpox, a mild disease related to smallpox that can be transmitted between species, did not show symptoms of smallpox. So he took material from a cowpox sore on a milkmaid's arm and injected it into the arm of his gardener's son, then months later exposed the boy to smallpox multiple times. The boy never got smallpox. After several similar experiments, Jenner published a paper in 1801 entitled 'On the Origin of the Vaccine Inoculation', for which he is credited with being the father of vaccination. Vaccination became widely accepted in the 1800s, and in 1959 the World Health Organization initiated a campaign to globally vaccinate everyone and eradicate smallpox. New high-quality freeze-dried vaccine technologies and laboratories increased production. The last known death from smallpox occurred in 1978, and in 1980 the World Health Assembly declared that the variola smallpox virus, which was one of the most deadly in human history, had been made extinct by vaccination, except for research samples kept in the USA and Russia. Vaccination saved hundreds of millions of lives and is now seen as the greatest achievement of international public health. Nevertheless there was opposition that had to be countered. Some clergy claimed that vaccination was against religious precepts because the serum was derived from animals. Civil libertarians opposed mandatory orders. Sceptics claimed smallpox was not caused by a virus but instead by bad air. Although different in some ways, the struggle against smallpox and against fossil-fuelled climate change share similarities. Both involve a global threat from runaway dynamics of nature. In both cases, science needs to be accepted and sceptics refuted. New technologies need to be innovated.

Social practices and cultures need to be changed. Governments need to enact new laws. International cooperation is necessary, including science-based global organizations like the World Health Organization (WHO) for pandemics, and the Intergovernmental Panel on Climate Change (IPCC), World Meteorological Organization (WMO), United Nations Environmental Programme (UNEP), and the United Nations Framework Convention on Climate Change (UNFCCC) for climate change. There are social justice issues such as a doctor trying out a vaccine on his gardener's son instead of his own. Despite demonstrable success of mitigating the threat there is ongoing opposition to measures of mitigation, such as unfounded claims that vaccines cause autism. Hopefully the struggle against fossil-fuelled climate change will be as successful as the fight against the deadly smallpox virus.

Societies have risen to meet enormous challenges in the past by being willing to make sacrifices and pay the cost. Wars are a primary example, with the sacrifices even involving dying. Another example consists of the first humans landing on the moon. NASA was given an open chequebook and told to hire the brightest people for fear the Soviets would win the race. The American population accepted to pay the huge expense for the necessary research and innovation, even though the USA was mired in the ill-conceived war in Vietnam. Far from expensive space-exploration research being a job killer, it stimulated economic growth in the long run. This could be a template for a race among countries to obtain the economic benefits and prestige of being first to develop non-carbon emitting, affordable sources of energy and storage.

Byers (2019: O3) argues that international cooperation is fostered where activities are dangerous and measures to ensure safety are expensive, hence giving an incentive for burden sharing. He points to space and the Arctic. Both are dark, cold, dangerous places where cooperation, even between enemy states, has been common. When big game hunters started using helicopters to kill polar bears and decimated their population, the USA, Soviet Union, Norway, Denmark, and Canada signed the 1973 Polar Bear Treaty banning the practice. Cooperation during the cold war led in 1979 to the creation of OSPAS-SARSAT, which pooled the capacities of satellites and ground stations of the USA, the Soviet Union, Canada, and France for search and rescue in the Arctic.

In 1987, Mikhail Gorbachev started international negotiations leading to the creation of the Arctic Council. The International Space Station is another illustration of cooperation between states, often rival ones, as is the use of Soyuz spacecraft to send American and other Western astronauts and equipment to it. These successes provide hope that interstate cooperation and burden sharing can be scaled up to deal with the expensive global climate threat.

Will Global-Warming Danger Elicit Solidarity?

Research has documented that the experience of disaster fosters solidarity, mutual aid, and increased willingness to obey leaders while it is occurring. 'The material context of grave danger augments support for leaders during a disaster or a war. In both cases, however, there is much criticism if things go wrong and much second-guessing when the threat ends and the situation returns to normal' (Murphy 2009: 239). Clark (2011: 156) wants 'to hold on to the idea that the disaster is an incitement to new alliances, practices, repertoires – and that the most important improvisation is community itself'. Nature's threatening constructions such as earthquakes have repeatedly acted as prompts that promoted mutual aid and enhanced a feeling of collective belonging: 'the rumbling of the earth is an ancient and unending imperative for human communing, not simply testing existing communities but reactivating, again and again, a primordial impulse for being together with others' (Clark 2011: 164). Claisse and Delvenne (2015) argued that, even on a global level, the anticipation of catastrophes prompts changes in discourse and policy that will prevent catastrophes. Will dangerous fossil-fuelled global warming elicit solidarity, social cohesion, and willingness to improve social practices and technologies to win the war against man-made climate change? That danger in a modern world tightly coupled with transportation and communication technologies has the potential to inspire a transition to realizing that we are all in this together everywhere on the planet.

Beck (2007: 9–10) presents a theoretical formulation of this argument that the anticipation of catastrophe leads to emancipation from harmful social practices. Material risk determines cultural perceptions of risk and actions: 'because this constant danger [of catastrophe] shapes our expectations, lodges in our heads and guides our actions, it becomes a political force that transforms the world'. It fosters the emergence of cosmopolitan perspectives regarding the consequences of decisions for others across space and time. The dangers of fossil-fuelled climate change will impel positive changes in social practices and technologies that mitigate it: 'it is not about the negative side effects of goods but the positive side effects of bads'. He refers to this hypothesis as 'emancipatory catastrophism' (Beck 2015: abstract). It is important to examine Beck's argument in detail, and evidence for or against it, because he was the pre-eminent social scientist of risk in his generation, and because he was a public intellectual who disseminated his arguments broadly, especially in Germany. Most important, the idea that the anticipation of global climate catastrophe will incite a change in orientations is relied on in society generally to assume that all will end well. Beck's hypothesis constitutes a window into society's thinking about this issue. Its logic and supportive evidence need to be rigorously evaluated.

Beck's argument can be briefly summarized as follows using his concepts. (1) The anticipation of global catastrophe violates sacred (unwritten) norms of human existence and civilization; (2) hence it causes an anthropological shock; (3) a social catharsis; (4) and a metamorphosis with new compasses for the twenty-first century and a cosmopolitan perspective. 'The global climate risk, far from an apocalyptic catastrophe, is instead –so far! – a kind of "emancipatory catastrophe"' (Beck 2015: 79). He (Beck 2015: 79) contends that 'global risks – like climate change or the financial crisis – have given us new orientations, new compasses for the 21-century world. … Climate change … is a reformation of modes of thought, of lifestyles and consumer habits of law, economy, science and politics. Global climate risk could usher in a rebirth of modernity'. Beck (2015: 81) claims that 'in this moment of catharsis the mind-walls of institutionally constructed side effects are breaking down and we can empirically study the cultural fact of how cosmopolitan horizons are emerging and being globalized'. A global risk like 'climate change induces fundamentally changing landscapes of social class and inequality created through rising sea levels which draw new maps of the world where the key lines are not traditional boundaries between nation-states and social classes, but rather elevation above sea – a whole different way of conceptualizing the world and the "life" chances, the chances of survival within it' (Beck 2015: 76). It induces a sense of ethical violation which creates new norms, laws, markets, technologies, understandings, and international cooperation. Global risks also expose the limitation of 'methodological nationalism'. Organized irresponsibility towards others is being transformed into a cosmopolitan outlook that will take into account the needs of others distant in space and time. Global cosmopolitanism consists of a moral responsibility to not engage in social practices which do harm to others and to the environment they need. Beck's theory that the anticipation of catastrophe will lead to a cosmopolitan orientation and emancipation from global warming risk is as optimistic as it gets.

Beck's analysis has a curious complementary-oppositional relationship with other prominent analyses of global environmental problems. The ecological modernization approach (Mol, Sonnenfeld, and Spaargaren 2009) shares the postulate that the anticipation of catastrophe is prompting adaptation, resilience, and mitigation by companies and governments. Giddens (2009: 70–71) agrees with ecological modernization authors who 'distanced themselves from the pessimism of the "limits to development" literature, and also from those in the green movement who set themselves against modernity and, to some extent, against science and technology more generally. … [I am] in general a supporter of the ecological modernization approach' to environmental problems. The theories of Beck, Giddens, and ecological modernization all assume that anthropogenic climate change is being mitigated. However, Giddens' analysis is based on a premise diametrically opposed to that of Beck. Beck's theory of catastrophism is about the motivating force of fear. Change is being brought about by dreading calamities scientifically predicted to bring harm to future generations and vulnerable populations in distant places. Giddens (2009: 12) argues, on the contrary, that 'Martin Luther King didn't stir people to action by proclaiming, "I have a nightmare". Fear and anxiety are not necessarily good motivators, especially with risks perceived as abstract ones, or dangers that are seen as some way off'. He presents '"Giddens paradox".

It states that, since the dangers posed by global warming aren't tangible, immediate or visible in the course of day-to-day life, however awesome they appear, many will sit on their hands and do nothing of a concrete nature about them. Yet waiting until they become visible and acute before being stirred to serious action will, by definition, be too late' (Giddens 2009: 2). There is not much fear of catastrophe in that quotation. What is needed is 'an emphasis on positives as much as on negatives, and on opportunities rather than on self-induced deprivations' (Giddens 2009: 106). The main impetus in Europe for increased taxes on fossil fuels and development of wind, solar, and nuclear energy was the opportunity to enhance energy security after the formation of OPEC in the nineteen-seventies, with the reduction of carbon emissions and mitigation of climate change being indirect beneficial side effects. The politics of hope, according to Giddens, trump the politics of fear. Hence this chapter involves examining whether societies are motivated to respond to threats through fear or opportunities, first with a detailed assessment of Beck's theory of fearful anticipation of global catastrophe and the remainder of the chapter consisting of hopeful examples of responses.

Nordhaus's (2013: 180) economic modelling research arrived at the following conclusion. 'Suppose we live in an ideal world – one where countries work together cooperatively to introduce emissions reductions, take care to ensure that all countries and sectors participate, and time their actions efficiently'. He calculated that this could slow global warming to about 2 °C at a cost of 1 to 2% of total world income annually, which is not high. This is Beck's optimistic world, having a cosmopolitan orientation everywhere willing to sacrifice some present income to emancipate humanity present and future from catastrophic fossil-fuelled climate change. It is a noble ideal, but does it correspond to the real world? If some countries or sectors refuse to participate or receive exemptions, or if this ideal is not started promptly, then the cost, atmospheric temperature, and consequences rise precipitously. The IPCC (2018) documented that even a rise of 1.5 °C could have devastating consequences.

Where does Beck find evidence to support his hypothesis? 'Analysing the discourses around Katrina makes apparent a paradigm shift, in fact, a social catharsis, that two formerly separate discourses came together:

ecological challenges and the history of racism in the US' (Beck 2015: 80). He claimed that Hurricane Katrina resulted in a paradigm shift in Louisiana. 'The traumatic experience produces a process of reflection in which things, which had not been thought of as being connected, are now connected – flooding of cities with racial inequality with questions of global justice. This is what I call "social catharsis"' (Beck 2015: 80). Thus 'anthropologic shocks provide a new way of being in the world, seeing the world and doing politics' (Beck 2015: 80).

Weaknesses in Presuming That the Anticipation of Catastrophe Will Incite Safe Practices

Beck's arguments are scholarly formulizations of presumptions in society that underpin optimism that scientific conclusions concerning global environmental problems like fossil-fuelled climate change will be heeded, social practices causing them will be changed and/or technological remedies developed, adverse consequences prevented, and sustainability enhanced. But will this happen in the real world, or will the contemporary period be seen later as the incubation of global catastrophe? Are such theories logical, valid, and well-founded empirically, or are they representative of wishful thinking? Will path-dependent social practices, perceived entitlements, and powerful economic interests override scientific conclusions of danger?

The response to threats is mixed. The COVID-19 pandemic elicited some cooperative solidarity but also inadequate reactions: belated acceptance of scientific evidence of danger and of community transmission, China blaming the USA and vice versa. Research documented that often disasters do not result in significant reductions in vulnerability when mitigation is costly or requires changed practices. Declaring an extreme-weather event a 'hundred year storm' or a 'thousand year flood' is frequent despite lack of evidence so far back. Such declarations are reassuring cultural practices so that costs of preparedness and changes in risky social practices can be postponed, resulting in 'man-made disasters' (Turner and Pidgeon 1978), 'repeat disasters' (Platt 1999), 'disasters by design' (Mileti 1999), and 'unnatural disasters' (Abramovitz 2001). If this occurred for the *experience* of disasters, then it is questionable whether the mere *anticipation* of catastrophes will result in mitigation. Much can be learned from the flaws of Beck's argument.

Distinguish Aspirations for Change from Changes in Real-World Practices

Beck's hypothesis conflates aspirations and facts. Thus he claims that (i) 'we *have to* attach central importance to the dangers that we have repressed until now', and that (ii) 'global climate risk *could* usher in a rebirth of modernity' (Beck 2015: 79). Is what 'has to' or 'could' be done, being done? Or will consciousness of dangers be discounted and the new ecological modernity be stillborn? Funk (2014: 285) argues convincingly that 'magical thinking is the fallacy that thoughts correspond to actions – that to think is to do, to believe is to act'. A clear distinction is required between what is and what is desired so that desires can be brought into being. Global risks like climate change are compasses indicating where to go, as Beck argues, but old orientations, habitual social practices, and economic interests predispose people not to go there.

A catastrophe does not always lead to positive change, and often 'bads' lead to more 'bads'. As Diamond (2005) documented, bad societal choices can multiple into positive feedback loops and pass a tipping point resulting in societal collapse. Diamond investigated traditional societies, and moderns can hope they are more rational. But that is uncertain: interests, predispositions, and power can trump reason when science brings inconvenient conclusions. Emancipation from harm by anticipating catastrophes refers to a hoped-for future state. Beck's concept of 'emancipatory catastrophism' is an aspiration and a goal, but the anticipation of catastrophe prompts a variety of responses, with a transformation of high-risk social practices being just one possibility. It can also result in doubling down on fossil-fuelled practices and discounting danger.

Beck (2015: 85) claims that global risk is a dystopian vision which 'has a significant power of mobilization because it is about the existence of humanity'. I share Beck's wish that this be so, but where is the evidence to confirm his assertion? The 'existence of humanity' is abstract and does not have as great a mobilizing power as immediate economic interests for oneself and one's family and friends. If carbon pollution immediately harmed polluters and their families, then this would likely mobilize changes of social practices. Invisible greenhouse-gas emissions typically cause harm distant in space and time, thereby diluting mobilizing possibilities because distant harm can be discounted in favour of immediate economic gain. Ongoing socioeconomic practices causing the increasing carbonization of the atmosphere demonstrate that, so far, the threatening dystopian vision has been a less potent source of mobilization than immediate economic interests. Important social relations are those with what sociologists have long called 'significant others'. Judging by the persistence of carbon pollution, 'the existence of humanity' constitutes up to now 'an insignificant other'.

Don't Mistake Emancipatory Discourse for Emancipatory Social Practices

This is a related problem. Beck (2015: 80) focusses on 'the discourse on climate justice' and that 'analysing the discourses around Katrina makes apparent a paradigm shift, in fact, a social catharsis'. The book (Stewart and Ray 2007) he uses as sole documentation of change prompted by the Hurricane Katrina catastrophe includes only changes in discourse. 'Norms and imperatives that guided decisions in the past are re-evaluated and questioned through the imagination of a threatening future. From that follow alternative ideas for capitalism, law, consumerism, science (e.g. the IPCC), etc.' (Beck 2015: 83). But it does not follow that those ideas necessarily lead to benign alternatives to present practices of capitalism, law, consumerism, and science. New discourses can be irrelevant or more devious legitimation of pollution, inequality, racist practices, and greenwashing. Beck does not analyse fossil-fuel, racial, and class practices resulting in vulnerability and catastrophe. Talk might change following a catastrophe, but polluting energy use, racial and class practices persist. Discourses and ideas only emancipate populations from material threats if they result in improved practices. Instead of being based on free-floating discourse and ideas, emancipation needs to be grounded in changes of social practices. Beck mistakes discourse advocating emancipation for emancipation itself. Simpson, Jaccard, and Rivers (2007) documented that many policies to mitigate global warming were 'hot air' and never implemented. Perverse consequences of well-intentioned discourse should not be ignored: talk about renewable energy freeing humanity from fossil fuels has sometimes led to faster rates of extraction before they are displaced by renewable energy. Failing to distinguish aspirations/discourse from changed practices results in a flawed understanding of fossil-fuelled climate change and fosters illusions of improvement even as global warming worsens.

Don't Assume Scientific Predictions of Danger Lead Necessarily to Anticipating Danger

The hypothesis that scientifically documented anticipation of catastrophe leads to safety requires all links in the chain between scientific prediction, anticipation of danger by the public, and corrective action be operative, but powerful economic and political forces act to disconnect the links. Studies in the sociology of science have shown that scientific conclusions cannot be straightforwardly equated with their understanding by the public and are often socially contested (Yearley 1992, 2004). Misrepresentations by fossil-fuel supported think tanks (Jacques, Dunlap, and Freeman 2008; Dunlap and Jacques 2013; Elsasser and Dunlap 2013; Dunlap, McCright, and Yarosh 2016), diversionary tactics (Freudenburg 2006), etc., result in discounting scientific conclusions of danger. An outlier scientist can be found to reject the scientific consensus, with outlier voices being amplified by proponents of fossil fuels, thereby transforming science into a contact sport (Schneider 2009). Rigorous empirical studies (Freudenburg et al. 2009) showed there were scientific predictions of catastrophe for Louisiana before Hurricane Katrina struck, but they were discounted by economic and political decision-makers. These studies provide evidence against hypotheses that scientific predictions of catastrophe necessarily lead to action promoting safety.

Specify What Is Being Transformed

Scientific measurements and modern means of communication are bringing knowledge of threatening biophysical transformations to all nations. In this sense a cosmopolitan perspective is indeed developing. But Beck is vague concerning what societies are being emancipated from through the anticipation of catastrophe. Beck's 'emancipatory catastrophism', 'metamorphosis of the world', and 'cosmopolitan perspective' remain 'highly abstract theoretical concepts' (Han 2015: 116) that are insufficiently specified (Blok 2015: 110) and inadequately grounded in empirical research. Where is the evidence that a sociocultural transformation is occurring in which a cosmopolitan worldview taking into consideration needs of all humans, including future generations, is becoming dominant in shaping present social practices? Restricted perspectives prioritizing near-term economic interests and lifestyles seem to prevail. To mitigate the fossil-fuelled climate crisis, the 'cosmopolitan perspective' has to lead to greenhouse-gas reducing social practices on a global scale promptly. An 'optimistic outlook must be qualified by the strong warning that it requires cooperative and efficient measures [such as placing a price on carbon pollution]' (Nordhaus 2013: 194). It may well be that cosmopolitanism and carbon pollution emancipation are occurring partially in some societies but much less so in others. They should be analysed not as a universal trend but rather as variables differentiating societies.

Nation-State and Social-Class Boundaries Remain Key

Beck's (2015: 76) contention that elevation above the sea is replacing social classes and nation-states as significant factors fails to recognize that there are important differences at the same elevation: between the prosperous Garden District in New Orleans and the poor Lower Ninth

Ward; between rich Switzerland and poor Nepal; between wealthy, well-protected Netherlands and underprivileged, vulnerable Bangladesh. The wealthy live in less vulnerable areas and the poor reside in more vulnerable ones. Elevation of residence above sea and other means of reducing risk are largely based on social class. Wealth gives the capacity to monopolize safety, whereas poverty excludes people from safety. This too is a dimension of social closure, which has been documented by environmental justice researchers (Bullard 2000, 2005; Bullard and Wright 2009; Roberts and Parks 2007). Risks from carbon pollution disproportionately affect more vulnerable social classes and poorer nations. It would be more accurate to argue that when new risks emerge, they are superimposed on old social class and nation-state boundaries.

Take into Account the Preponderance of the Evidence, Not Just What Supports the Hypothesis

Beck selected cases indicating emancipatory possibilities and ignored disconfirming evidence: 'the agility with which the Chinese are promoting the boom in the trade in renewable energy sources' (Beck 2015: 79). This is a strange example of emancipation involving the Communist Party dictatorship, a country whose principal agility is its increasing use of the most polluting fossil fuel (coal), contamination of the air in China's biggest cities, etc. Since it is the excess of emissions over carbon withdrawals that causes global warming, praise for China and Louisiana should be tempered.

The Test Case of Louisiana

As supporting evidence, Beck cites the <u>experience</u> of catastrophe when Hurricane Katrina struck Louisiana. This is not a valid test of whether the scientific <u>anticipation</u> of global catastrophe leads to a cosmopolitan outlook taking the interests of humans distant in space and time into consideration. Nevertheless, if scientific <u>predictions</u> of catastrophe are to incite action, then the actual experience of catastrophe, which confirmed scientific predictions, should result in action. If not, then serious doubt is cast on the theory that the mere prediction of catastrophe would have that effect. Louisiana is an interesting case for investigating whether the experience of environmental catastrophes, and anticipation of recurrence, prompt change of social practices causing catastrophes.

i. It suffered recurring catastrophes resulting from hurricanes (Betsy in 1965, Katrina and Rita in 2005) and from oil spills (2010 BP Deepwater Horizon two-month underwater oil gusher in 2010).
ii. New Orleans is vulnerable to extreme weather and sea level rise. It is located at sea level, with some wards below sea level, and is jammed between the Gulf of Mexico, the Mississippi River, and huge Lake Pontchartrain (1600 km^2). Hence it is plausible that anticipation of recurring catastrophes is high.
iii. The importance of the oil industry and the fishery and tourism to its economy leads to conflicting interests.

The catastrophes of Hurricane Katrina and the BP blowout incited talk about the dangers of hurricanes and fossil fuels and about the problems of poverty and race relations. But did they change social practices? Blumenthal (2014) provides data after the Katrina catastrophe and after the BP blowout catastrophe to answer the fossil-fuel dependence question. 'More than 300,000 people in the state are employed in the oil and gas sector, which provides tens of billions in tax revenue. The state hosts 19 oil refineries, second only to Texas, and just offshore in the Gulf of Mexico lie 4,000 drilling rigs'. The Hurricane Katrina catastrophe in 2005 nine years before these data were gathered, and the 2010 BP oil blowout four years prior, were not leading to emancipation from economic dependence on the fossil-fuel industry by 2014. Discourse about the 2005 catastrophe of Hurricane Katrina did not bring emancipation from danger is indicated by the 2010 BP oil catastrophe five years later, which poured oil over Louisiana's seacoast for two months threatening its fishery (Freudenburg and Gramling 2011). This too was followed by renewed dependence on fossil fuels. This confirms the present book's premise that the focus needs to be on socioeconomic practices, with discourse only being important to the extent it influences those practices. Despite the experience of catastrophes, Louisiana remains heavily dependent on fossil fuels, whose combustion results in global warming, extreme weather, and sea level rise.

I visited New Orleans in 2015 to assess whether its catastrophes emancipated it from fossil fuels, inequality, and racism. I was invited to speak at Tulane University in New Orleans, which is a prestigious university that has a billion-dollar endowment fund invested in Louisiana's fossil-fuel industry. A few students started discourse of divestment, but it was rejected by the university administration. In Louisiana, talk about global warming has been drowned out by discourse about economic growth, which is more influential for socioeconomic practices, leaving greenhouse gases to accumulate in a failure of foresight and the incubation of disasters. As a state regularly struck by hurricanes and vulnerable to rising sea level, Louisiana is trapped between the biophysical and geographical impossibility of adapting to global warming and the sociopolitical nonstarter of replacing fossil fuels with low-carbon energy because of powerful fossil-fuel interests and path-dependent, fossil-fuelled practices.

Did these catastrophes result in emancipation from inequality? By 2010, half a decade after the Katrina catastrophe, Louisiana ranked as one of the states where income inequality is the highest (McNichol et al. 2012) in a country with among the highest income inequality of modern developed nations (Butler 2012; Piketty 2014). The catastrophe of Hurricane Katrina striking New Orleans resulted in an existential shock, but not one that prompted emancipation from inequality. Nor did Katrina result in reducing inequality of educational opportunity. The fees for 2014–2015 at New Orleans' private university, Tulane, were \$48,305 tuition per year per student plus \$12,556 for room and board. The public universities—University of New Orleans, University of Louisiana at Lafayette and Monroe, and Louisiana State University at Baton Rouge—with tuition fees of \$6000–8000 for in-state students, are struggling because taxpayers demand reduced taxes. The offspring of the wealthy enjoy small class sizes at Tulane University whereas those of the poor and middle classes face large classes at public universities, having to earn money while studying, and debt burdens upon graduation.

Hurricane Katrina demonstrated the oversimplification of Beck's contention that the key lines are not between social classes. In New Orleans, wealthy former plantation mansions are located on safe, high ground, whereas ten years after the Hurricane Katrina catastrophe the vulnerable lower ninth ward remains populated by the poor, mainly African Americans, where city tour buses are prohibited to avoid showing the misery of the inhabitants. I found no evidence that the catastrophe was liberating the state from inequalities and divisions between races. Despite its oil riches, Louisiana remains, after Katrina as before, one of the states where poverty is highest. African Americans remain the poorest group with the lowest opportunities. This catastrophe did not result in emancipation from fossil-fuel dependence, from racism, or from inequality. There was increased talk about emancipation immediately following the catastrophe, but little sustained change in energy, racial, and class practices. It produced an existential shock, but not one that prompted transformations of practices. Far from having an emancipatory effect, reversion to pre-catastrophe fossil-fuelled practices was the outcome of this disaster. The traumatic experience resulted only in learning to have more robust defenses (improved end-of-pipe remedies like levees, pumps, blowout protectors, etc.) against nature's forces.

Hurricane Sandy did not transform climate politics in New York either: 'our interviews found that for the most part, people who were already concerned about climate change continued to be so, and those who were not, continued not to be even if they were persuaded that climate change played a role in the storm. … Hurricane Sandy exacerbated crises which existed before the storm and continued afterwards in heightened form, including poverty, lack of affordable housing, precarious or low employment, and unequal access to resources generally' (Superstorm Research Lab 2014).

Do Scientific Predictions of Catastrophe Incite Prevention?

If the actual experience of catastrophe failed to prompt change, then how can the mere scientific prediction of one whose principal consequences will be distant in time and space incite a transformation? The presumption that the scientific anticipation of catastrophe will emancipate societies from social practices causing it is not an accurate portrayal of what is occurring in the real world. Economic interests and attachment to normality can override scientific warnings and lead danger to be discounted. It is necessary to be vigilant that aspirations not slide down the slippery slope into wishful thinking and false hopes, not to mention greenwashing, as fossil-fuelled climate change worsens. Furthermore, the mobilizing effect of the anticipation of catastrophe is not universal. It is contingent and differs between societies and social groups, even when the anticipation is well grounded in scientific understanding available to all cultures and societies. Groups respond to the scientific anticipation of catastrophe either by (i) emancipating themselves from social practices incubating catastrophe, or (ii) doubling down on fossil-fueled normality. The real world exhibits these two different tendencies. Theories that the anticipation of catastrophe will result in emancipation from fossil-fuelled practices extrapolates from changes that are occurring. Investigations of discounting danger shine light on tendencies unresponsive to scientific warnings, on what is not being done but is needed to prevent global warming, and how the problem is worsening. These two capture opposite important elements and have different weaknesses. The first runs the risk of failure to perceive the depth of the problem. The second risks falling into beliefs that everything must change to change anything, that fossil-fuelled global warming is too big to solve, and of fostering fatalism and inaction.

Beck's theory of emancipatory catastrophism is pitched at a high level of abstraction and focusses on discourse. Generalities like that are of dubious value but are common. After a study of the human impact on nature, including fossil-fuelled climate change, the Chair of the Intergovernmental Panel on Biodiversity and Ecosystem Services concluded as follows. 'The Report also tells us that it is not too late to make a difference, but only if we start now at every level from local to global,… Through "transformative change", nature can still be conserved, restored and used sustainably – this is also key to meeting most other global goals. By transformative change, we mean a fundamental, system-wide reorganization across technological, economic and social factors, including paradigms, goals and values' (IPBES 2019). This is also pitched at a general level.

Identification of Needed Measures

It is necessary to examine concrete socioeconomic practices and be inspired by specific cases of changed practices fostering safety. The most important reason for hope is that impact natural science has provided an understanding of the processes involved and suggested what must be done to prevent disaster. Measures needed to mitigate the greenhouse effect and global warming have already been identified. Countries need to reduce their 'emissions by 80 percent by 2050 through a combination of steps: (1) energy efficiency gains, both in electricity generation and use and in transportation, including fuel-efficient vehicles; (2) renewable energy development, especially wind and solar energy (3) other energy efficiency gains including improvements in residential and commercial buildings; (4) shifting to low-carbon fuels; (5) geological disposal (sequestration) of carbon dioxide; (6) reducing emissions of greenhouse gases other than carbon dioxide; and (7) enhanced forest and soil management practices' (Speth 2009: 29–30). Many books and articles have suggested multiple solutions to mitigate fossil-fuelled global warming and slow-onset unsustainability: Barnes 2008; Jaccard 2005; Dauncey 2009; Speth 2009, 2012; Jackson 2009; Bullard and Wright 2009; Giddens 2009; Lomborg 2010; Pielke 2010; Rand 2010, 2014; Nordhaus 2013; Moser and Boykoff 2013; Hawken 2017; Flannery 2015; Klein 2017; Suzuki and Taylor 2009; Suzuki and Hanington 2017; Harvey and Orbis 2018; Ramish 2018; Pink 2018; Klenert et al. 2018. Some of these (Harvey and Orbis 2018); Rand 2010, 2014) have made excellent contributions in terms of detailed and specific policy guides for designing solutions and moving towards low-carbon energy, typically using already available clean technologies.

However, suggesting solutions is easy. The hard part in a socioeconomic context of fossil-fuelled social practices, interests, and power is gaining acceptance of such measures. The problem is not lack of solutions nor even lack of technology, but whether proposed solutions will be implemented or discounted and left to gather dust on library shelves.

Foresight Prevails: Reconciling the Economy with the Environment

There are cases where leaders had the foresight to tackle the threat of fossil-fuelled climate change. On the international level, the best known case is the 2015 Paris Agreement, where all nations committed to reduce emissions and limit global warming to below 2 °C, to increase adaptation, build resilience, reduce vulnerability, and developed nations agreed in principle to help developing countries finance these goals. A less known case consists of the Kigali Amendment to the Montreal Protocol, which is especially significant because enforcement practices were accepted. That amendment dealt with refrigeration, which includes both keeping foods from spoiling and air conditioning. Science developed an understanding of the chemical processes of refrigeration and production science applied this knowledge and innovated the early chemical refrigerants, namely CFCs and HCFCs. However, impact science discovered they were depleting the ozone layer which shields the Earth from harmful ultraviolet radiation. Despite a campaign of disinformation by companies like Dupont, which produced these chemicals, a counter-campaign resulted in political leaders agreeing to the 1987 Montreal Protocol to phase them out. This resulted in the innovation of replacement chemicals HFCs, which do not deplete the ozone layer. Hence, air conditioning and refrigeration could be maintained and expanded. Unfortunately, it was then found that HFCs are a thousand times more powerful causes of a greenhouse effect than carbon dioxide. This is extremely threatening because air conditioning, which previously was seen as a luxury, is now viewed as a necessity in hot climates. As global warming intensifies, there will be even more demand for air conditioning. In 2016 in Kigali, 170 countries negotiated an amendment to the Montreal Protocol. They agreed that high-income countries would start phasing out HFCs in 2019, low-income ones would do so between 2024 and 2028, rich countries would finance the transition even in poor ones, and trade sanctions would be used to enforce compliance. This Kigali amendment has been hailed, notably by then US Secretary of State John Kerry, as 'the biggest thing we can do [on climate] in one giant swoop' (Hawken 2017: 164). Replacement chemicals have already been developed, and no deleterious consequences of their use have been found, so far. Hawken (2017) rates this sociotechnical innovation in refrigeration as the most beneficial and cost-effective remedy for the avoidance of carbon in the atmosphere and the greenhouse effect for the years between 2020 and 2050.

It should be noted that refrigeration, such as air conditioning, constitutes a double jeopardy global warming social practice. The chemicals used, such as HFCs, cause a greenhouse effect when released into the atmosphere, usually at disposal. But also refrigeration is powered by electricity, which in most cases uses fossil fuels as its primary energy source, which when combusted cause carbon to combine with oxygen to produce the greenhouse gas of carbon dioxide. Air conditioning in vehicles, planes, and ships is powered directly by fossil fuels. The Kigali agreement, valuable as it is, only solves the HFC part of refrigeration's contribution to global warming, but not the fossil-fuel combustion part. Nevertheless, it is a hopeful template for other international agreements to deal with the fossil-fuelled climate crisis. The Paris Agreement and the Kigali Amendment are examples of the interaction of sociopolitical practices with nature's properties and dynamics as actants, of the positive relationship natural science can have with political leadership, and of a global threat being mitigated by cooperation among nations.

In 1991, Sweden implemented a carbon tax of US$133 a metric ton. By 2008, emissions there had fallen 40% compared to 1990 levels, and its economy had grown 44%. Since carbon pollution continued increasing globally, Sweden did not merely point to other polluting countries as an excuse for climate inaction, but instead raised its carbon tax in 2014 to $168 (Suzuki and Hanington 2017: 232). Sweden illustrates the fact that countries which do the most to mitigate fossil-fuelled climate change through high gasoline prices are not thereby prevented from being prosperous. Similarly, Switzerland has high gasoline prices, the best environmental performance of 180 countries, and the second best climate and energy performance (Yale University 2018), yet this has not prevented it from being one of the world's most affluent countries without having the windfall of oil deposits. Environmental performance and near-term economic performance can be complementary.

Norway is as good as it gets for exploiting fossil fuels. First, its North Sea oil has the lowest life cycle emissions of any type of oil (see Table 2.1 Chapter 2). Second, it extracts maximum royalties and benefits for the Norwegian people from private oil companies by having a state-owned oil company to leverage royalties from competing private-sector companies. Third, it implements a high-carbon tax throughout its economy to reduce carbon-polluting activities and stimulate low-carbon innovation. Fourth, it saves a large proportion of the oil benefits in a sovereignty fund for future generations and economic downturns. Parenthetically, it borrowed this idea from Alberta, but whereas Norway saved massive amounts in it, Alberta didn't because its right-wing leaders refused to form its own state oil company and government involvement. Norway's economic practices result in taking maximum benefits for its citizens so that the least amount of oil need be extracted and combusted. This contrasts with other crude oil exporting states whose prosperity depends on volume: extracting as much crude oil as possible. They then suffer the boom-and-bust cycles of the market. Fifth, Norway built efficient systems of low-carbon rapid public transportation and heating. Sixth, it spends part of its oil revenue to offset emissions, principally by financing reforestation in Indonesia, Brazil, and other countries. Some environmentalists complain that Norway's measures do not go far enough, contending Norway should abandon North Sea oil entirely as the perfect solution to climate change. If the only goal is eliminating Norway's contribution to climate change, they are right. However, such an environmental utopia has been rejected by Norwegians as threatening to provoke an economic dystopia. Norway's approach is the best practices template for exploiting fossil fuels in an imperfect world.

Germany has huge reserves of lignite coal, which has been the source of its energy but is a particularly polluting type of coal. The population voted strong support for its Green Party, which gave it significant influence in an electoral system of proportional representation. Germany then promoted a green energy transition, called *Energiewende*, away from fossil fuels towards low-carbon energy. It developed feed-in tariffs whereby consumers accept higher electricity costs to transform generation from fossil fuels to low-carbon energy, with the latter increasing from 5% to 40% of electricity production (Turner 2019). The low-carbon facilities are 40% owned by farmers and citizens and 340,000 Germans now work in the renewable energy industry, five times more than in the coal industry. The transition was opposed by the coal industry and the powerful automotive industry. After the nuclear disaster at Fukushima, Germany decided to phase out its nuclear reactors, which resulted in pressure to return to coal. Nevertheless, the energy transition marches on with the support of the population and without damaging Germany's performance as an economic powerhouse. The country is on track to shutter its remaining coal-fired power plants and leave its valuable but polluting coal reserves underground.

California, which is the most populous American state and one of the most prosperous but which has been devastated by drought, wildfires, and floods, is implementing its own strict fuel-efficiency standards to cut emissions and other pollutants in cars, pickup trucks, and SUVs, which will also reduce air pollution and smog, and save money for drivers in fuel costs. Governor Newsom argued that 'we're not a small, isolated state. California moves markets' (McCarthy and Lewis 2019: B2). Thirteen American states and the government of Canada indicated they will follow California's lead. Nevertheless, not only the Trump Administration but also some automakers are opposed, arguing it will add US$1800 to the price of vehicles, will split the market, and disrupt highly integrated supply chains. States like California are willing to pay that cost and solve the split-market problem by promoting a single, ambitious standard to diminish the danger. Many automakers complied with California's fuel-efficiency standards. But it would have been better to have a federal administration apply rigorous fuel-efficiency standards to all states rather than the Trump Administration fighting to suppress California's right to implement fuel-efficiency standards.

Some other social practices are changing and technologies are being innovated with the climate challenge in mind. Cities are attempting to densify and develop dedicated bicycle lanes for the home-to-work commute. Deforestation has decelerated and afforestation has begun, which withdraws carbon from the atmosphere. The practice of draining wetlands has largely stopped, and some are being expanded because they are effective protecting against floods and in withdrawing atmospheric carbon. The birth rate is dropping everywhere and the education of girls is rising. The environmental movement is strong in many countries. Young people are becoming more involved in politics because they realize it is they who will be most affected by slow-onset problems like global warming.

Private-Sector Solutions

There are private-sector solutions for mitigating the fossil-fuelled climate crisis, unsurprisingly often found in Nordic countries with a history of social democracy. One case occurred in Denmark (see Reguly 2019). Before 1974, energy in Denmark was almost entirely based on oil imported largely from Saudi Arabia. The OPEC oil embargo of 1974 sent prices soaring, resulting in Denmark having to pay almost 400% more, immediately, for Saudi oil. This would make Denmark's energy bill ruinous, so the country converted its oil-burning electricity generating plants to coal, which was less expensive and less vulnerable to foreign volatility. A state-owned company called DONG, an acronym for Danish Oil and Natural Gas, was founded with the mission of discovering and extracting oil and natural gas in the Danish part of the North Sea. Over the following decades, DONG expanded its oil and gas portfolio in the North Sea, and merged with other companies and expanded into coal. It built an enormous coal-fired power station in northeast Germany and became one of the most coal-intensive companies in Europe.

By 2006, however, scientific findings concerning the danger of global warming became accepted by the Danish population. Al Gore's *Inconvient Truth* had a major impact. The European Union introduced its carbon dioxide reduction goals for 2020. Sustained anti-coal protests erupted in Denmark. Debates within DONG's leadership began to occur. One group argued that Denmark had high carbon dioxide emissions, that more investing in fossil fuels would result in being locked into carbon emissions, that offshore wind is the energy of the future, and that the company should replace fossil fuels and especially coal with wind energy. An opposing leadership group claimed that the company's core competence was in fossil fuels, that there was almost no wind energy being produced in Denmark, and that changing from one of Europe's most coal-intensive companies to wind energy would be risky and unprofitable, hence fossil fuels should remain the priority. The social change to wind leadership group won this internal struggle over the risk averse, path-dependent coal group. Despite major expenses to decommission coal plants and construct offshore wind farms, the decision was made to dismantle the fossil-fuel infrastructure and replace it with wind energy. High-carbon taxes in Denmark provided the push, and the pull came from the conviction that remaining a carbon polluter was morally wrong. There was nevertheless nervousness in the company and the country concerning whether the transformation would involve a substantial financial sacrifice. External factors also came into play. In 2012, European natural gas companies, such as DONG, were clobbered by the plummeting prices because of fracking in the USA. Vast amounts of low-price American coal were dumped onto Europe. DONG's huge fossil-fuel business lost money that year, which prompted it to sell those assets and focus on offshore wind, which was growing in profitability. Furthermore, the technology and installation costs of offshore wind were decreasing dramatically. In 2017, DONG sold its North Sea oil and gas business, and changed its name to Orsted. By 2018, Orsted's wind energy output was 75% of its total energy on the way to 99% by 2025. Its carbon dioxide emissions fell 98% by 2015 compared to the level in 2009. It now markets itself as the greenest energy company in Europe. The transition from fossil fuels to wind energy was made far more rapidly than expected. Not only did it result in a major reduction in Denmark's carbon dioxide emissions, it also meant cleaner air in Copenhagen and heat for homes and businesses as a by-product of electricity generation. The transformation has also been exceptionally profitable, making Orsted one of Europe's most valuable energy companies. 'In 2009, Orsted was largely a domestic Danish company. Today, it is the leader in offshore wind power, with control of 30% of the global market. Orsted has more than two dozen offshore farms in Denmark, Britain, Germany, Netherlands and Taiwan, and has several in development off the U.S. east coast' (Reguly 2019: 40). This demonstrates how the economy can be reconciled with the environment if leaders do not discount future harm in favour of short-term benefits and if they have the foresight to innovate out of carbon-polluting fossil fuels towards clean renewable energy.

There are many other examples of market-driven technological innovations reducing emissions. 'New LEDs [light emitting diodes] use about one-eight as much energy as the incandescent bulbs they replace and last about 20 times longer. As a result, more than 80 million LED bulbs have been installed in the United States today, which have avoided millions of tons of CO_2 emissions and saved billions of dollars' (Harvey and Orbis 2018: 11). This despite the fact that upfront costs are greater for LEDs than for incandescent bulbs. Hawken's (2017: 222) research team estimates that over the thirty year period from 2010 to 2050 replacing incandescent lighting globally with LEDs would cost $324 billion but would save $1730 billion and would avoid emitting 7.8 gigatonnes (= 7.8 trillion kilograms) of carbon dioxide equivalent into the atmosphere. Resistance to this simple remedy results from aesthetic claims that incandescent bulbs give a psychological feeling of warmth, which is because they waste energy as heat.

Solar, wind, and energy storage, particularly in batteries and electric vehicles, are becoming more efficient and cheaper as they scale up, pushed especially by China seeking to leap-frog Western countries into clean energy of the future (Rand 2018: B4). A variety of electric vehicles are being developed and their prices are diminishing. Electric vehicles 'are the perfect storage system for renewables, and the synergy created with their widespread deployment may well provide the momentum for a decisive end to the fossil-fuel era' (Flannery 2015: 131). Fossil fuels are becoming less attractive to investors pursuing long-term profits, especially carbon-laden ones like coal and heavy oil. One energy consultancy forecasts 'that global demand for oil will peak in 2022 owing to a surge in electric-vehicle adoption and a bearish outlook on petrochemicals. The demand for natural gas could surpass global oil demand by

2026' (quoted in Quinlan 2019: B4). The investment firm BNP Paribas calculated that to be competitive over the next 25 years with renewable energy, oil will have to be priced between US$9 and US$20 a barrel (Lewis 2019). Many oil reserves cannot meet those requirements, hence will be left underground. Lewis (2019: 3) states that 'economic and environmental benefits [are] set to make renewables in tandem with EVs [electric vehicles] irresistible. …We conclude that the economics of oil for gasoline and diesel versus wind- and solar-powered EVs are now in relentless and irreversible decline'. Divestment from fossil fuels is increasing. Renewable energy, especially wind, solar, and hydro, is coming on stream and their costs are decreasing as they begin to be mass produced. Flannery (2015: 120) points out that 'renewables are now successfully competing with fossil fuels. For two years running more renewable energy, including wind and solar, has been installed globally than fossil fuel-based generation'. After an exhaustive study, Hawken (2017: 1) concludes that humanity is 'squarely in the middle of the greatest energy transition in history. The era of fossil fuels is over, and the only question now is when the new era will be fully upon us. Economics make its arrival inevitable. Clean energy is less expensive'.

The Anglo-Australian mining company BHP is the world's biggest exporter of coking coal used for steelmaking, the third largest iron-ore miner, and a major producer of liquefied natural gas, oil, and copper. It announced in 2019 it will invest US$400 million over five years to reduce emissions and, more significantly, will take into consideration scope 3 emissions (Russell 2019). Whereas scope 1 and 2 include an organization's direct and indirect emissions from its own activities, scope 3 refers to emissions generated by the use of its products, and is always excluded from consideration by resource extractors. Scope 3 emissions of BHP would cover emissions from burning its coal, oil, and natural gas in whatever country they are combusted, as well as fossil fuels combusted by ships and trains bringing them to their destination. Most of BHP's fossil fuels are extracted in Australia, the world's largest extractor of coal, iron ore, and liquefied natural gas, so taking scope 3 into consideration would mean more emissions included in Australia's already high total. Taking scope 3 into account could potentially attract ethical investors and pressure downstream users of fossil fuels to increase efficiency and reduce emissions. Whether these grounds for hope are fulfilled depends on whether it is more than a public relations coup, on how it is implemented, on whether investors reward or punish the company's foresight in mitigating the fossil-fuelled climate crisis, and on whether its initiative inspires other resource companies to follow suit.

The insurance industry will be most affected by global warming, hence tends to be in the vanguard of transforming its practices to heed warnings of scientists. Swiss-based Zurich Insurance Group Ltd, with US$190 billion investments worldwide, committed to using only renewable power by 2023, eliminated coal used for electricity production from its investment portfolio in 2017, is creating standards to set targets and measure the carbon footprint of companies' underwriting and investment portfolios, and will divest from Alberta's bituminous sands extractors, their pipeline facilities and their crude-by-rail companies within two years unless they produce business plans to reduce emissions (McCarthy 2019).

Green bonds are being issued by banks to raise capital to mitigate environmental problems like climate change (Flannery 2015: 111) by financing wind and solar power, energy efficient buildings, and clean transportation. They are being driven by demand from institutional investors. The Green Bond market taking into account environmental impacts has grown from US$10 billion in 2013 to US$170 billion in 2018 (Le Hanerou and Lamontagne 2019). This growth followed the construction of the Green Bond Principles organization in 2014, which assesses and monitors green bonds to ensure they finance climate-friendly projects and are not greenwashing. That organization also promotes disclosure. This is significant for reconciling the economy and the environment rather than obsessing about only the financial bottom line.

Some other examples of private-sector innovation towards lower carbon energy are the following. The USA reduced its greenhouse-gas emissions, not intentionally by government mandate to switch away from fossil fuels, but rather by fossil-fuel companies innovating new cost-effective technology to extract natural gas from shale by hydraulic fracturing, which made natural gas as inexpensive as coal and resulted in transitioning from high-emissions coal to lower emissions natural gas. A flat-roofed shopping mall took the initiative of installing solar panels, which provide electricity to run air conditioning especially in the hottest, sunniest days of summer when demand is highest (Immen 2019). This reduces the need for fossil-fuelled electricity, saves the mall money, and could be a template for all malls, big-box stores, and other large flat-roofed buildings. Patents for innovation in renewable energy have recently become far greater than patents for innovation in fossil fuels, whereas they were roughly equal before 2000 (Flannery 2015: 125).

However, the pursuit of profit in the market leading to innovations is a double-edged sword, with one side having been sharper than the other to date. It invented fracking, deepwater drilling, Arctic drilling, etc., which increased fossil-fuel extraction and resulted in more emissions worsening global warming. And hydraulic fracturing deflects investment away from clean, renewable energy. Profit-seeking entrepreneurs vastly expanded fossil-fuelled cruises and jet-fuelled intercontinental tourism. They developed air conditioning and computer servers which enable social media and data storage to function, but these worsen global warming by requiring huge amounts of electricity frequently provided by fossil-fuel combustion. These innovations are all desired, but have harmful effects of increasing emissions and worsening global warming. The market's innovation of low-carbon energy, carbon capture and storage (CCS), and other means of bringing emissions into line with carbon withdrawals trails way behind its success in developing technologies to extract and use more fossil fuels. Dependence on the market by itself to solve the climate change crisis is contradicted by its overall failure to bring cost-effective innovations to decrease emissions without government regulations. Reliance on profit-seeking companies to implement solutions is exceedingly risky. It constitutes dangerous brinkmanship by making global warming worse before hoped-for solutions are implemented.

References

Abramovitz, J. 2001. *Unnatural Disasters*. Washington: World Watch Paper.

Barnes, Peter. 2008. *Climate Solutions: What Works, What Doesn't, and Why*. White River Junction, Vermont: Chelsea Green Publishing.

Beck, U. 2007. *World at Risk*. Cambridge: Polity Press.

Beck, U. 2015. Emancipatory Catastrophism. *Current Sociology* 63 (1): 75–88.
Blok, Anders. 2015. Towards Cosmopolitan Middle-Range Theorizing. *Current Sociology* 63 (1): 110–114.
Blumenthal, P. 2014. Mary Landrieu's Fossil Fuel Stance Both a Blessing and a Curse in 2014 Election Bid. *Huffington Post*, 2 October. http://www.huffingtonpost.com/2014/10/02/mary-landrieu-2014-election_n_5921010.html. Accessed 26 August 2016.
Bullard, R.D. 2000. *Dumping in Dixie: Race, Class and Environmental Quality*, 3rd ed. Boulder: Westview Press.
Bullard, R.D. 2005. *The Quest for Environmental Justice: Human Rights, and the Politics of Pollution.* San Francisco: Sierra Club Books.
Bullard, R.D., and B. Wright (eds.). 2009. *Race, Place and Environmental Justice After Hurricane Katrina.* Boulder: Westview Press.
Butler, R. 2012. Global Gini Index. *All Data from the CIA World Factbook.* http://www.mongabay.com/reference/stats/rankings/2172.html. Accessed 2 July 2012.
Byers, Michael. 2019. Watch This Space. *Globe and Mail*, 27 July: O3.
CDC (Centers for Disease Control). 2016. *History of Smallpox.* Washington: Centers for Disease Control and Prevention. https://www.cdc.gov/smallpox/history/history.html. Accessed 10 May 2020.
Claisse, Frédéric, and Pierre Delvenne. 2015. Building on Anticipation: Dystopia as Empowerment. *Current Sociology* 63 (2): 155–169.
Clark, N. 2011. *Inhuman Nature.* London: Sage.
Dauncey, Guy. 2009. *The Climate Challenge: 101 Solutions to Global Warming.* Victoria, BC: New Society Publishers.
Diamond, Jared. 2005. *Collapse: How Societies Choose to Fail or Succeed.* New York: Viking.
Dunlap, Riley E., and Peter J. Jacques. 2013. Climate Change Denial Books and Conservative Think Tanks: Exploring the Connection. *American Behavioral Scientist* 57: 699–731.
Dunlap, R., A. McCright, and J. Yarosh. 2016. The Political Divide on Climate Change: Partisan Polarization Widens in the U.S. *Environment Science and Policy for Sustainable Development* 58 (5): 4–23.
Elsasser, Shaun W., and Riley E. Dunlap. 2013. Leading Voices in the Denier Choir: Conservative Columnists' Dismissal of Global Warming and Denigration of Climate Science. *American Behavioral Scientist* 57: 754–776.
Flannery, Tim. 2015. *Atmosphere of Hope: Searching for Solutions to the Climate Crisis.* New York: Atlantic Monthly Press.

Freudenburg, W. 2006. Environmental Degradation, Disproportionality, and the Double Diversion. *Rural Sociology* 71 (1): 3–32.

Freudenburg, W., and R. Gramling. 2011. *Blowout in the Gulf*. Cambridge, MA: MIT Press.

Freudenburg, William, Robert Gramling, Shirley Laska, and Kai Erikson. 2009. *Catastrophe in the Making*. Washington: Island Press.

Funk, McKenzie. 2014. *Windfall: The Booming Business of Global Warming*. New York: Penguin.

Giddens, Anthony. 2009. *The Politics of Climate Change*. Cambridge: Polity Press.

Han, Sang-Jin. 2015. Emancipatory Catastrophism from an East Asian Perspective. *Current Sociology* 63 (1): 115–120.

Harvey, Hal, and Robbie Orbis. 2018. *Designing Climate Solutions: A Policy Guide for Low-Carbon Energy*. Washington: Island Press.

Hawken, Paul. 2017. *Drawdown: The Most Comprehensive Plan Ever Proposed to Reverse Global Warming*. New York: Penguin.

Immen, Wallace. 2019. Solar Panels Help Retail Property Owners Tap into Savings. *Globe and Mail*, 6 August: B6.

IPBES (Intergovernmental Science-Policy Platform on Biodiversity and Ecosystem Services). 2019. *Global Assessment Report on Biodiversity and Ecosystem Services*. Bonn: IPBES. https://ipbes.net/sites/default/files/2020-02/ipbes_global_assessment_report_summary_for_policymakers_en.pdf. Accessed 10 April 2020.

Jaccard, M. 2005. *Sustainable Fossil Fuels*. Cambridge: Cambridge University Press.

Jackson, Tim. 2009. *Prosperity Without Growth: Economics for a Finite Planet*. London: Earthscan.

Jacques, Peter, Riley E. Dunlap, and Mark Freeman. 2008. The Organization of Denial: Conservative Think Tanks and Environmental Scepticism. *Environmental Politics* 17: 349–385.

Klein, Naomi. 2017. *No Is Not Enough*. Toronto: Alfred A. Knopf.

Klenert, David, Linus Mattauch, Emmanuel Combet, Ottmar Edenhofer, Cameron Hepburn, Ryan Rafaty, and Nicholas Stern. 2018. Making Carbon Pricing Work for Citizens. *Nature Climate Change* 8: 669–677.

Le Hanerou, Philippe, and Paul Lamontagne. 2019. Canadian Firms Must Adopt a Common Standard for Impact Investing. *Globe and Mail*, 3 August: B6.

Lewis, Mark. 2019. Wells, Wires, and Wheels—EROCI and the Tough Road Ahead for Oil. *BNP Paribas Asset Management*, August. Paris:

BNP Paribus. https://docfinder.bnpparibas-am.com/api/files/1094E5B9-2FAA-47A3-805D-EF65EAD09A7F. Accessed 15 April 2020.

Lomborg, Bjorn. 2010. *Smart Solutions to Climate Change: Comparing Costs and Benefits*. Cambridge: Cambridge University Press.

McCarthy, Shawn. 2019. Zurich Insurance to Pivot Investment Away from Oil Sands. *Globe and Mail*, 26 June: B3.

McCarthy, Shawn, and Jeff Lewis. 2019. Ottawa, California Target Vehicle Fuel Efficiency as Part of New Clean-Transportation Partnership. *Globe and Mail*, 27 June: B2.

McNichol, Elizabeth, Douglas Hall, David Cooper, and Vincent Palacios. 2012. Pulling Apart: A State-by-State Analysis of Income Trends. *Center on Budget and Policy Priorities*, 15 November.

Mileti, D. 1999. *Disasters by Design*. Washington: James Henry.

Mol, A.P.J., D. Sonnenfeld, and G. Spaargaren. 2009. *The Ecological Modernisation Reader: Environmental Reform in Theory and Practice*. London: Routledge.

Moser, Susanne C., and Maxwell T. Boykoff (eds.). 2013. *Successful Adaptation to Climate Change: Linking Science and Policy in a Rapidly Changing World*. London: Routledge.

Murphy, Raymond. 2009. *Leadership in Disaster: Learning for a Future with Global Climate Change*. Montreal: McGill-Queens University Press.

Nordhaus, William. 2013. *The Climate Casino: Risk, Uncertainty, and Economics for a Warming World*. New Haven: Yale University Press.

Pielke, Roger Jr. 2010. *The Climate Fix*. New York: Basic Books.

Piketty, T. 2014. *Capital in the Twenty-First Century*. Cambridge, MA: Belknap Harvard.

Pink, Ross Michael. 2018. *The Climate Change Crisis: Solutions and Adaptations for a Planet in Peril*. Cham, Switzerland: Palgrave Macmillan.

Platt, R. 1999. *Disasters and Democracy*. Washington: Island Press.

Quinlan, Kevin. 2019. Why Canada Needs Climate Stress Testing. *Globe and Mail*, 3 October: B4.

Ramish, Mridula. 2018. *The Ultimate Solution: India's Climate Change Crisis and What We Can Do About It*.

Rand, Tom. 2010. *Kick the Fossil Fuel Habit: Ten Clean Technologies to Save our World*. Toronto: Eco Ten Publishing.

Rand, Tom. 2014. *Waking the Frog: Solutions for Our Climate Change Paralysis*. Toronto: ECW Press.

Rand, Tom. 2018. A Sound Debate About Canada's Emissions, Brought on by Fury Over Trans Mountain. *Globe and Mail*, 3 May: B4.

Reguly, Eric. 2019. From Black to Green Energy. *Corporate Knights*, Spring: 39–47.

Roberts, J.T., and Bradley Parks. 2007. *A Climate of Injustice: Global Inequality, North-South Politics, and Climate Policy*. Cambridge, MA: MIT Press.

Russell, Clyde. 2019. Will BHP's Bid to Cut Emissions Inspire Others to Follow Suit? *Globe and Mail*, 30 July: B4.

Schneider, S. 2009. *Science as a Contact Sport*. Washington: National Geographic.

Simpson, Jeffrey, Mark Jaccard, and Nic Rivers. 2007. *Hot Air: Meeting Canada's Climate Change Challenge*. Toronto: Emblem McClelland & Stewart.

Speth, James Gustave. 2009. *The Bridge at the Edge of the World: Capitalism, the Environment, and Crossing from Crisis to Sustainability*. New Haven: Yale University Press.

Speth, James Gustave. 2012. *America the Possible: Manifest for a New Economy*. New Haven: Yale University Press.

Stewart, Q.T., and R. Ray. 2007. Hurricane Katrina and the Race Flood. *Race, Gender and Class* 14 (1–2): 38–59.

Superstorm Research Lab. 2014. A Tale of Two Sandys. *White Paper*, December. New York. https://superstormresearchlab.files.wordpress.com/2013/10/srl-a-tale-of-two-sandys.pdf. Accessed 13 February 2015.

Suzuki, David, and Ian Hanington. 2017. *Just Cool It: The Climate Crisis and What We Can Do*. Vancouver and Berkeley: Greystone Books.

Suzuki, David, and Dave Robert Taylor. 2009. *The Big Picture: Reflections on Science, Humanity, and a Quickly Changing Planet*. Vancouver: Greystone Books.

Tenner, E. 1997. *Why Things Bite Back*. New York: Vintage.

Turner, Chris. 2019. "Energiewende", or Energy Transition, Is the Word on Everyone's Lips in Germany. *Globe and Mail*, 1 June: O1, O9.

Turner, B., and N. Pidgeon. 1978. *Man-Made Disasters*. London: Wykeham.

Yale University. 2018. *EPI Environmental Performance Index 2018*. https://epi.envirocenter.yale.edu/epi-country-report/CAN. Accessed 22 September 2019.

Yearley, Steven. 1992. Green Ambivalence About Science. *British Journal of Sociology* 3: 511–532.

Yearley, Steven. 2004. *Making Sense of Science*. London: Sage.

9

Faith 2.0 in the Mastery of Nature

Humans are embodied beings immersed in dynamics of nature that provide them with sustenance but also threaten them with danger. Graphs portrayed as a hockey stick curve, namely a long slow increase followed by an inflection point at the time of the industrial revolution and thereafter becoming an exponential increase, have been debated in climate science. Nonetheless, that curve is clearly applicable to the evolution of human societies and activities, whether it be population increase, scientific discoveries, technological innovations, life expectancy, use and combustion of fossil fuels, global anthropogenic environmental problems, etc. The amazing success of our species led to belief in the human capacity to master nature's dynamics and thereby rely on innovating technological solutions to the fossil-fuelled climate crisis. This chapter explores the evolution of such beliefs.

R. Murphy, *The Fossil-Fuelled Climate Crisis*,
https://doi.org/10.1007/978-3-030-53325-0_9

Emergence of Faith 1.0 in the Mastery of Nature

The conviction that nature could be mastered emerged as science, technology, organizational efficiency, and markets brought major benefits. They gave people the capacity to travel over land and oceans and fly much more rapidly and further than any other species. Human ingenuity seems to shrink time and space. Deadly viruses like smallpox were made extinct by developing vaccines. Others like polio, measles, and diphtheria only exist where communities could not or would not be vaccinated. Antibiotics were developed which killed bacteria without harming the host humans. Modern methods of agriculture and food delivery vanquished famine and replaced it with abundance. Nuclear physics provided potentially inexhaustible energy from nuclear reactors, and extremely lethal bombs. Satellites enhanced communication over vast distances, and with other technological innovations enabled the emergence of smart mobile phones that revolutionized communication. Astronauts landed on the moon, a previously unimaginable technological achievement. After the deadly flood in Galveston Texas in 1900 and the fatal storm surge in the Netherlands in 1953, seawalls were built that protected these low-lying places ever since (Zebrowski 1997; Larson 2000). Fatalities from natural disasters were highest where technological protections had not been developed, such as when the 1556 earthquake in Shaanxi China killed 830,000 people (Zebrowski 1997). The development of applied science improved robustness of defences and reduced fatalities. Air and water pollution were worse at the beginning of industrialization; now air and water have become cleaner because of purification and filtration plants. Life expectancy has lengthened dramatically. These successes fostered beliefs in the primacy of human desires over nature's limits. Tenner (1997: 348–349) argues that capitalists and communists shared this technological optimism that they could reconstruct nature.

Faith 1.0 believes in the teleological ascendency of reason over nature in which human social constructions are abolishing nature's constructions. 'For generations technological development had progressed on the premise of transforming, even replacing, the natural world' (Worster

1994: 350). Nature is mastered in the sense of being socially re-engineered and even eradicated. There is an embedded zero-sum presumption: as human activities affect all of nature in the biosphere, there are fewer autonomous forces of nature left until eventually there will be none. Autonomous nature is assumed to exist only in nature's pristine form. The key concept is 'autonomous'. There is another either/or presumption: as knowledge of nature's dynamics advances, ignorance recedes.

This faith 1.0 in the mastery of nature emerged because of success in exploiting nature's resources and technologically manipulating its properties and dynamics. Countries with minimal natural resources are often the most technologically advanced and prosperous—Japan, South Korea, Singapore, Switzerland, the United Kingdom, Netherlands, Germany, Sweden—because they rationally developed their human capital to add value to raw materials. Countries rich in natural resources—South Africa, Nigeria, Brazil, and Russia—are not prosperous because they failed to develop their human capital. Simon (1981) gave an economist's formulation of the widespread belief that prosperity depends on developing the ultimate resource: the unique human capacity of reason. The horse-and-buggy age did not end because of a scarcity of horses. Human reason innovated more efficient transportation based on fossil fuels. The asbestos age did not end for lack of asbestos. It is now safely underground because market-based research innovated a cost-effective alternative when it was proven dangerous. Where there is demand and financial rewards for innovating supply through high prices, market competition will technically innovate more of the resource or better substitutes. Beckerman (1974) claimed that man gives nature its characteristics; therefore as reason develops, resources grow rather than become depleted. The industrial editor of the *New Scientist* claimed that humans are no longer subject to nature's constraints: 'Technology can achieve practically anything today if we spend enough on it. It gives Man unprecedented powers over his environment and himself' (Hamilton 1973: 41). Such claims are based on extrapolations: previous problems were solved by technological innovation, so environmental problems will be similarly solved. Prior to the 1840s, whale oil was the means of lighting and lubrication, but whaling fleets were hunting whales to extinction. Butts (2019) argues that whales were saved by the technological innovation of kerosene (coal oil) in 1846 and then petroleum in the 1850s, which provided better lighting and lubrication. Extrapolation from past successes is firmly embedded in cultural beliefs and is used to plan for the future.

The US Surgeon General believed in the mastery of nature in 1967 when stating it was 'time to close the books on infectious disease' (quoted in Tenner 1997: 74). The 'chief executive of Merck declared in 1988 that the company expected to find an anti-AIDS drug within five years' (Tenner 1997: 81). It was assumed that applied science is enabling humanity to escape from nature's forces. The presumption nature could be reconstructed without limit was especially prominent during the Reagan era in the USA in the anti-environmental backlash against ecological limits (Mitchell 1990; Dunlap and Scarce 1991; Dunlap and Catton 1994).

In the 1990s many social scientists believed in the mastery of nature. Grundmann (1991a, b) postulated that technology provides almost unlimited powers to overcome nature's constraints and master it. Giddens (1991: 224) contended that the 'invasion of the natural world by abstract systems brings nature to an end as a domain external to human knowledge and involvement'. Beck (1995: 37–38) postulated that the 'process of interaction with nature has consumed it, abolished it. … it no longer exists'. Book titles were revealing: *The social Creation of Nature* (Evernden 1993); *The Social Construction of Nature* (Eder 1996); *The Social Construction of Oceans* (Steinberg 2001). These titles did not refer to discourse about nature and oceans or to affecting them, but to nature and oceans per se. Dunlap and Catton (1994) critically referred to this as the human exemptionalist paradigm in social science and society, which postulates that humans are making themselves exempt from nature's constraints.

Often the mastery of nature presumption is deployed when it suits vested interests and avoided when it doesn't. In order to avoid government regulation to prevent environmental problems, companies claiming they are innovative risk-takers state they will find solutions when problems arise. But they refuse to make technological solutions mandatory now, for example, requiring carbon capture and storage reducing emissions as a prerequisite for combusting fossil fuels. This demonstrates that professed faith in mastering nature's forces is often rhetoric deployed to promote economic interests. There is a huge difference between valuing science (i) as an evidence-based foundation for social practices in order to prevent environmental and health problems and (ii) as an ideology of mastering nature's forces to innovate last-minute solutions so that costly or/and annoying regulations to prevent problems need not be adopted. The first constitutes scientific thinking based on the preponderance of the evidence and admitted fallibility; the second amounts to scientism and magical thinking founded on illusory assumptions of infallible certitude concerning the capacity to discover cost-effective technical solutions in the future. Relying on technological innovation to master nature's forces requires certainty it will succeed in the future, even for unforeseeable forces that could emerge. Such wishful thinking is engaged in because of reluctance by decision-makers, the powerful, and the population to make expensive, inconvenient changes in social practices to prevent environmental and health problems.

The Collapse of Faith 1.0 in the Mastery of Nature

This belief in human mastery of nature's forces was, however, confronted by powerful disconfirming phenomena. Problems emerged after a time lag. Many seemingly miraculous technological innovations turned out to be mirages that receded as use increased. Using asbestos as insulation and fireproofing seemed amazing, but later it became evident that asbestos promotes the uncontrollable growth of fatal cancers. No technology was found to make it safe, so it had to be left underground. The creation of DDT reduced malaria, but when it manifested carcinogenic properties the only solution was to abandon it. PCBs turned out to have similar perverse effects. The invention of CFCs enabled refrigeration and air conditioning to make buildings climate-controlled, which resulted in economic development in hot zones. Impact scientists discovered, however, that CFCs depleted the ozone layer, and the only solution was to abandon them. Although HCFC substitutes were devised, they too have perverse properties. The innovation of Thalidomide had potential to help pregnant women, but its catastrophic reconfiguration of fetal growth only became visible after babies were born, so it too had to be abandoned. It was becoming evident that technological innovations often unleash the emergence of second-order harmful constructions of nature.

Moreover, social constructions continue to be overpowered by nature's constructions. In 2005, Hurricane Katrina killed 1836 inhabitants in New Orleans where it destroyed parts of the city in the world's most powerful country. The city was resilient in that it pumped the water back out and rebuilt most of the city, but it remains in harm's way for the next big storm. Moving the city to a safer location was rejected because of cost and tradition. Both alternatives—remaining in harm's way or moving—demonstrate that nature's powerful forces are not being conquered. Forces like hurricanes are being rendered more powerful and frequent by fossil-fuelled global warming. A nuclear reactor was constructed at Fukushima Japan with claims it was well protected against tsunamis. But a tsunami produced a wall of water that flowed over the protective seawall flooding the reactor, resulting in a nuclear meltdown and drowning 20,000 people (Hasegawa 2012, 2015, 2016). BP's risk analysis given to regulators claimed its blowout protector was failsafe from the enormous pressures in deepwater oil drilling. But it failed to be safe and a 2010 blowout produced an oil gusher lasting two months contaminating waters and beaches in lucrative fishing and tourist areas of Louisiana (Freudenburg and Gramling 2011).

Clark (2011: 30) describes the unleashing of nature's autonomous forces: 'it is incautious human agents who engineer the conditions under which dangerous elements are likely to be accidently released into the world. But the backstory tells of a disconcerting *willingness* on the part of these absconders to break loose, disperse and pursue their own agendas'. Those agendas of nature's forces defy human mastery. Researchers (Turner and Pigeon 1978) documented that humans often ignore the power of nature's autonomous forces, resulting in catastrophes. Even the success of technological defences often has adverse consequences. Galveston learned from the drowning of 7000 residents in 1900 by a storm surge, and built an enormous seawall robust enough to withstand surges ever since, but it ruined its beautiful beaches, hence the city's growth stagnated (Zebrowski 1997).

It became evident that faith in the mastery of nature had myopically focussed on the benefits of technological innovations and ignored failures and adverse side effects, especially slow-onset dangers. Science and technology have failed to increase the life span, and by lengthening life expectancy produced a growing proportion of the population near its upper limits with untreatable, chronic maladies of the very old, including dementia, Alzheimer's, etc. The US Surgeon General who closed the books on infectious disease in 1967 had to reopen them shortly thereafter when AIDS appeared, and the books have had to be repeatedly reopened ever since with the emergence of SARS 2002–2004, H1N1 (swine flu) 2009, MERS 2014, Ebola 2014–2016, COVID-19, etc. The successful development of the internal combustion engine and other motors using fossil fuels resulted in carbon taken from safe storage underground and emitted into the atmosphere causing chain reactions of global warming, melting permafrost and methane emissions, reduced reflective capacity of the Arctic Ocean, and perhaps runaway climate change.

There has been no technical solution to the harmful consequences of the vast amount of garbage being produced. Fortunately sociocultural innovations have been moderately successful: regulations, household triage, and recycling. Antibiotics that appeared miraculous are leading to the emergence of harmful antibiotic-resistant super bacteria because of overuse. So efforts have begun to reduce overuse. These cases illustrate that technical solutions of mastering nature's forces are often illusory, and that the problems can be managed by modifying social practices to emphasize prevention.

Tenner (1997) concludes that 'things bite back' and the only solution is to monitor, attempt to prevent, and bounce back after being bitten because it is impossible to dominate nature's forces completely. Acute problems aren't solved but instead are transformed into chronic ones that must be continually managed. The belief that nature's global forces will be mastered is misleading, and a more accurate conception is that of the problematic manipulation of nature. Harnessing nature is merely a trial-and-error endeavour, and nature often escapes its leash

(Turner and Pidgeon 1978). As technology and markets develop and population grows, pristine nature vanishes in the biosphere as everything becomes affected by human activities. But nature's autonomous primal dynamics (e.g. hurricanes) are not eliminated; instead becoming increasingly internalized into societies (Murphy 2002). Innovative sociotechnical constructions recombine nature's dynamics to achieve particular goals, with some cases succeeding but others failing, and still others unleashing threatening forces of nature.

Even that part of the social sciences that took a cultural turn leading it to presume nature can be socially constructed is relinquishing that one-sided virage. It is now taking a more inclusive turn acknowledging that human social constructions are caught up in nature's powerful, autonomous biophysical dynamics which shape them, and that this material context should not be excluded from the social sciences (Clark 2011: 105). Collins (2011: 5) concludes that 'science is still the best thing we have where knowledge about the natural world is concerned'. In one of his last works, Beck (2015: 123) concluded that 'the technocratic vision of finding a technological answer to the climate risk stands against the empirical facts'. He repudiated his earlier assumption that human interaction with nature has abolished it: 'the idea that we are the masters of the universe has totally collapsed and turned into its opposite' (Beck 2015: 75). Giddens' (2009) book on the politics of climate change refutes his earlier scepticism about anthropogenic climate change and his 1991 contention that science and technology bring nature to an end as a realm beyond human knowledge. Natural scientists have long recognized that the more they know, the more they become aware how much more there is to learn about nature's dynamics. Scientific uncertainties demonstrate that nature persists as a domain external to human knowledge and that its dynamics continue to interact with the human invasion of the natural world by technological manipulations. The interaction of technological constructions and social practices with nature's forces has become more intense in modern societies than traditional ones (Murphy 2002). Fortunately, the environmental social sciences never excluded nature's autonomous biophysical dynamics from the analysis, instead integrating impact natural science assiduously as they researched the interaction between sociocultural dynamics and biophysical dynamics (see Benton

1994, 2001; Dickens 2004; Freudenburg and Gramling 1993; Freudenburg, Frickel, and Gramling 1995; Dunlap and Catton 1994; York and Rosa 2003; Foster, Clark, and York 2011; Dunlap and Brulle 2015).

Society's assumption that technological solutions can always be constructed in a timely, cost-effective way to master nature's forces safely is just a speculative belief based on extrapolating from past successes and ignoring past failures. The continuing surprises and unforeseeability of nature's earthquakes, tsunamis, hurricanes, wildfires, floods, etc., especially their location, timing, and force, have prompted a reformation of this faith in the mastery of nature.

A Reformation of Belief: Faith 2.0 in the Mastery of Nature

'Defending the right of U.S. citizens to buy semi-automatic rifles or carry concealed weapons is akin to denying any human responsibility for climate change. Rational arguments are not the point. No matter how many schoolchildren are gunned down or what the scientific evidence may be for the effects of carbon-dioxide emissions, people will not change beliefs that define their identity' (Buruma 2018: O11). Beliefs are difficult to change when they define identities and when based on interests and long-standing social practices. Sclerosis maintaining fossil-fueled social practices goes deep. Rational arguments must confront all of these.

Belief in the mastery of nature was demonstrably wrong, which shows it is misleading to blur the difference between the world as it exists and beliefs about it. The overwhelming contradictory evidence and experiences have not eliminated this belief because it is so deeply rooted in modern social and economic interests, path-dependent practices, and identities. A new, more subtle formulation has emerged. Preventing fossil-fuelled climate change by following the counsel of impact natural science—slowing down extraction of carbon-emitting fossil fuels to limit global warming to 2 °C—threaten to require the sacrifice of consumption and profits and to modify social practices and identities. Resistance to making necessary changes results in falling back on mastery of nature presumptions, but they are given a new form. To use an analogy, the Protestant reformation did not abandon faith in God and Christianity, but instead reformulated those beliefs. So too inadequacies of faith in the mastery of nature, which became visible as nature's dynamics struck back, have not led to abandonment of such faith, but rather to a modification of its content. Belief 1.0 in the mastery of nature has collapsed, but it has been reformed rather than eliminated, much like the Protestant reformation renewed Christianity. The persistence of faith in mastering nature by technological innovation, albeit a reformed faith, is evidenced by reliance on technological solutions to global warming by politicians, companies, and people who oppose placing a price on carbon pollution.

Although there are commonalities between faith 1.0 and faith 2.0 in mastering nature, there are significant differences. Faith 2.0 acknowledges that nature's autonomous constructions will always occur. It has no illusions about replacing the natural world and is not blind to nature's powerful forces that will continue to threaten humans, including those inadvertently unleashed by technology and market competition. These forces constitute primal nature (Murphy 2002). Floods, drought, heatwaves, infectious diseases, etc., are not being eliminated. Fossil-fuelled global warming is making them worse. Dams built to protect some areas redirect water to other areas making floods there worse. The novelty of this new belief in mastering nature consists of faith that technological innovations will always be able to prevent catastrophic global damage in a timely manner when needed, to robustly defend against nature's forces, to adapt to them, and resiliently bouncing back after a disaster. There is assumed certainty that technological innovation gives societies the capacity to mitigate, adapt to, defend against, and ride out (Pellizzoni 2016) everything nature throws at us and be resilient. If there were no certainty, then it would be reckless to continue fossil-fuelled social practices demonstrated dangerous by impact science and to assume nature's forces will always be mastered in the future even though they cannot be mastered now. Precaution and prevention would be required, not a reliance on last-minute mitigation, adaptation, robustness, and resilience come what may. But precaution and prevention are costly and require changes of present fossil-fueled lifestyle practices. Just-in-time future mitigation, adaptation, construction of robustness, and resilience constitute paths of least political and economic resistance, and they rely on faith that nature's forces will be mastered by future innovation. When proponents of fossil fuels and opponents of taxes on carbon pollution are asked for their solution to fossil-fuelled climate change, they almost always answer future technological innovation, usually prompted by market dynamics. This faith legitimates present fossil-fuelled practices. Future 'innovation' is the recurrent buzzword used to solve anthropogenic climate change and justify present combustion of fossil fuels. This requires faith that technological innovation will master nature's forces.

In faith 2.0, warnings by impact science of dangers like climate change are not denied but are discounted so as to not require expense or immediate change of fossil-fuelled practices. It is believed that, when these problems become visibly serious, future societies will develop technological innovations to mitigate them, adapt to them, and build robustness and resilience. The reformed faith refuses to believe it could be technically and economically impossible to defend against nature's forces. It assumes that the elimination of services that nature does for free now, such as the Arctic ice cover reflecting the sun's rays back into space, is not a significant loss because either (i) benefits of an ice-free Arctic Ocean will exceed dangers, or (ii) market-incited technological innovation will find a substitute, even placing an artificial sunscreen in the sky. The possibility of encountering irreversible tipping points into new steady states of nature providing inferior services to humanity is excluded, as is the possibility some of nature's essential services are irreplaceable.

Beliefs that applied science, market competition, and efficient organization will always enable humanity to mitigate, adapt and be resilient are faith-based assumptions in trial-and-error interactions between socioeconomic constructions and nature's constructions. Beliefs that nature's forces will always be mastered sufficiently enabling adaptation and resilience consist of speculation, as does opposite beliefs in inadaptability and incapacity to bounce back after catastrophes, because there is no definitive evidence yet about tipping points in the future. The question is what to do concerning indicative scientific evidence about possible tipping points but current lack of definitive evidence. Do warnings lead to caution, or does lack of definitive evidence result in continuance and acceleration of carbon polluting practices? In this context of uncertainty and unforeseeability, belief fills the void of lack of definitive evidence. Mastery of nature faith in future technological solutions justifies present path-dependent, fossil-fuelled practices even when good scientific evidence indicates danger and where uncertainties about specifics abound. This reformed faith consists of a cultural framing that guides practices concerning threats for which there is indicative scientific evidence of danger but where prevention would be expensive, disruptive to fossil-fuelled consumption practices, and confronts powerful vested interests.

Pellizzoni (2016) argues convincingly there is an emergent belief in a new mastery of nature for human enhancement, whereby risks and uncertainties of manipulating nature's dynamics are not ignored, nor feared, nor even managed, but instead exuberantly ridden. I would add, for better or for worse. The CEO of Exxon declared that climate models predicting calamity may be wrong, and solutions will present themselves as those challenges become clear. When carbon accumulation in the atmosphere clearly caused by fossil-fuel emissions produces disasters in the future, market incentives will promptly incite the development of technologies to capture and store carbon, suck it out of the sky, and innovate clean energy that is cheaper than fossil fuels. It is presumed that the market and technology can adapt to any of nature's dynamics and constructions. The possibility that risk makers are unleashing dangerous forces of nature that societies will not be able to master or even ride is treated as a non-problem (Freudenburg 2005, 2006; McCright and Dunlap 2010) which distracts from pursuing near-term economic benefits. This is precisely what exacerbated vulnerability to Hurricane Katrina in New Orleans in 2005 (Freudenburg et al. 2009). Prevention of long-term threats from carbon emissions is pushed to the back of the mind.

Danger is discounted by the belief that market dynamics will bring wealth and technological innovations to solve the climate problem. 'People living in the future will probably be wealthier than our contemporaries and will be better buffered from the shocks that weather and climate produce' (Rayner and Malone 1998: 106), hence preventive action by changing present fossil-fuelled practices is deemed unnecessary. Lomborg (2001, 2007) argues the market will make all countries, including developing ones, wealthy by the end of the century, that these wealthy countries will be able to protect themselves at little cost, that there is little need to worry about environmental problems like climate change, and that the precautionary principle must be strictly circumscribed to enable the market to create wealth. Giddens (2009) agrees that the precautionary principle should be strictly circumscribed. 'Risk adverse' becomes a pejorative expression and 'risk-takers' are sought, or more accurately concerning the fossil-fuelled climate crisis, risk makers for everyone.

A recent case reinforced this faith in exuberantly riding out problems by relying on market-driven technological innovation. There was worry about 'peak oil' because societies are so dependent on oil and because conventional reserves in wells are approaching their limit (Hughes 2009). Scarcity threatened and the price increased; oil companies innovated ways to meet this profitable demand. Hydraulic fracturing was invented to extract oil and natural gas from shale where they are abundant. Methods were devised to extract oil from bituminous sands and to upgrade heavy oil as a substitute for light, sweet oil. Drilling techniques were improved to drill in deepwater and in the frigid Arctic. These made oil and natural gas abundant and prices dropped. These technological innovations solved the peak oil problem, but worsened the fossil-fuelled global warming problem (Davidson and Andrews 2013). Solving one problem but creating another is typical of market-driven technological solutions.

By extrapolating from such cases, it is assumed that when adverse consequences of carbon pollution are widely experienced, demand for clean, cheap energy will incite cost-effective innovations to render fossil fuels low-carbon or find replacements. Carbon capture and storage (CCS) and direct air capture (DAC) are oft-mentioned solutions for constructing emissions-free fossil-fuel industries. Claims of these solutions strengthen beliefs that market-based technological innovation will solve any problem nature throws up. Faith that wind and solar technology will replace fossil fuels is widely shared, as expressed by former

American President Barack Obama (2017) who claimed they will inevitably drive fossil fuels out of the market.

Microsoft pledged to reduce its carbon emissions more than half by 2030, and remove as much carbon from the atmosphere as it emitted during its 45-year history by investing US$1-billion over four years to develop appropriate technology (Nellis and Dastin 2020). If all companies and countries removed their past carbon, it would respond to criticisms by developing countries that only considering present emissions allows early-bird polluters to get away with already fouling the global nest. The International Energy Agency (IEA) proposed reducing the environmental impact of petrochemicals through technological 'measures that would cut air pollutants from chemicals production by 90 per cent over the next three decades, reduce greenhouse gas emissions by 60 per cent and halve ocean-bound plastic waste' (Willis 2019: B5). Whether these technological measures would effectively reduce emissions on the scale needed will only be known when these refineries operate. Until then, the proposals remain in the realm of faith in technological solutions.

Alert commentators admit that relying on technology to solve global environmental problems is a matter of faith. Concerning the need for battery storage of energy for electrical vehicles, Gorrie (2017: 45) argues that 'technological advancements are key, but uncertain. The optimistic forecasts assume more efficient battery chemistries will supplant lithium-ion. But no one yet knows what those might be, or when. It's simply an article of faith'. The economist Piketty (2014: 11) concludes that economists' 'overly developed taste for apocalyptic predictions gave way to a similarly excessive fondness for fairy tales, or at any rate happy endings'.

A reformed faith in mastering nature is also prevalent in some social science analyses. The Hartwell group provides a particular case for managing anthropogenic climate change (Rayner 2010, 2012). Pielke (2010: 230) puts his faith in technologically mastering nature's dynamics to develop accessible sources of energy cheaper than fossil fuels thereby leaving them safely in the ground: 'progress on accelerating decarbonisation of the global economy will be a consequence of technological innovation'. Lomborg (2019: O11) agrees. After criticizing electric vehicles, vegetarianism, wind and solar energy as empty, trivial gestures, he presents his solution to make mitigation compatible with economic growth: 'a vast increase in spending on green energy research and development, so that these energy sources eventually become cheap enough to out-compete fossil fuels'. Pielke (2010) refers to this as a 'climate fix'. Part of his proposal involves developing technologies of direct air capture (DAC) to suck carbon out of the atmosphere. Mastery of nature faith 2.0 consists of both dreams of (i) upstream technical miracles to produce abundant, clean, cheap energy and (ii) end-of-pipe geoengineering solutions to suck carbon out of the atmosphere. Faith in the mastery of nature has been reformed, but it has not collapsed and remains a pillar of modern thought.

Every oil extraction state is threatened by the conclusion of impact science that fossil fuels must be left in the ground in order to mitigate the greenhouse effect. They are vulnerable to the requirement that clean, renewable energy (wind, solar, hydro, tidal, geothermic, etc.) and nuclear energy must replace fossil fuels rather than merely add to them. For fossil fuels to continue to be used, they have to wish for technological innovations (i) that develop emissions-free combustion, with the only existing one being carbon capture and storage (CCS), or (ii) direct air capture (DAC) that sucks carbon out of the atmosphere, or (iii) geoengineering solutions that screen sunlight from getting in.

Surveys found Canadians want solutions to climate change, but do not want to pay for them, for example through higher prices for gasoline, oil, electricity, heating, plane fares, etc. The wealthy Canadian province of Alberta extracts 3 million barrels of oil daily from its tar sands, with ensuing carbon emissions, and wants to extract more. Its conservative government admits fossil-fuelled climate change is a threat. To avoid carbon taxes, its solution is technological innovation, as its environment minister put it 'our focus on climate change is around technology, … We're not in crisis mode. We're focussed on actually being able to address emissions where the previous government taxed people' (Giovannetti 2019: A4). That this will lower Alberta's disproportionate input to global warming is a matter of faith. The Canadian Conservative party strongly supports Alberta's tar sands, is opposed to carbon taxes, to significant environmental assessments, and fuel-efficiency regulations. It proposes instead requiring big polluters to invest in technological innovation to find solutions, thereby having faith in polluters to find technologies to combust fossil fuels without carbon pollution. Success of its policy depends on technologies which haven't been invented. This requires a leap of faith.

Uncertain Technological Success Equals Faith in the Unforeseeable

Since danger is evident under both lay and scientific assessments of global warming and since fossil-fuel restraint is unpopular everywhere, particularly in oil extraction states, faith in an almost miraculous technological innovation to solve global warming is relied upon, even when there is little or no basis for such a belief. Faith in a utopian technical fix attracts many adherents because it would enable humanity to avoid the unpleasant task of changing fossil-fuelled practices and paying the full cost of harm they cause. Beck (2015: 123) refers to this as 'taking "the easy way" of technocratic catastrophism', namely the belief that market dynamics and technology are capable of enabling societies to avoid the unleashing by fossil fuels of catastrophic forces of nature. No major change of social practices and cost would be needed. Thanks to the mastery of nature's dynamics, people will be able to fly the world using abundant carbon-free fuel because it will be cheaper than jet fuel, take cruises powered by carbon-free energy cheaper than bunker fuel, and import food from the other side of the planet. Such are the implications of Pielke's belief, which is optimistic that technological innovation financed by government-imposed escalating carbon taxes will succeed in reducing emissions to withdrawal rates by inventing carbon-free energy cheaper than fossil fuels, but pessimistic that the population would accept carbon taxes. This constitutes a contradiction if ever there was one.

The continuation of fossil-fueled social practices relies on faith that the danger of global warming will prompt the emergence of market-based technologies to economically capture and store carbon underground or suck it out of the atmosphere. However, the 'change is so vast, so universal, that it seems to test the limits of human reason. So it should not be surprising that the ideologies that led us here, those that have guided the postindustrial age – techno-lust and hyperindividualism, conflation of growth with progress, unflagging faith in unfettered markets – are the same ones many now rely on as we try to find a way out. Nowhere is humankind's mix of vision and tunnel vision more apparent than in how we're planning for a warmed world' (Funk 2014: 6).

Where will the technologies come from? Pielke (2010: 226) admits that fundamental breakthroughs developing new energy sources, such as nuclear fusion, are unlikely. Nevertheless, the 'alternatives to fossil fuels are well known and include various technologies of wind, solar, biomass, nuclear, hydropower carbon neutral fossil fuels' and energy storage such as better batteries. None are currently ready on the massive scale needed, hence he advocates 'technological agnosticism, since we do not presently know where advances might lie' (Pielke 2010: 226–227). Agnostic though he is, he has faith that technology will succeed. He excludes a priori the possibility that unforeseeable technological advances could be inadequate to mitigate the foreseeable fossil-fuelled climate crisis. But we know that needs do not necessarily result in the emergence of technologies and social structures that meet those needs (the old structural-functionalist fallacy).

Empirical evidence (York 2012) contradicts Obama's belief that wind and solar energy is driving fossil fuels out of the market. Despite increased carbon-free energy, fossil-fuel use continues to grow to power global economic growth. Microsoft's pledge quoted earlier is only an aspiration since the necessary direct air capture technology and its economics are unproven. It relies on faith that the US$1-billion investment will translate aspiration into achievement, but money has often failed to develop hoped-for technologies. Meanwhile, Microsoft expects to emit 16 million tonnes of carbon into the atmosphere in 2020. Whether the IEA's proposal quoted previously of technological measures to reduce the environmental impact of petrochemicals amounts to effective mitigation or discursive greenwashing could be determined by a simple test: will these measures, which would increase costs of refineries, oil, and plastics, thereby decreasing profitability for investors and increasing prices for consumers, be implemented as a precondition for operating those refineries? Profitability and cheap prices depend on future harms not being included in the price and externalized to be paid later by victims.

Lomborg's claim about research and development quoted earlier sounds good, but he fails to deal with two key problems. (i) Where will the money for research come from? Hopefully, it will be paid by carbon polluters who are causing global warming. (ii) Although research resulted in amazing, unanticipated innovations, it has often failed to yield needed inventions, cures, etc. Research to discover technologies to out-compete fossil fuels can be hoped for, but it is reckless to rely on it. Since presently low-carbon energy is more costly and less flexible and reliable than polluting fossil fuels therefore providing a small fraction of global energy, belief in this solution requires a leap of faith in the capacity to re-engineer nature's dynamics to construct such technological breakthroughs. The ongoing accumulation of atmospheric carbon pollution means that there must also be faith that the technological solution will be implemented quickly.

In 2000, emissions-free hydrogen fuel cells were expected to revolutionize the global automotive industry by replacing gasoline-powered internal combustion engines. Ballard, the company that developed the technology, partnered with Daimler-Benz and Ford. Its share price rose to $192. But hydrogen fuel cells proved too expensive for mass adoption, and Ballard's share price fell to 59 cents by 2012 (Shufelt 2020). To succeed, innovation has to be both technically effective and cost-effective. Other putative solutions similarly failed, for example, iron seeding of part of the Canadian ocean coastline to withdraw carbon from the atmosphere, and is laden with unknowns and risk.

Nordhaus describes how long it would take to implement technological innovations to capture and store carbon dioxide. 'A technology like CCS might require a decade of research and development (R&D), another decade of pilot plant testing, continuous public and environmental and boardroom scrutiny, perhaps another decade of roll-out of large-scale plants in many countries, and only then – if it passes all the tests along the way – would it be ready for deployment on the scale needed to capture and store billions or tens of billions of tons of CO_2 every year' (Nordhaus 2013: 282). CCS only works for large stationary sources of carbon dioxide, not moving ones, like planes and vehicles. It requires 20–25% of the energy produced by a power plant to operate, making it prohibitively expensive (Flannery 2015: 179). CCS has been costlier and more difficult to build than projected, and only captures 65% of the carbon dioxide produced; 35% escapes polluting the atmosphere. Demonstration projects of CCS have been promoted and tried in Canada but have been found wanting. Fossil-fuel companies won't pay for them, hence they require massive public tax subsidies. This is not very promising for urgent cumulative problems, hence many specialists conclude CCS is a dead end. Researchers developing carbon capture (Keith et al. 2018) and economists (Nordhaus 2013) argue that such technological attempts should be additional to carbon taxes, never used to avoid them.

Faith in technological innovation to solve the climate crisis amounts to faith in the unforeseeable. Extrapolation from past successes is a non-sequitur when there are discontinuities, as foreseen with global climate change. It could well be that no non-carbon energy source cheaper than fossil fuels and as flexible will be found, that geoengineered sunscreens in space will have unforeseen harmful side effects, that air capture of carbon will not work, that carbon capture and storage will be prohibitively expensive or applicable to only a small portion of carbon emissions, and that the greenhouse effect will result in passing through tipping points into an irreversible habitat less beneficial to humanity for which no adaptation will be possible. Current faith in these technological innovations may well be based on ignorance of the physical obstacles to their development, of their limitations and side effects, and of their cost. Modern social practices of the knowledge society are constructing uncertainty and ignorance, paradoxical as it may seem. The reformed faith in mastering nature involves riding the risk of known unknowns and unknown unknowns and of possible technological impotence in the face of global biophysical forces. Even those who favour technological climate fixes admit their success is uncertain. 'Although some scientists believe that there may be "tipping points" or thresholds in the climate system where catastrophes occur, there inevitably remains much that is unknown. … the impacts of increasing carbon dioxide are already occurring, and no one knows if or when there might be a threshold effect' (Pielke 2010: 12). Belief in technologically mastering nature's dynamics to solve global warming is not based on evidence. It is faith.

Efficiency as Mastery of Biophysical Dynamics

There are different kinds of reformed faith in the mastery of nature concerning the fossil-fuelled climate crisis. One is 'strong faith 2.0' in future technological innovation, like the development of sunscreens in space, withdrawing carbon from the atmosphere, carbon capture and storage, etc. Another is 'faith 2.0 light' in technologically mastering nature's forces. It believes that technological innovations to increase efficiency, mitigate, adapt, be robust, and resilient will enable societies to ride out problems their fossil-fuel use cause. In some quarters, this is seen as complementing prevention, even facilitating it, and should occur now. In many, however, it is relied on to avoid carbon taxes and restraints on fossil-fuelled practices.

One belief is that increased efficiencies will enable societies to do more with less fossil fuels, hence decrease their consumption and reduce emissions. This involves technologically reconfiguring and recombining properties and dynamics of biophysical materials to attain the same or greater output using decreased input of fossil fuels. Technological improvements in fuel efficiency would enable travel of a specified distance with fewer litres of jet fuel or petrol (gasoline) and less greenhouse gases-emitted. Maximizing economic activity for each unit of emissions is necessary. However, improvements in efficiency confront biophysical limits and diminishing returns to doing more with each litre of petrol or jet fuel. Achieving high efficiency can be very costly, so it also meets economic limits. It takes time to innovate and implement efficiency improvements for industry and the global vehicle fleet, plane fleet, and ship fleet, which contrasts with the rapid accumulation of greenhouse gases caused by ongoing fossil-fuel combustion.

Increased efficiency also has rebound effects. Jevons (1865) noted a paradoxical association: when technical innovations made coal-fired steam engines more efficient, coal consumption rose instead of declining. This correlation between efficiency improvements and greater resource consumption has been documented recently for many resources and technologies: 'The main point that we draw from this analysis is that we should not assume that efficiency is necessarily a sufficient solution to resource consumption problems, and that it is at least plausible that, in some contexts, efficiency may, counter-intuitively, contribute to growth in resource consumption' (York and McGee 2016: 85). Efficiency gains reduce prices and stimulate consumption thereby aggravating path-dependent, polluting social practices. This is 'where efficiency leads to a structural transformation of production/consumption processes, setting a pathway to future reliance on high levels of resource consumption' (York and McGee 2016: 79). After OPEC raised oil prices in the 1970s, it prompted fuel efficiency gains: 'When oil prices fell in the 1980s the efficiency gains made in the prior decade were translated into making vehicles more powerful and better equipped (heavier) while possessing the same fuel costs per mile as earlier models. Enter sports utility vehicles – aka SUVs' (Carolan 2014: 150).

York and McGee (2016: 82–83) present the following thought experiment to illustrate how ineffective improved efficiency can be to reduce resource consumption and pollution. In one scenario, cars are awfully inefficient, needing 50 litres of petrol to go one kilometre. In the other they are incredibly efficient, capable of going 50 kilometres on one litre. Hence in the second, cars are 2500 times more fuel-efficient than those in the first. Imagine also that planes, boats, trains, etc., differ similarly between the two scenarios. It would be simplistic to presume there would be less fuel used in the highly efficient scenario. In the inefficient scenario, fuel tanks on cars would be huge and fuel costs would be high because of all the fuel needed. Therefore, few people would use cars. Ditto for planes, boats, etc. Infrastructure and innovation would develop with fuel inefficiencies in mind, namely dense, compact cities instead of urban sprawl. The highly efficient use of fossil fuels would stimulate the use of cars, urban sprawl and steer innovation towards car-centric technologies. Technological innovations have made fossil-fueled travel much more efficient and cheaper, so people commute alone by car from homes in distant suburbs to work in centre town, shuttle on weekends to cottages beside lakes, fly on aeroplanes to watch sporting events in distant places, and the affluent fly and take cruises to see the other side of the planet. Efficient fossil-fuelled transportation has propelled intercontinental tourism, and thereby increased carbon emissions. Even the less affluent drive motorhomes across continents on vacation. Improved efficiency has, paradoxically, been the catalyst to the increased consumption of fossil fuels and worsening of greenhouse-gas emissions. York and McGee (2016: 83) conclude that 'it is quite possible that there would be less fuel consumed in the low fuel-efficiency world than in the high fuel-efficiency world', and lower greenhouse-gas emissions.

Studies vary in concluding how big this effect is. 'Although the rebound effect is not trivial, it is more than offset by savings from the fuel economy standards' (Harvey and Orbis 2018: 127). Increasing efficiency is unlikely by itself to solve environmental problems of resource depletion and pollution. This must not be misinterpreted as against efficiency. Getting as much bang for each unit of energy is essential, especially for fossil fuels to restrain carbon emissions. Rigorous fuel-efficiency regulations are crucial. They are necessary but not sufficient, and if they alone are promoted, those problems could perversely be worsened. The point is that improvements in efficiency need to be coupled with a high price on carbon pollution that takes the full cost of fossil fuels into account to dampen their consumption. This is necessary to avoid perverse effects of efficiency incentivizing more fossil-fuel consumption and emissions.

Faith 2.0 in Robustness, Adaptation, and Resilience

The dogged attachment to current fossil-fueled practices is based on faith that societies will be capable of adapting to whatever dynamics of nature are being unleashed by present socioeconomic practices, and being resilient and able to bounce back or up to a better state after adverse consequences of fossil-fuelled climate change. Efficiency improvements, geoengineered sunscreens in space, capturing atmospheric carbon and sequestering it, and innovating cheap, non-carbon sources as energy-dense and flexible as liquid fossil fuels will necessitate major technological advances. Storage of wind and solar energy will need a quantum leap of technical improvements. Robustness, adaptation, and resilience will also likely require significant technological improvements, otherwise they will fail. Defences against hurricanes, drought, and wildfires are already overwhelmed, and will need significant technological advances to confront more extreme weather as global warming intensifies. Modern societies are dependent on reliable electrical grids, which must be made more robust to withstand extreme weather and designed to be rebuilt promptly if destroyed so that societies can be resilient (Murphy 2009). Likewise for other essential infrastructures such as water and transportation. Agriculture will need to technologically adapt and become biologically resistant to climate changes, perhaps with new seeds or developing appropriate genetically modified ones. A remarkable example of adaptation requiring major technological advances involves current research to discover how to make coral adapt as climate change renders ocean waters warmer and more acidic, and even grow coral and restore coral reefs. Researching technologically enhanced robustness, adaptation, and resilience is worthwhile, but it is uncertain whether they will work on the scale needed, avoid having harmful side effects, and be cost-effective. Faith in these technologies attempting to manipulate and master nature's dynamics will be severely tested when confronted by massive forces of nature like extreme weather, drought, wildfires, flooding, ocean level rise and acidification, changing jet stream and Gulf Stream patterns.

Although fossil-fuelled global warming has been scientifically well documented, the specifics of what needs to be adapted to in the future are largely unknown. Construction of worse-case scenarios helps prepare for disasters, but research (Murphy 2009) has documented that nature's catastrophic dynamics often surpass these make-believe scenarios that were socially constructed. Faith 2.0 light will have to confront known unknowns: e.g. whether tipping points into irreversibility will be reached. It will also have to confront unknown unknowns that can't even be imagined at the present time as humanity travels along this dangerous new trajectory. Because it is uncertain what societies will have to be robust against, adapt to, and bounce back from as fossil-fuelled climate change intensifies, the possibility of constructing robustness, adaptation and resilience should never be used to avoid preventive measures. Nordhaus (2013: 150) describes 'the siren song of relying solely on adaptation or geoengineering. … They may be part of a strategy of risk management, but even the best geoengineering and adaptation will still leave significant and unacceptable risks to the planet'. Prevention is easier, albeit costly and inconvenient, in the sense that the means of doing it are known, namely carbon taxes, exercising restraint about using fossil fuels, and replacing them with low-carbon energy.

Nature's Services

Both faith 1.0 and 2.0 in the mastery of nature tend to ignore services to humanity done by nature, for example the Albedo effect of the reflective property of the white Arctic ice cover and the carbon storage by the Earth's permafrost. These services are so omnipresent they are taken for granted. If the rate of anthropogenic carbon emissions continues in excess of the rate of nature's reabsorption, melting the Arctic ice cover and the permafrost, then to prevent dangerous carbon accumulation and to be safe humans will have to implement ways to reflect sunlight and remove carbon themselves through geoengineering and direct air capture (DAC), which could prove to be a prohibitively expensive, impossible task.

Technological Innovations Worsening Global Warming

Faith in last-minute technological innovations to solve global warming by mastering nature's dynamics has a major flaw. Innovations have been much more abundant at worsening fossil-fuelled climate change than solving it. Hydraulic fracking, extracting oil from tar sands, from deepwater, from the frigid Arctic, upgrading and refining heavy oil, and liquefying natural gas have been successfully innovated. They accelerated the treadmill of fossil-fuel extraction and greenhouse-gas emissions. On the other hand, technological innovations to withdraw carbon from the air or to capture it from combustion sources and store it underground have either failed or not been tried for economic reasons on the scale needed to combat emissions.

Technological innovations which increased the demand for fossil fuels, with resulting emissions, have been prolific. Heavier, gas-guzzling sports utility vehicles (SUVs) with all the comforts, protections, and gadgets imaginable are crowding out light, fuel-efficient vehicles. Advanced jet-fuelled planes make discount flying and long-distant flights cheap. Enormous fossil-fuel-powered ships transport huge quantities of goods and make possible inexpensive mass cruise tourism. Market-based technological innovations produced air conditioning for hot climates that makes indoors comfortably climate-controlled. This requires electricity, whose primary energy source is usually fossil fuels. Similarly for smartphones capable of manipulating enormous amounts of data, using huge data servers powered by electricity often from combusting coal. The storage site of this data is called a 'cloud'. A more literal referent would be a global carbon cloud in the atmosphere that results in global warming. Crematoriums have been made less expensive than burial, and are becoming significant sources of emissions. Concrete, held together by cement whose production results in massive emissions, has been used to construct ever higher skyscrapers. Although advertising promotes all these, the very availability of such previously inexistent means of comfort and pleasure made possible by technological innovation, stimulates demand.

The treadmill of these technological innovations overwhelmingly based on carbon-emitting fossil fuels cancels out the benefits of technological innovations to replace fossil fuels with low-carbon energy of wind, solar, hydro, geothermal, etc. The latter has been additional to, rather than replacing, fossil fuels. Hence carbon emissions continue to accumulate in the atmosphere. Market-based technological innovation has its own dynamics based on the pursuit of profit and can't be relied upon to solve health and environmental problems. It has hitherto been ineffective in bringing emissions in line with carbon withdrawal. Far from being the solution, it has been part of the problem. Technological innovation is not coming to the rescue to resolve the fossil-fuelled climate crisis, and instead is worsening it by developing appealing new ways to emit more carbon. Therefore the belief that the crisis will be solved by technologically mastering nature's dynamics has little supporting evidence. York (2017: 2) concludes that 'it is likely the case that transitioning to a carbon-free economy will not be accomplished by technological developments alone' (see also Sim 2012). Faith 2.0 in technologically mastering nature's dynamics to solve the climate crisis is misplaced. Shifting the present dangerous trajectory to safety will require a purposeful transition away from fossil-fuelled social practices as well as technological innovations.

Hope for Technological Solutions or Reliance on Them?

The reformed faith in mastering nature involves not just hope for technological solutions to fossil-fuelled climate change, which justifiably everyone shares. It relies on technological solutions in order to avoid annoying, costly changes of current dangerous fossil-fuelled practices. Faith 2.0 depends on nature's constructions not overwhelming sociotechnical constructions on a global scale, even though that occurs regularly on a smaller scale. It involves scaling up to the global level of the faith BP had in its supposedly failsafe blowout protector to master extreme pressures of deepwater. Nature's dynamics have repeatedly usurped the hubris of claims that market-driven technological innovation will inexorably master nature's forces (Freudenburg et al. 2009). If disasters occur frequently for small scale phenomena, there is no assurance they could not happen globally. Faith 2.0 risks the scaling up of failure of foresight, discounting danger, and the incubation of global disaster.

Attempts to reconstruct nature's dynamics through technological innovations, usually propelled by the pursuit of profit, involve trial and error. Some succeed—such as hydraulic fracturing, wireless communication, etc.,—but many don't—like fuel cells for transportation, nuclear fusion, lung cancer cures, etc. Others accomplish their objectives but bring threatening new dynamics of nature, as when the innovation of DDT and asbestos fireproofing caused cancer, and CFCs depleted the ozone layer. Hence they had to be abandoned. Since technological solutions often do not come when needed, or come with unforeseen harmful side effects, they can be wished for but relying on them is reckless. Faith-based dependence on future technological fixes mastering nature's dynamics—to avoid paying the full cost of combusting fossil fuels or restraining fossil-fuel use—enables societies to discard the restrictive precautionary principle, and legitimates dangerous fossil-fuelled economic growth. The distinction between hope for technological solutions and reliance on them is crucial. Pielke's iron law of the population's priority to near-term economic goals over long-term ones must be contextualized and balanced against an iron law of the uncertainty of technological success: needed, expected innovations often fail to materialize, but unexpected innovations do, sometimes worsening problems. The success or failure of technological innovation to find solutions to global problems is largely unforeseeable, and history is littered with promised solutions that failed. However, neither are iron laws. Both are contingent on socioeconomic conditions and nature's dynamics.

Steering Technological Innovation

If harnessing nature through future technological innovations is the solution relied upon for the fossil-fuelled climate crisis, why isn't it being implemented now as a pre-condition for fossil-fuel combustion to prevent carbon accumulation in the atmosphere, for example by making carbon capture and storage mandatory? The answer is because it is prohibitively expensive and largely ineffective. What fails locally now is relied upon globally in the future. This contradiction shows that technological innovation, especially if it is market driven free of government regulation, as a solution for the fossil-fuelled climate crisis is not evidence-based but rather faith-based. Moreover, the belief that societies will innovate their way out of the climate crisis with new technologies as they become more wealthy is contradicted by the evidence that wealthy societies are the principal per capita carbon emitters, for example the USA, Canada, and Australia. Depending on technological solutionism could prove to be perilous magical thinking because of nature's powerful reaction.

Faith 2.0 holds the belief that market-driven technological innovations and improving efficiencies, technically adapting, building robustness and resilience can be counted on to ride out whatever nature's forces throw at society when problems become visibly serious. These are important, but will be extremely difficult without prevention. Protecting against fossil-fuelled climate change, adapting to it, and bouncing back after its shocks will be easier if it is at least partially prevented. But they are often chosen instead of prevention because prevention is costly and requires modifications of lifestyle social practices using fossil-fuel energy.

There are many weaknesses of the belief that nature's dynamics can be technologically mastered by market-driven innovations to solve the fossil-fuelled climate crisis, whether in terms of a strong program of sunscreens in space or more modest technical innovations in terms of efficiencies, robustness, adaptation, and resilience. The success in extracting new sources of fossil fuels and innovating uses for them has had no equivalent success in capturing and safely storing underground the carbon emitted when combustion occurs. Carbon capture and storage has hitherto been expensive and ineffective and is unprofitable unless a significant cost for carbon pollution has been enforced. Carbon air capture that would be cheap, effective, and innocuous is nowhere to be found. Hence faith in future market-based technological solutions founded on mastering nature's dynamics has to be particularly strong in the context of its present evidence-deprived underpinning. Profit-driven innovation rarely solves environmental problems by itself, which requires purposive environmental-driven innovation promoted by government policy. The Nobel Prize-winning economist Nordhaus (2013: 33) concluded that the best analyses of economic and energy experts 'indicate that the CO_2 problem is not going to disappear or be magically solved by unrestrained market forces'. He adds that 'for the foreseeable future there are no mature technologies that can meet ambitious emissions reduction targets economically. …[but he also concludes] that a rapid decarbonisation will require substantial changes in our energy technologies' (Nordhaus 2013: 283). Developing the needed low-carbon technologies requires a government imposed high price on carbon pollution, government subsidies for basic scientific research, and doing this promptly. The economist Leach (2019) concludes that 'technology is not, as it's often portrayed, a substitute for carbon pricing. The value of emissions reduction technology is increased through carbon pricing, thereby providing incentives to innovate ... The notion that we can meet our GHG reduction goals by focusing on large industry alone and can do so with new, costless technological solutions is a convenient but unrealistic safety blanket. But it's more appealing to many than a carbon price'.

It is important to avoid misunderstanding. The alternative to faith 2.0, namely the belief that market-based technological innovation will succeed in mastering nature to solve global environmental problems, is not to stop technological innovation. Rather it is to avoid relying solely on future market-based technological innovation, which currently is incubating a fossil-fuelled climate disaster. To wish for technological solutions, while supporting government mandated full-cost pricing of carbon pollution to create incentives for zero- and low-carbon energy and exercising restraint on fossil-fuelled social practices, is a reasonable hope. This would replace blind faith in market-driven technological solutions with an eyes wide open steering of technological innovation to mitigate the climate crisis.

Take an example from another area. A century of investment in research to find a cure for lung cancer failed and the technological invention of light cigarettes did not prevent lung cancer. Nature's dynamics are often recalcitrant to innovations to solve problems. Fortunately, regulations, taxes, etc., were accepted by the population, decreased smoking from 50 to 15% in wealthy societies, and mitigated cancer. Where technological rationality and market rationality fail, social rationality implementing changes in social practices can succeed. But market-based technological innovation can undo positive social innovations. The development of e-cigarettes and vaping, usually legitimated to help smokers stop smoking, has been pitched to young non-smokers with flavours like cookies-and-cream, and is hooking them on smoking. Technological innovation failed to solve the cigarette–lung cancer causal relationship and made it worse by inventing a new delivery system for nicotine, other harmful chemicals, and lung cancer. A second round of social rationality will be required to promote health.

York (2017: abstract) concludes there is 'limited potential for technological developments to help overcome environmental problems without concurrent political, economic, and social change that supports conservation'. Technological innovation needs to be purposefully steered towards mitigating the fossil-fuelled climate crisis, typically by government. This will be examined in the next chapter.

References

Beck, U. 1995. *Ecological Politics in an Age of Risk*. Cambridge: Polity.

Beck, U. 2015. Emancipatory Catastrophism. *Current Sociology* 63 (1): 75–88.

Beckerman, Wilfred. 1974. *In Defence of Economic Growth*. London: Jonathan Cape.

Benton, Ted. 1994. Biology and Social Theory in the Environmental Debate. In *Social Theory and the Global Environment*, ed. M. Redclift and T. Benton, 28–50. London: Routledge.

Benton, Ted. 2001. Why Are Sociologists Naturephobes? In *After Postmodernism*, ed. J. Lopes and G. Potter. London: Athlone.

Buruma, Ian. 2018. When Laws Compete with Myths, the Gunslinger Triumphs. *Globe and Mail*, 17 March: O11.

Butts, Ed. 2019. The Cautionary Tale of Whale Oil. *Globe and Mail*. 5 October: O8.

Carolan, Michael. 2014. *Cheaponomics: The High Cost of Low Prices*. Abingdon, UK: Earthscan.

Clark, N. 2011. *Inhuman Nature*. London: Sage.

Collins, H. 2011. *Gravity's Ghost*. Chicago: University of Chicago Press.

Davidson, D., and J. Andrews. 2013. Not All About Consumption. *Science* 339 (6125): 1286–1287.

Dickens, P. 2004. *Society & Nature*. Cambridge, UK: Polity Press.

Dunlap, Riley, and Robert Brulle. 2015. *Climate Change and Society: Sociological Perspectives*. New York: Oxford University Press.

Dunlap, R., and W. Catton. 1994. Struggling with Human Exemptionalism. *The American Sociologist* 25 (1): 5–30.

Dunlap, R., and R. Scarce. 1991. The Polls-Poll Trends. *Public Opinion Quarterly* 55: 651–672.
Eder, K. 1996. *The Social Construction of Nature*. London: Sage.
Evernden, N. 1993. *The Social Creation of Nature*. New Haven: Yale University Press.
Flannery, Tim. 2015. *Atmosphere of Hope: Searching for Solutions to the Climate Crisis*. New York: Atlantic Monthly Press.
Foster, John Bellamy, Brett Clark, and Richard York. 2011. *The Ecological Rift: Capitalism's War on the Earth*. New York: Monthly Review Press.
Freudenburg, W. 2005. Privileged Access, Privileged Accounts. *Social Forces* 84 (1): 89–114.
Freudenburg, W. 2006. Environmental Degradation, Disproportionality, and the Double Diversion. *Rural Sociology* 71 (1): 3–32.
Freudenburg, William R., and Robert Gramling. 1993. Socioenvironmental Factors and Development Policy. *Sociological Forum* 8: 341–364.
Freudenburg, William R., and Robert Gramling. 2011. *Blowout in the Gulf*. Cambridge: MIT Press.
Freudenburg, William R., Scott Frickel, and Robert Gramling. 1995. Beyond the Nature/Society Divide. *Sociological Forum* 10: 361–392.
Freudenburg, William R., Robert Gramling, Shirley Laska, and Kai Erikson. 2009. *Catastrophe in the Making*. Washington: Island Press.
Funk, McKenzie. 2014. *Windfall: The Booming Business of Global Warming*. New York: Penguin.
Giddens, Anthony. 1991. *Modernity and Self-Identity in the Late Modern Age*. Stanford: Stanford University Press.
Giddens, Anthony. 2009. *The Politics of Climate Change*. Cambridge: Polity Press.
Giovannetti, Justin. 2019. UCP's Climate Change Plan Puts Focus on Technology, Industry, Minister says. *Globe and Mail*, 22 May: A4.
Gorrie, Peter. 2017. Crossroads. *Corporate Knights* 16 (2) (Spring): 42–45.
Grundmann, Reiner. 1991a. The Ecological Challenge to Marxism. *New Left Review* 187: 103–120.
Grundmann, Reiner. 1991b. *Marxism and Ecology*. Oxford: Oxford University Press.
Hamilton, David. 1973. *Technology, Man and the Environment*. London: Faber and Faber.
Harvey, Hal, and Robbie Orbis. 2018. *Designing Climate Solutions: A Policy Guide for Low-Carbon Energy*. Washington: Island Press.

Hasegawa, Koichi. 2012. Facing Nuclear Risks: Lessons from the Fukushima Nuclear Disaster. *International Journal of Japanese Sociology* 21 (1): 84–91. https://doi.org/10.1111/j.1475-6781.2012.01164.x.
Hasegawa, Koichi. 2015. *Beyond Fukushima*. Melbourne: Trans Pacific Press.
Hasegawa, Koichi. 2016. *Sociology of Climate Change Policies* (co-edited, in Japanese).
Hughes, J.D. 2009. The Energy Issue. In *Carbon Shift*, ed. T. Homer-Dixon, 58–95. Toronto: Random House.
Jevons, W.S. 1865. *The Coal Question*. London: Macmillan.
Keith, David W., Geoffrey Holmes, David St. Angelo, and Kenton Heidel. 2018. A Process for Capturing CO_2 from the Atmosphere. *Joule* 2 (July 6): 1–22. https://doi.org/10.1016/j.joule.2018.05.006.
Larson, Erik. 2000. *Isaac's Storm*. New York: Vintage.
Leach, Andrew. 2019. How the Commuter vs. Polluter Narrative Could Backfire on Alberta. *Policy Options*, 15 July. https://policyoptions.irpp.org/magaxines/july-2019/how-the-commuter-vs-polluter-narrative-could-backfire-on-alberta/. Accessed 13 November 2019.
Lomborg, B. 2001. *The Skeptical Environmentalist*. Cambridge: Cambridge University Press.
Lomborg, B. 2007. *Cool It*. Alfred A. Knopf: New York.
Lomborg, Bjorn. 2019. Empty Gestures Trivialize the Very Serious Challenge of Climate Change. *Globe and Mail*, 28 December: O11.
McCright, A., and R. Dunlap. 2010. Anti-Reflexivity. *Theory, Culture, and Society* 27 (2–3): 100–133.
Mitchell, R.C. 1990. Public Opinion and the Green Lobby. In *Environmental Policy in the 1990s*, ed. N.J. Vig and M.E. Kraft, 81–99. Washington: CQ Press.
Murphy, R. 2002. The Internalisation of Autonomous Nature into Society. *The Sociological Review* 50: 313–333.
Murphy, Raymond. 2009. *Leadership in Disaster: Learning for a Future with Global Climate Change*. Montreal: McGill-Queens University Press.
Nellis, Stephen, and Jeffrey Dastin. 2020. Microsoft Announces Plan to Erase Carbon Footprint. *The Globe and Mail*, 17 January: B2.
Nordhaus, William. 2013. *The Climate Casino: Risk, Uncertainty, and Economics for a Warming World*. New Haven: Yale University Press.
Obama, Barack. 2017. The Irreversible Momentum of Clean Energy. *Science* 355: 126–129.
Pellizzoni, L. 2016. Catching Up with Things. *Environmental Sociology*. https://doi.org/10.1080/23251042.2016.1190490.

Pielke, Roger Jr. 2010. *The Climate Fix*. New York: Basic Books.

Piketty, T. 2014. *Capital in the Twenty-First Century*. Cambridge, MA: Belknap Harvard.

Rayner, Steve. 2010. How to Eat an Elephant: A Bottom-Up Approach to Climate Policy. *Climate Policy* 10: 615–621.

Rayner, Steve. 2012. Uncomfortable Knowledge. *Economy and Society* 41 (1): 107–125.

Rayner, Steve, and Elizabeth Malone (eds.). 1998. *Human Choice and Climate Change*. Columbus: Battelle Press.

Shufelt, Tim. 2020. As China Looks to Cut Emissions, Ballard Finds New Life. *The Globe and Mail*, 16 January: B10.

Sim, Hans-Werner. 2012. *The Green Paradox: A Supply-Side Approach to Global Warming*. Cambridge, MA: MIT Press.

Simon, Julian. 1981. *The Ultimate Resource*. Princeton: Princeton University Press.

Steinberg, P.E. 2001. *The Social Construction of Oceans*. Cambridge: Cambridge University Press.

Tenner, E. 1997. *Why Things Bite Back*. New York: Vintage.

Turner, B., and N. Pidgeon. 1978. *Man-Made Disasters*. London: Wykeham.

Willis, Andrew. 2019. The Future Is Plastics: Murray Edwards, Li Ka-Shing Add to Oil Patch Holdings as Others Flee. *Globe and Mail*, 7 July: B1, B5.

Worster, Donald. 1994. *Nature's Economy*. Cambridge: Cambridge University Press.

York, R. 2012. Do Alternative Energy Sources Displace Fossil Fuels? *Nature Climate Change* 2 (6): 441–443.

York, R. 2017. Why Petroleum Did Not Save the Whales. *Socius: Sociological Research for a Dynamic World* 3: 1–13.

York, R., and E. Rosa. 2003. Key Challenges to Ecological Modernization Theory. *Organization & Environment* 16 (3): 273–288.

York, R., and J.A. McGee. 2016. Understanding the Jevons Paradox. *Environmental Sociology* 2 (1): 77–87. https://doi.org/10.1080/23251042.2015.1106060.

Zebrowski, Ernest, Jr. 1997. *Perils of a Restless Planet: Scientific Perspectives on Natural Disasters*. Cambridge: Cambridge University Press.

10

Technological Solutions and Social-Technological Solutions

A generalized faith in mastering nature through technological solutions to the fossil-fuelled climate crisis is empty without knowing what they could be, how feasible they are, and how they could be implemented. Hawken's research team proposes an effective plan to reverse global warming, which they call 'drawdown'. 'In atmospheric terms, drawdown is that point in time at which greenhouse gases peak and begin to decline on a year-to-year basis. … by reducing the amount of carbon in the atmosphere' (Hawken 2017: x, xiii). Note that this refers to the inflection point where carbon withdrawals exceed emissions thereby reversing global warming. Although very demanding, their extensive modelling demonstrates it can be done in the next thirty years by 2050 and that it will benefit the economy and save money compared to continuing current social and technological practices. But how will it be done? This chapter assesses some of the most likely possibilities. The issue is not one of choosing either sociocultural change or technological innovation. Since the fossil-fuelled climate crisis is caused by the interaction of socioeconomic constructions with nature's dynamics, remedies almost invariably require hybrid innovations of both technology and social practices. Technological innovations are needed and have to be prompted by

R. Murphy, *The Fossil-Fuelled Climate Crisis*,
https://doi.org/10.1007/978-3-030-53325-0_10

social innovations, for example feed-in tariffs prompted development of wind and solar energy in Germany.

No energy source is perfect: all have some adverse consequences. Thus it is important to discriminate between them and be selective when confronting fossil-fuelled climate change. Some environmentalists oppose nuclear energy, others oppose large-scale hydroelectric generation, and still others oppose wind farms especially if they are visible from their favourite backyard. In most cases where initiatives to implement low-carbon energy are opposed, the default options are fossil fuels which cause far more long-term harm. 'With the growing urgency of climate change, we can't have it both ways. We can't shout about the dangers of global warming and then turn around and shout even louder about the "dangers" of windmills' (Suzuki and Hanington 2017: 176). Discernment is required to select the energy source that avoids the most danger, even at the cost of minor inconveniences, such as seeing a windmill in the distance or hearing a faint swish. Technological solutions can be viewed as belonging to two broad categories: (i) those that are claimed to mitigate global warming while combusting fossil fuels, and (ii) those that result in fossil fuels being left safely in the ground.

Mitigating Global Warming While Combusting Fossil Fuels

The proposed solutions to global warming that do not require leaving fossil fuels in the ground unsurprisingly are favoured by that industry. Fossil fuels could continue to be extracted and combusted without causing global warming. These could in principle either (i) bring emissions in line with carbon withdrawals or (ii) prevent global warming despite an excess of atmospheric carbon. Countries that base their economies on continuing fossil-fuel extraction must presuppose one or all of these technological climate fixes will be successful, cost-effective, and promptly implemented everywhere to avoid runaway global warming.

Replacing Coal and Heavy Oil with Natural Gas

Pielke (2010: 98) documents that the combustion of coal emits 94.4 million metric tonnes of carbon dioxide per quad of energy, petroleum emits 70.0 and natural gas 53.1. Nordhaus (2013: 162) concludes that 'natural gas is the cleanest of the fossil fuels, emitting about half as much CO_2 per kilowatt hour (kWh) as coal when burned for electricity generation. Shifting a greater fraction of electricity to natural gas is an important way of reducing CO_2 emissions'. And natural gas can be liquefied to be shipped long distances. It is a bridge for transitioning away from coal and petroleum to low-carbon wind or solar energy, providing that methane flaring and leaking be eliminated. Under the American Obama administration, greenhouse-gas emissions were reduced by replacing coal with natural gas for generating electricity. Much natural gas is now extracted by the hydraulic fracturing of shale, but that has caused other environmental problems including some small earthquakes. Other natural gas deposits are located in remote places far from where it will be used, therefore it has to be pumped in pipelines, liquefied, and shipped, which requires energy, usually fossil-fuel energy and associated emissions. Natural gas is a fossil fuel whose combustion causes emissions thereby worsening global warming. Hence it must itself be replaced by non-emitting sources to mitigate climate change.

Carbon Capture and Storage [Sequestration] (CCS)

This would consist of capturing CO_2 emissions from large stationary sources and storing the carbon dioxide underground in appropriate geological formations, like saline aquifers or depleted oil or gas wells or in deep ocean waters. Globally there are over 8000 large stationary sources of carbon dioxide (e.g. fossil-fuel-fired electricity generation plants) emitting about half of global emissions, so this remedy has potential. However, the capture of CO_2 emissions is well below 100%, hence there would still be significant amounts of emissions. Moreover, it is difficult to find secure storage sites nearby, so the carbon dioxide has to be pumped long distances. CCS is prohibitively expensive making operations uneconomical, hence proponents demand subsidies from the state. The rare attempts typically involve enhanced oil recovery and enhanced coal-bed methane recovery to counterbalance the cost, which results in extracting even more fossil fuels. The scale of CCS needed for 8000 large stationary sources of emissions is enormous. CCS does not work for sources that are not large nor stationary, namely enormous numbers of small emitters like vehicles and planes. Nevertheless, CCS is the great hope of the fossil-fuel industry and states where there are large emitters because capturing and storing carbon emissions would enable fossil-fuel extraction and combustion to continue.

Direct Air Capture (DAC) of Carbon Storing the Carbon Underground

Direct air capture is already being done successfully through afforestation using the growth of trees to capture carbon from the air and storing it in trees and roots. Planting trees has, however, been no match for the scale of carbon emissions; globally it has not even kept up with deforestation, and forest fires resulting from global warming caused drought emit the carbon back into the atmosphere. Another method of carbon withdrawal would seed parts of the ocean with iron to promote phytoplankton growth thereby drawing carbon dioxide from the atmosphere and depositing it at the ocean bottom. This was tried in Canada, was unsuccessful (Suzuki and Hanington 2017), and brings risks of serious side effects for oceans. Other experiments demonstrated that 'most of the carbon captured by the algal bloom fails to sink to the sea floor and quickly returns to the atmosphere and ocean' (Flannery 2015: 144). Flannery (2015: 93 states that 'when I wrote *The Weather Makers* [in 2006], algae was showing great promise as a source of renewable fuel. Today [in 2015], most innovations remain in the lab or operate at very small scale'. Yet it is the huge scale that is crucial for mitigating global warming.

The technical type of direct air capture strips CO_2 from the air, then concentrates it into a solid or liquid which can be buried and sealed off

(Keith et al. 2018). This requires compression of gas, massive amounts of clean energy, and is expensive. The principal investors in this start-up technology are the oil companies Occidental Petroleum and Chevron and their executives (McCarthy 2018a: B6). They seek to inject carbon dioxide into depleted wells to extract more oil, referred to as enhanced oil recovery, and store the carbon there. Most rely on futuristic technological innovation. It could be overwhelmed by the amount of carbon emissions, especially if done belatedly after melted permafrost releases methane into the atmosphere and melting of the Arctic Ocean ice cover weakens its reflective capacity leaving more sunlight to heat the atmosphere. Hence it seems fanciful to depend on this method to bring carbon withdrawal into balance with ongoing emissions. Furthermore, it becomes cost-effective only where there are carbon sequestering credits, as in California (Littlemore 2019). Both carbon capture from large stationary sources and from the air need a price on carbon pollution and/or low-carbon fuel standards to make them price-competitive. As Keith and Parson (2017) state forcefully, they could be complements to carbon pricing, not a way to avoid it.

Seaweed

Seaweed cultivation is better at carbon dioxide absorption from the atmosphere than land-based crops and faster growing. It would provide the co-benefits of food and medicine, would reduce ocean acidification, increase sustainable fish production, and is already used on a miniscule scale in China. Seaweed farming could capture carbon dioxide from the air, then store it deep on the ocean floor. Why isn't this simple solution used to solve the urgent climate crisis? The problem is the amount of seaweed needed to withdraw all the carbon dioxide being emitted. 'Covering 9 per cent of the world's oceans with seaweed farms, and then processing the voluminous product yielded, is far beyond our current capabilities' (Flannery 2015: 166). It would require covering the equivalent of half the Indian Ocean with a cultivated seaweed forest, then processing all the stuff, continually and permanently. Better to prevent carbon dioxide emissions than to adapt by being burdened with the sisyphusian task of pulling the carbon dioxide out of the atmosphere.

Cement

Every tonne of cement emits one tonne of carbon dioxide. Globally four billion tonnes of cement are produced annually (Flannery 2015: 170–171), emitting four billion tonnes of carbon dioxide into the atmosphere. New ways are being researched to make and cure cement using less water and energy, thereby making concrete produced from cement more durable, flexible, stronger, and less expensive. These new ways could even absorb carbon dioxide from the atmosphere and store it in the concrete. Instead of concrete production being an important contributor to emissions, concrete would withdraw carbon dioxide and store it. This seems too good to be true, and maybe it is. It is unproven, and would have to be tested rigorously to be certain tall concrete buildings would not collapse.

Where to Store the Carbon Dioxide?

The amount of carbon dioxide that would have to be captured is huge. A difficult issue 'is what to do with all that carbon dioxide, as twenty years' worth of human carbon dioxide emissions, captured and liquefied, would fill Lake Michigan' (Dyer 2008: 212). After capture from the air, it could be stored under the huge ocean pressures at depths over 3000 metres where it would remain trapped as a liquid. Or under the Antarctic ice cap where the ultra-cold temperatures would store it in solid form (Flannery 2015: 181–184). Transporting it to these places would be difficult and expensive. Sucking carbon out of the atmosphere might have worked if undertaken at the Rio Conference in 1992, but now after thirty more years of enormous emissions, the task of pulling it out and storing it by whatever means would be immense.

A Geoengineered Sunscreen in Space

To reduce global temperature despite high levels of atmospheric carbon, a geoengineered solar sunscreen proposal involves placing a film of sulphur dioxide or titanium dioxide high in the atmosphere around the planet. 'The most plausible method of achieving this is to use aircraft to produce a fine mist of reflective material in the stratosphere, where it would act like a thin cloud reflecting a little sunlight back to space' (Keith and Parson 2017: O5). This mimics what happens when a volcano erupts cooling the planet. Research modelling for this proposal has begun. Geoengineering is a form of adaptation, in that it seeks ways to adapt to carbon emissions rather than prevent them. The sunscreen proposal would let carbon dioxide accumulate in the atmosphere but prevent global warming by blocking some of the sun's rays.

Attempts like these have many unknowns in an extremely complex system, and perhaps dangerous consequences, because of interactions with other dynamics of nature. It would be difficult to cool the atmosphere by the right amount uniformly over all our planet, and could cause a patchwork of cooling and warming, flooding and droughts. Too little screening would leave the problem unsolved; too much could produce catastrophic cooling as occurred with the Mt. Tambora volcanic eruption in 1815 that resulted in crop failure and worldwide famine (Zebrowski 1997: 208–211). Flying aircraft all over the globe to maintain sufficient atmospheric sunscreen would be expensive, and if flights were stopped for whatever reason (refusal to pay the cost, wars, recessions, etc.) and the sunscreen depleted, then the atmosphere would overheat quickly because of all the carbon in it, which would be worse than slow warming due to its gradual accumulation of carbon. If sunscreen geoengineering is started, it must not be stopped. Moreover, implementing geoengineering remedies, like sunscreens in the sky, would mitigate global warming but not the acidification of oceans when the carbon descends from the atmosphere.

The complexity and unknowns of the global planetary biophysical system could result in geoengineering inadvertently unleashing even worse problems than global warming, perhaps irreversible ones. If fossil-fuelled practices weaken services that nature's dynamics do freely for humans, such as the Arctic ice cover reflecting the sun's rays out to space, and humans have to do it themselves technically, then it will be costly and dangerous. A study by meteorologists estimated that twelve atmospheric services essential for the habitat for humanity have a monetized value between 100 and 1000 times greater than the gross world product (Thornes 2010).

The possibility of geoengineering solutions, however improbable, could reinforce an illusory faith in last-minute technological solutionism and give decision-makers and populations excuses to continue unrestrained use of low-cost fossil fuels. But since geoengineering solutions carry the threat of unimaginable global harm, they are viewed by geoengineering scientists as 'at best, supplements to emissions cuts, not substitutes for them. Emissions must still be cut' (Keith and Parson 2017: O5). They argue that the longer carbon emissions exceed withdrawals and accumulate in the atmosphere, the more likely measures of desperation like geoengineering will be needed to limit global warming to 2 °C. Hence they advocate a research programme of modelling and small field experiments. This can be interpreted as disaster preparedness, namely as last resort remedies if societies refuse to reduce their destructive dependence on fossil fuels and if no other technological remedy like carbon capture and storage works, all of which seems likely.

Mitigating Climate Change and Leaving Fossil Fuels in the Ground

Solar, Wind, Hydro, Tidal, Biofuel, and Geothermal Energy

There are huge amounts of clean, renewable energy available: the sun's solar energy striking Earth is enormous, wind energy is abundant, tidal energy resulting from the moon's rotation around Earth is plentiful, geothermal energy is so powerful that it causes volcanos and earthquakes, etc. Iceland is one of the few countries using geothermal energy. Pielke (2010: 220, 230) contends that 'to reach low stabilization targets implies a massive transition to a nearly carbon-free global energy supply.

... progress on accelerating decarbonisation of the global economy will be a consequence of technological innovation'. Technological innovation could be steered towards making low-carbon energy viable and cost-effective, instead of extracting more fossil fuels. Wind and solar energy requires development of storage capacity for times when the wind doesn't blow or the sun doesn't shine. Electric vehicles require development of lightweight energy storage for long-distance travel. These renewable energy solutions to fossil-fuelled climate change, and others like hydro, geothermal, tidal, etc., consist of improving and increasing the use of existing technologies which have proven their worth in varying degrees. Technological improvements require socioeconomic changes to incentivize their development. Feed-in tariffs in Germany propelled the development of wind and solar energy for electricity. There have been major efforts to augment the amount of electricity generated from wind and solar, but this increase failed to decrease the amount of electricity produced from fossil fuels, which remains at two-thirds on the world level, and failed to stem the rise in carbon dioxide emissions because the demand for electricity is increasing rapidly. The electricity sector currently results in over 40% of global energy-related carbon emissions. The use of electricity will increase more rapidly if electrically powered vehicles replace gasoline-powered internal combustion engines. Electricity is a carrier of energy from primary sources, hence it is crucial that the primary sources of electricity be low- or zero-carbon.

Since windmills and solar farms can be built close to where energy is used, spill-prone pipelines and ships are not needed to transport energy long distances. Mass public transportation and electric cars powered by low-carbon energy would be privileged. Just as coal was eliminated from powering trains and ships by the advent of more malleable oil, so too fossil fuels could be surpassed by low-carbon energy. The fossil-fuel industry would shrink, thereby leaving dangerous fossil fuels safely underground. And just as the replacement of coal by oil stimulated economic growth, so too would the replacement of fossil fuels by renewable energy.

Unlike fossil fuels, wind and solar energy does not result in externalities of costly fossil-fuelled global warming, hence there is no belated cost of intensified wildfires, floods, etc., to pay, nor environmental deficits, nor accumulating environmental debt. Hawken's (2017) research team assessed proposed solutions to reverse global warming from 2020 to 2050, with their detailed estimates being a quantitative example of foresight guiding social practices. A technically effective and cost-effective way to reverse global warming without hindering economic growth was replacing fossil-fuel energy with wind energy, especially onshore turbines. They estimated onshore turbines could reduce carbon dioxide emissions by 85 gigatonnes and the offshore ones by 14 gigatonnes. Although onshore turbines would cost $1.23 trillion, they would save $7.4 trillion compared to fossil-fuel energy. In the long run, they would result in financial savings. Wind turbines don't interfere with farming, can be built quickly, and electrical interconnections over a big geographical base can transfer electricity from where the wind is blowing to where it isn't. Unlike fossil-fuel extraction and nuclear reactors, wind turbines do not use massive amounts of water nor contaminate it. No tailings impoundment areas, euphemistically called 'ponds', are left to leak or burst their retaining structures. Claims that turbines are noisy and kill birds have been resolved by designing slow-moving blades and siting practices that avoid bird migration routes. Although windmills kill 45,000 birds and bats yearly in the USA, housecats allowed to roam outdoors kill 30 billion (Diamond 2019: 402). The main impediment to deploying this cost-effective source of clean, renewable energy is the cultural claim that wind turbines are aesthetically unpleasant and will lower property values. The 'not-in-my-backyard sentiment – from the British countryside to the shores of Massachusetts – remains an obstacle' (Hawken 2017: 3). The issue is whether people will learn to prefer seeing windmills in the distance to scenes of fossil-fuel smokestacks, polluted tailings ponds, and eventually to destructive flooding, hurricanes and wildfires resulting from fossil-fuelled global warming.

Solar energy is another valuable replacement for fossil fuels. The main collection method consists of photovoltaic solar farms, namely a vast array of solar panels. They are cheap to install and operate. Over a thirty year-period $81 billion would be saved compared to building fossil-fuel power plants, and $5 trillion would be saved in operational costs. Most importantly, 37 gigatonnes of carbon dioxide emissions would be avoided (Hawken 2017: 9). So why aren't solar farms replacing fossil-fuel power plants? Fossil-fuel power plants have the advantage of being built first and their fixed costs have long been paid, the cost per kilowatt hour of solar is presently slightly higher than for fossil fuels because externalities of fossil fuels are excluded from their price, and solar farms need much land. The latter problem can be resolved by placing solar panels on roofs so that households can collect their own energy: 'they transform generation and its ownership, shifting away from utility monopolies and making power production their own' (Hawken 2017: 11). In Freiburg Germany, a 59-home community called Solar Settlement has all roofs made of solar panels. Each energy-efficient home has a positive energy balance and yields $5600 per year in solar-energy benefits to homeowners (Hawken 2017: 5). In Australia, 16% of homes have photovoltaic rooftops. Solar panels could become the roofs of big-box stores. Rooftop solar panels can be installed in remote places without electricity lines, hence contribute to reducing poverty. They would result in a clean, sustainable future if people can escape the cultural habit of desiring shingles on rooftops. Hawken's (2017: 11) research team estimates that by 2050, rooftop solar will increase from 4% of electricity generation to 7% globally and that will avoid 24.6 gigatonnes of emissions. Although photovoltaics cost money to produce and install, they save seven times more in electricity bills.

Hybrid renewable energy projects combining wind, solar, and hydroelectric are being innovated to solve energy storage problems. When the sun doesn't shine, wind power is used. When wind doesn't blow, solar power is used. The best-known example of hybrid renewable energy is El Hierro in the Canary Islands where wind power pumps water uphill to a reservoir in a volcanic crater where it is stored and released as needed to provide hydroelectric power (Healing 2019: A6).

Electricity is just one component of energy use. Fossil fuels are used for about 80% of global energy because aviation is entirely powered by fossil fuels, as is cement production and much industry, and ships and most vehicles are fossil-fuelled. Cars and trucks powered by gasoline in internal combustion engines consume two-thirds of the world's oil and produce 23% of emissions. Presently there are one billion vehicles, and estimates predict the number will exceed two billion in 2035 as developing countries motorize. Each gallon of gasoline combusted in vehicles produces 25 pounds of carbon dioxide emissions. Given the deep embeddedness of cars and pickup trucks in sociocultural practices, decreasing their use is difficult. The alternative is to transition them to low-carbon energy. Emissions can be reduced by 95% if powered by electric motors using low-carbon energy. This improvement requires electric vehicles, carbon-free electricity generation, and a network of charging stations similar to gasoline service stations. Hawken's (2017: 142–143) team estimates that if the use of electric vehicles could be increased from the present miniscule amount to 16% of passenger miles travelled in the next thirty years, almost 11 gigatonnes of carbon dioxide emissions would be avoided. Aviation is more difficult. 'The first non-petroleum based biofuels were used in commercial jet aircraft in 2008. Yet, as of 2014, demand from airlines remains low, … The incumbent fossil fuels industry, which has already built plants that make jet fuel from crude oil, have a great advantage' (Flannery 2015: 93).

The main challenge for low-carbon energy sources will be to scale them up to replace fossil fuels. That is difficult because liquid energy like oil has a dominant place in energy consumption. It is energy dense and flexible, hence difficult to replace in a cost-effective way, and the infrastructure for oil, gasoline, diesel, and natural gas is already in place. Most importantly, the price of fossil fuels does not represent their full cost, which would require a carbon tax to pay for their externalities.

The development of hydroelectric, solar, wind, tidal, geothermal, nuclear energy, and efficiencies result in energy with little if any greenhouse-gas emissions. The electric car is coming, albeit slowly. The media tell us about these innovations regularly, fostering the inference that climate change is being solved. The media have just reported the first long-distance flight propelled by 10% biofuel from Los Angeles to Melbourne by the airline Quantas, with a resulting 7% reduction in greenhouse-gas emissions. Biofuels have been criticized for using pesticides and crowding out agriculture on good land. But this 'jetoil' is derived from carinata which does not need much water, fertilizer, or pesticides and can be grown on land not used for agriculture (McCarthy 2018b: B1, B7). The jetoil example consisted of one flight, but even if all flights used jetoil thereby reducing emissions 7%, rapid growth in international aviation would have easily overpowered this technological improvement.

Nuclear Energy

A renaissance of nuclear-powered electricity is occurring, with 55 nuclear reactors under construction mostly in Asia, particularly China (Reguly 2019: B4). The interest is nevertheless much broader, motivated by increasing recognition that wind and solar energy will not suffice to reduce carbon emissions enough to keep the global average temperature from rising more than 2 °C. 'Despite the strong growth [of wind and solar energy], the percentage of emissions-free electricity in the world has not increased in 20 years. It's stuck at 36 percent' (Gorman 2019: B4). The reason is because the global demand for energy keeps increasing, nuclear reactors are being shut down, and wind and solar are backed up by fossil fuels when there is little sunshine or wind. Paradoxically, wind and solar are in this sense prolonging the use of natural gas, with attendant emissions. Thus John Gorman, who previously was the chief executive of the Canadian Solar Industries Association and who continues to support solar and wind, has now concluded that nuclear energy is a necessary part of solving the climate crisis. 'People fail to realize that nuclear is the only proven technology that has decarbonized the economies of entire countries, including France and Sweden' (Gorman 2019: B4). He means the electricity supply, not energy for vehicles, planes and boats, but he still has a point. He argues that nuclear can replace fossil fuels as the back-up for wind and solar, is cost-effective over the lifetime of a reactor despite high upfront costs, meltdowns are rare, radioactive waste is small in quantity and can be managed when well regulated.

The World Nuclear Association (2019) claims that nuclear energy can be deployed on a large scale in time to meet emission-reduction goals, that it is emissions-free during the operation of a reactor, is reliable and affordable, and therefore its expanded use is essential to deal with the fossil-fuelled climate crisis. The Association gives the example of France, which generates more than 70% of its electricity from nuclear power, resulting in its electricity industry having one-sixth of the European-average emissions, and has been safe. Presently the world's 430 nuclear reactors in 30 countries provide 11% of the world's electricity, and the Association argues this should be increased to 25% by 2050. The International Energy Agency (2018), the Massachusetts Institute of Technology Energy Initiative, the US Energy Information Administration, the World Energy Council, and many independent energy analysts agree that increased nuclear power is essential for achieving deep decarbonization. Rhodes (2018) argues that nuclear's worst accidents have been less destructive than other major industrial accidents, that nuclear waste disposal has been solved technologically, and that 'a full accounting of the external costs [such as fossil-fuelled climate costs] of different energy sources would find nuclear cheaper than coal or natural gas. … It's a valuable, even an irreplaceable, part of the solution to the greatest energy threat in the history of humankind', namely the climate crisis. Furthermore, small Modular Reactors (SMR) are being developed that proponents claim will consume and eliminate radioactive waste. Electricity generation will become even more massive and essential when electric vehicles replace the fossil-fuelled, internal combustion engine. If societies fail to replace fossil fuels with renewable energy, then emissions will continue to exceed carbon withdrawals, the consequences of global warming will become worse, and societies will increasingly have to turn to nuclear energy.

Nuclear power is, however, the most contentious energy source to replace fossil fuels. Critics argue it consists of trading one risk for another rather than decreasing risk: risk displacement instead of risk decrease (Hasegawa 2012, 2015, 2016). One intractable problem is that of scale. Pielke (2010: 113–116) calculated how many regular-size nuclear reactors would be needed to accomplish a 50% reduction of the world's carbon emissions between 1990 and 2050. To have eliminated all coal and natural gas consumption by 2006 would have required 2800 reactors; to power economic growth of 1.5% between 2006 and 2050 would necessitate another 5000 reactors; to eliminate oil, perhaps using electrical vehicles, would require 750 more reactors; and to bring electricity to the 1.5 billion people without it would require thousands more reactors. The total would be about 12,000 additional reactors, or opening a new one every day between 2010 and 2050. That massive deployment of nuclear would only reduce emissions by 50%. This also demonstrates how dependent the world is on the enormous extraction and combustion of fossil fuels. The radioactive waste that would accumulate from such an immense deployment of reactors would be huge, which would constitute a radioactive debt left by each generation to the following ones. Imagine the resistance to building and operating all those reactors. There is already lack of trust in nuclear authorities to dispose of radioactive waste safely and to prevent nuclear accidents. Nuclear energy, promised to be used for only peaceful purposes, could be redirected to nuclear weapons, as occurred in India. There are risks of antagonistic states acquiring nuclear weapons, of terrorists obtaining reactors or using them as enticing targets for mass destruction. Nuclear reactors have much more inherent danger than wind and solar farms and hydroelectric power. They take longer to build and deploy, rendering them a slow remedy to an urgent problem. Constructing reactors invariably involve cost overruns and there is the cost of decommissioning old reactors, which undermines proponents' claims of cost-effectiveness. The price of nuclear energy is competitive with that of coal and natural gas only if the latter's costly externalities are included in their price, which has never been the case. Moreover, there is the risk of running out of fissile uranium, and having to turn to other unproven sources.

Nuclear fission energy must not be overestimated as the solution to fossil-fuelled climate change, but it can have important but minor roles under certain conditions. It could be used judiciously: not in earthquake-tsunami zones like Japan but instead to diminish high industrial emissions in remote areas for cement and aluminum factories. Northern Alberta's massive reserves of tar sands oil could be transformed from having high emissions to low emissions if extraction, upgrading, and pipelines were powered by nuclear energy instead of natural gas, but this is not occurring because nuclear energy's higher cost would depress profits. Most important, a precondition for widespread deployment of nuclear energy is a binding, enforced international agreement to avoid its weaponization. Such agreements are extremely difficult to negotiate, implement, and enforce. Admittedly they are needed for almost every solution to the fossil-fuelled climate crisis, such as pricing carbon pollution and equitably distributing the cost among nations (Nordhaus 2013).

Decision-makers and populations will have to realize that the longer it takes to include the full cost of carbon pollution in the price of fossil fuels and to exercise restraint in fossil-fuelled social practices, the greater the accumulation of carbon in the atmosphere, the worse global warming becomes, and the more likely costly, risky remedies will become the only ones remaining. If one generation refuses visible wind farms, carbon taxes, etc., then the harms of fossil-fuelled climate change will force their grandchildren to accept dangerous alternatives. From a risk comparison perspective, Flannery (2015: 184) argues that 'we are on a worst-case emissions trajectory at present, so if nothing changes, in coming decades the idea of storing CO_2 as snow at the South Pole, or deep in the ocean's sediments, may not look so risky after all', or so costly compared to living with massive amounts of CO_2 in the atmosphere. Similarly, in this trial-and-error process, nuclear energy may eventually appear less risky than fossil fuels.

Direct Air Capture (DAC) of Carbon Producing Energy

Another version of DAC would be to capture carbon dioxide from the atmosphere, and then convert it into fuel. This could capture both ongoing emissions and legacy emissions (greenhouse gases emitted since industrialization). The carbon capture could be done using seaweed, then existing technologies such as methane digesters could use it to make fuels. 'This would produce enough biomethane to replace all of today's needs in fossil fuel energy, while removing 53 gigatonnes of CO_2 per year from the atmosphere, thus more than offsetting all human CO_2 emissions' (Flannery 2015: 41). Instead of getting the fuel from underground where it's stored safely, it would be captured by seaweed from the air where it otherwise causes a greenhouse effect, global warming, and climate change. No new technologies would be necessary. However, as stated previously for DAC storing the carbon underground, the limitation is one of scale. The seaweed needed to extract all this carbon from the atmosphere for energy would cover half the Indian Ocean.

The carbon dioxide could also be captured from the air technically and transformed into energy instead of extracting oil from underground (Keith et al. 2018). This technology would 'mix the CO_2 with hydrogen to form something very similar to the hydrocarbon fuel that got us into this trouble in the first place. This might seem perverse at first blush, but the CE [Carbon Engineering = company name] fuel would be endlessly renewable, perfectly clean-burning ... and carbon neutral – it wouldn't create a net addition of CO_2 to the atmosphere because CE's feedstock is atmospheric CO_2' (Littlemore 2019: 32). This fuel does not require a change of infrastructure: it's easy to refine, blendable to 100%, can be poured into a vehicle's gas tank, is portable, being a liquid is 30 times more efficient for carrying energy than batteries, and works in vehicles, ships and planes. The company promotes it as a world-changing fuel without needing the world to change. It does require large amounts of electricity, so this hydrocarbon fuel extracted from air presupposes an abundant supply of low-cost, zero-emissions electricity. Instead of pulling even more fossil fuels from the ground, this technological innovation of reusing the carbon dioxide emitted into the air since the industrial revolution, a very abundant supply indeed, seems too good to be true. And it might be. It remains to be seen whether it works on the large scale needed to replace all fossil fuels, whether that can be accomplished promptly for an urgent climate crisis, and whether it is cost-effective. Its inventor, David Keith of Harvard University, warns it is not reasonable to depend on this technology to remove CO_2 from the atmosphere while still emitting massive amounts, and that carbon pollution taxes, regulations, and other means of reducing emissions are needed.

Other Technological Fixes to Solve Global Warming

These would involve more extensive changes. Examples could be replacing the internal combustion engine with hydrogen-based fuel cells, changing fossil-fuel based electricity production to electricity based on nuclear fusion, or some other technical innovation that currently exists only in the imagination. Cement might be replaced by another chemical needing little fossil-fuel combustion and releasing little carbon dioxide.

These solutions presume the success of technological innovation on demand to solve specific problems promptly and bring benefits quickly, which is often not the case. Hence these imaginative technical fixes remain aspirations rather than factual remedies.

Note that all these alternatives would involve leaving fossil fuels in the ground and replacing them with other forms of energy. Therefore technological innovation producing alternative sources of energy, even if it is market generated, is as much a threat to the fossil-fuel industry as are governments imposing high prices on carbon pollution, or regulating the use of fossil fuels downward, or the development of a degrowth culture dramatically reducing consumption of fossil fuels. After all, the technological innovations of the internal combustion engine and of fossil fuels resulted in the demise of the horse as an energy source for transportation. The same could happen to fossil fuels.

Other Possibilities

All of the Above

These remedies are not necessarily opposed, although they may appear so in the minds of their proponents. They can be combined and all of them implemented to varying degrees, which may become necessary as fossil-fuelled climate change worsens. Moreover, technical solutions to the crisis and socioeconomic ones to be examined next are not mutually exclusive. 'Together with a steep reduction in emissions and seaweed farming, CO_2 storage in deep water marine sediments might just be planet saving' (Flannery 2015: 181).

A Presently Unimaginable Technological Solution

It can be hoped for, but it would be reckless to count on it in order to keep on combusting fossil fuels at a rate vastly exceeding carbon withdrawals.

None of the Above

It is easy to imagine solutions, but more difficult to implement them. Presently solutions of CCS, DAC, and aerosol sunscreens injected into the sky don't work, or are unproven at the necessary scale, or haven't found where to store the massive amounts of carbon, or are prohibitively expensive, or threaten to have dangerous side effects (Suzuki and Hanington 2017: Chapter 6). Hence their development as safe and cost-effective solutions can be hoped for, but it would be irresponsible to depend on them. Researchers agree that because fossil-fuelled climate change is so advanced and urgent, geoengineering measures should never be used to avoid reducing greenhouse-gas emissions (Keith and Parson 2017). The possibility that humanity will be unable to technologically reduce global warming or unwilling to reduce their fossil-fuelled practices cannot be ruled out. What can be done?

Fossil-Fuel Energy: Paying Its Full Cost Upon Use

Many of the technologies are available right now. Wind, solar, hydro, geothermal, and tidal energy are all in use already, but not on a scale that reduces global emissions. Why not? There is an immediate obstacle to implementing even the most feasible, technologically available, energy sources to replace fossil fuels. The obstacle consists of the cheap price of fossil fuels because the costs of global warming they cause are not included in their price when purchased, and are paid belatedly by victims, as explained in Chapter 4. To convince decision-makers and populations to take action against climate change, it is important to explain the belated costs of fossil fuels, such as the destruction caused by wildfires and the expense of fighting them, the damage caused by droughts, floods and hurricanes, and the high costs of delayed prevention and adaptation if atmospheric greenhouse gases are allowed to accumulate. But more is needed than explaining those deferred costs to nudge those who combust fossil fuels, big and small, to modify their carbon polluting practices. Economists have a practical solution based on the premise that price has a major influence on consumer choices and on technological innovation, an influence built into every market transaction. This is confirmed by experience. 'As energy prices have risen, measures such as LED lighting, increased use of insulation and improvements in appliance efficiency have become widespread. … This has had a huge impact on demand in the developed world, which was long profligate with electricity because its price was so low. … It's a great illustration of seemingly small changes cumulatively having a big impact' (Flannery 2015: 81).

Policies based on carbon taxes, cap-and-trade, feed-in tariffs, etc., seek to shift economic growth away from fossil fuels and towards low-carbon energy (solar, wind, hydro, tidal, geothermal). 'Pigou suggested that external effects be corrected by taxes or subsidies, which are now known in his honour as Pigouvian taxes or subsidies. If the private costs of action [or price of a commodity] are less than its total costs because of an external cost, then we should levy a charge that will raise the private cost [price] by the amount of the external cost so the company [and consumer] then faces the correct cost' (Heal 2017: 16). The environmental economist Geoffrey Heal (2017: 28) notes the success in dealing with sulphur dioxide emissions that produced acid rain, then concludes the following. 'And while there will be no silver bullet, no single solution for all these diverse environmental problems, there is a common principle uniting the various solutions: making people pay the full costs of their actions'. Economists argue that the least costly measures to curb fossil-fuelled climate change consist of governments placing a fee on carbon pollution through carbon taxes, favoured by Nordhaus (2013), or cap-and-trade, advocated by Wagner and Weitzman (2015), to include the cost of externalities in the price of fossil fuels. Carbon taxes give certainty about the price of fossil fuels but uncertainty about the amount of emissions, whereas cap-and-trade yields certainty about the upper limit of emissions but uncertainty about the price. Both result in pricing carbon pollution to compensate its belated costs, which would otherwise remain unpaid by fossil-fuel users, and are effective in restraining emissions. Low-carbon energy like wind and solar have much less global warming costs than fossil fuels and would become more competitive. It is important to incorporate into the price of any type of energy its long-term costs so as to avoid accumulating an environmental debt of harm to the environmental commons (Raymond 2016) of the atmosphere, oceans, and land. If the full cost of fossil fuels is paid upon purchase through a fee for carbon pollution, then the environmental debt will be reduced.

Jaccard (2005, 2009), Nordhaus (2013), Heal (2017) and Leach (2019) contend that the mitigation of fossil-fuelled climate change is, from their economics perspective, simple, relatively inexpensive, and would improve economic efficiency by paying the full cost of fossil fuels—including the cost of emissions—in their price. It would attack the global-warming problem at its source, whether through a carbon tax or cap-and-trade. 'Whenever a firm or person burns fossil fuels, and the CO_2 enters the atmosphere, the firm or person must pay an additional price that is proportional to the quantity of CO_2 emitted' (Nordhaus 2013: 222). This is based on the carbon-polluter pays principle: the more firms or individuals pollute; the more they would pay. The price of carbon pollution would function as a proxy for the externalized costs resulting from the combustion of fossil fuels. Everyone who uses fossil fuels pollutes to some extent, but built into this carbon pricing mechanism is the principle of proportionality (Freudenburg 2006): big polluters would pay more and very big polluter would pay much more. The price of energy would reflect the quantity of emissions. Since there are emissions at every stage of the fossil-fuel life cycle, as it is called, there would be a charge at each stage. It takes energy to extract coal from the ground, oil from a well, from deepwater, and from tar sands, and natural gas by fracturing shale. This initial energy is supplied by combusting fossil fuels, with corresponding emissions. Energy is needed to upgrade it and transport it to refineries, and to refine it into gasoline, diesel, jet fuel, bunker fuel, or transform it into petrochemicals like plastics. More energy is required to transport it to gasoline or diesel pumps, and to liquefy natural gas to ship it, etc. At every stage of doing this, fossil-fuel energy is combusted. Combustion at the end in vehicles, planes, ships, etc., is only part of the emissions. Vehicle drivers, plane and cruise passengers, etc., are only small carbon polluters compared to the principal polluters: companies (and their investors) that extract, upgrade, transport, refine, and transform fossil fuels. Although all would pay a price on carbon to compensate the otherwise unpaid costs of their emissions, payment would be proportional to their emissions; hence drivers and passengers would pay much less than the above companies. Note there is no double counting: these are different emissions at each stage in the fossil-fuel cycle from underground to sky.

Social justice researchers also focus on externalities and the polluter pays principle. 'Having polluters pay is efficient, since it puts formerly externalized costs back on them, which should inspire their cleanup. … That polluters should pay the costs of dealing with their pollution reflects the most fundamental principles of justice and responsibility' (Kahn and Roberts 2013: 139–140). Making all polluters pay through including costly externalities in the price of fossil fuels would make low-carbon energy more competitive, provide economic incentives for firms and states to avoid excessive extraction and combustion, spur technological innovation to eliminate carbon pollution, and provide funds for prevention, adaptation, and research into improving low-carbon energy, storage of electrical energy, etc. For individuals it would give an incentive to insulate homes, buy fuel-efficient vehicles, take public transport, enjoy non-fossil fuelled activities (row boating instead of motor-boating, skiing, and snowboarding instead of snowmobiling), etc. Placing a price on carbon pollution would drive home that carbon pollution is not cost-free and has a cost that someone has to pay, either polluters or victims.

Nordhaus (2013: 19 concludes that 'governments must step in and regulate or tax activities with significant harmful externalities'. The proposal to include the cost of carbon pollution in the price of fossil fuels makes sense since consumption, production, and innovation are price sensitive. When gasoline prices decrease, purchases of gas-guzzling vehicles usually increase and fuel-efficient vehicles decrease. Diamond (2019: 405) notes that the American energy consumption per person is almost double that of Europe despite Europe having a higher average standard of living than in the USA. He explains the difference by higher European taxes on vehicles and gasoline, which can then be used to pay for electrified public transit. This proposal gives incentives to polluters to reduce pollution, either by changing polluting practices or changing technologies to reduce pollution. It has the potential to diminish the monopolization of the atmosphere as a carbon pollution sink by present polluters to the exclusion of latecomers, particularly future generations. It is more equitable to price carbon pollution so that the polluter pays it rather than externalizing its costs to future generations as innocent victims.

Placing a price on carbon pollution recognizes that every company and person using fossil fuels contributes to carbon pollution and global warming. The necessary distinction is between amounts of pollution and hence between minuscule polluters, average polluters, and enormous polluters. Huge polluters would pay a huge price for their carbon emissions, whereas tiny polluters would pay a tiny price. Conservative political parties, which originally denied fossil-fuelled climate change but now confront the growing evidence, have advanced the nice-sounding slogan: 'tax polluters not commuters'. They refuse to recognize that commuters who combust gasoline or diesel in their vehicles are also polluters, and because of their enormous numbers and aggregation of their emissions, are major contributors to the greenhouse effect. The important factor is proportionality: big carbon polluters should pay big pollution costs, and small polluters small pollution costs. Economists argue that every company or individual polluting the atmosphere with carbon, big and small, should pay for it in proportion to their carbon emissions. Climate change remediation is a zero-sum contest. If any group is allowed to be a free-rider, then other riders have to do more than their share of emissions reduction. If the huge number of small polluters, such as commuters, are allowed to pollute freely, this increases demand for fossil fuels, and to meet an emissions reduction target, big polluters would have to pick up the slack and pay even more than their pollution would warrant. In some cases, this may be mathematically impossible because of the huge number of small polluters. So economists call for a broad-based price on carbon pollution, proportionately big and small.

When carbon pollution becomes costly for the polluter instead of being free, it gives strong incentives to decrease pollution. Nordhaus (2013: 287) concludes that a 'high price of carbon is necessary to induce profit-oriented business to undertake research, development, and investments in new low-carbon technologies. ... if the carbon price is zero, then projects to develop promising low-carbon technologies like CCS will not get to the boardroom of a profit-oriented company'. The Danish coal company DONG was an enormous industrial polluter, but with the financial incentives of carbon taxes as well as moral incentives to do what is best for future generations, it transformed itself into a prosperous offshore wind energy company and a small carbon polluter. Similarly, carbon taxes give incentives to individuals to take public transportation, switch from gas-guzzling vehicles to light fuel-efficient ones and eventually to electrical vehicles. Carbon taxes don't work optimally when exceptions are made. If they are reduced for cement factories because of competition with companies where pollution is free, for example in Trump's Republican America, then market competition drives everyone to pollute. This illustrates the need for an international agreement, such as Paris 2.0 or Kigali 2.0. All companies and societies need to pay the full cost of fossil fuels (i) because those that do not reduce their fossil-fuel pollution affect all the others through the medium of the environment and (ii) because those that do not pay the full cost of their fossil fuels have a cost advantage over those that do, which incites high-carbon emissions by everyone and a race to climate catastrophe.

Many decision-makers and consumers in society are willing to 'talk the talk' of environmentalism, in particular talk of doing something to solve the problem of climate change. They are, however, unwilling to 'walk the walk' by modifying their fossil-fuelled social practices (Kennedy 2015; Kennedy et al. 2009). Including a price on carbon pollution would build mitigation into the price of fossil-fuelled social practices, and test whether the talk is sincere or just greenwashing.

All environmental economists advocate a price on carbon pollution to diminish the use of fossil fuels, to help develop emissions-free, renewable energy they compete with, and to incite technological innovations. This could be carbon taxes or cap-and-trade, where the government sets a ceiling on carbon emissions and then companies trade within that cap to reduce cost and achieve efficiencies. Whichever form the price on carbon pollution takes is less important than having a price and setting it high enough to reduce emissions. These are market based solutions founded on the principle that polluters should pay in proportion to their pollution. Harvey and Orbis (2018: 263) argue that the carbon price should steadily rise, that it should be part of a long-term policy of at least a decade to influence investors, that it should be placed on all the market to prevent leakage to freely polluting sectors, and that it should affect especially pinch points, namely places where energy flows are concentrated. 'Economists and public finance experts universally agree there are efficiencies to raising funds through charges on pollution or other socially harmful activities' (Harvey and Orbis 2018: 277). They point out that Nordic countries have used the revenue to lower taxes on labour, and far from the price on carbon being a job killer, it is contributing to those countries being particularly prosperous. Suzuki and Hanington (2017: 230–234) also show that placing a price on carbon pollution has been beneficial to the economy. Harvey and Orbis (2018) stated that the California–Quebec–Ontario agreement is the best designed cap-and-trade system and the proposed Canadian carbon tax is a game changer. Unfortunately a new conservative government in Ontario withdrew from that agreement and Canadian conservatives oppose carbon taxes. This raises the issue of gaining acceptance by the population. Emphasizing co-benefits could help, for example, a fee for carbon pollution will decrease the combustion of coal and natural gas which will decrease air pollution and smog harmful to health.

Placing a price on carbon pollution leads to the question of where the revenue raised would go. The least effective place would be general government revenues, which would cause a backlash as just another tax grab. Some of it would have to be targeted to help the poor and particular groups that would have difficulty paying for the higher prices of fossil fuels and no way to avoid them. To most effectively combat fossil-fuelled climate change, ideally most of taxes on carbon pollution should go to fund prevention, adaptation, and especially basic research to innovate better low-carbon technologies, energy storage, and processes that could capture and sequester carbon (Pielke 2010). It could be used to finance low-carbon alternatives, such as electric cars, wind and solar energy generation, etc. This would give the most climate-change prevention bang for each buck of carbon tax. Unfortunately, this does not take into account the difficulty of selling a carbon tax to the population and implementing it. A price on carbon pollution could be sold to a reluctant population through immediate benefits of reducing taxes elsewhere, that is, by being revenue-neutral, or given out as a tax credit. Likely the best way to gain acceptance would be to give the revenue to citizens as a visible cheque.

These practices to use the raised revenue compete with one another: to deal with climate change most effectively, quickly, and at least cost, directly funding climate action and research would be best; to gain acceptance by the population, a visible cheque would be best. Which one is chosen will likely depend on whether decision-makers are (i) desperate to gain acceptance because of powerful resistance to a price on carbon pollution, or (ii) desperate to combat the ravages of fossil-fuelled climate change. 'One of the greatest challenges facing this approach [of a carbon tax] would be the ability of nations individually and collectively, to direct funds raised by a carbon tax to energy innovation' (Pielke 2010: 230). Basic research to lay the foundation for developments in storage of electrical energy, miniaturization of batteries, more efficient deployment of solar, wind, geothermal, and tidal energy must be paid for. Although there is resistance by carbon polluters to paying carbon taxes even for such research, there would likely be much more resistance to increasing general taxation to pay for it.

Undermining Pielke's Iron Law of Climate Change

Imposing a price on carbon pollution, whether in the form of a carbon tax or a cap and then trading pollution within the cap, is a proxy to cover the otherwise unpaid costs of fossil fuels that will harm the habitat of future generations and bystanders. It is a logical, cost-effective solution that will prompt changes in technologies and social practices. But it won't work if it is not accepted by populations and decision-makers and never implemented. It will increase the price of gasoline, plane and cruise tickets, goods from afar, and decrease profits. So resistance is likely to be fierce. Carbon taxes and/or cap-and-trade have been rejected by the Trump Republican administration in the USA, yellow jackets in France, conservatives in Canada and Australia, and many more countries. Many people want to solve global warming, but want others to pay to do it and do not want to change their fossil-fuelled practices. They seem to believe, to rework John F. Kennedy's famous phrase, what's mine is mine, and what's yours is to pay for solutions or to locate windmills. Pielke (2010: 46–50) generalizes this rejection into an 'iron law of climate change', namely the premise that populations will never accept to pay a price on carbon pollution anywhere near what is needed to mitigate global warming.

Formulated as an iron law, this risks becoming a self-fulfilling prophecy. If believed, then it is pointless trying to get populations to accept carbon taxes, the costs of cap and trade, or even the escalating carbon taxes suggested by Pielke to finance research to develop carbon-free technologies, which could be expensive. However, one important reason why populations resist the costs of mitigation is because they do not understand how their ordinary fossil-fuelled social practices add up to cause global warming, as illustrated by Nordhaus' example given at the beginning of this book. Carbon dioxide cannot, after all, be perceived by the senses, and so too its global accumulation in the atmosphere, which can only be assessed by science. If there is such an iron law, why bother trying to dissipate this ignorance by translating the scientific language of gigatonnes into units the population can understand? Just rely on wishing for last-minute, end-of-pipe technological solutions.

Exceedingly infrequent are media presentations about how a few pounds of carbon combusted during the vehicle drive from suburban home to work, during rides in motorboats, four wheelers, and snowmobiles interact with oxygen in the air to produce a far greater number of pounds of carbon dioxide emitted into the atmosphere thereby contributing to the greenhouse effect. Rare indeed have been the attempts to inform populations about how jet fuel, bunker fuel, and other fossil fuels combusted during flights and cruises, during transportation of goods in the global market, and powering social media servers, are chemically transformed into huge amounts of carbon dioxide polluting the atmosphere. One action in a huge planet would have a negligible effect, so each person assumes he or she is not responsible for climate change. The media rarely elucidate how billions of these fossil-fuelled practices every day worldwide cumulate into a huge impact. It is little wonder these practices are popular when the population remains uninformed about the damage they cause to the atmosphere, oceans, and the habitat future generations of humans will need. In addition to the aggregation of harm from individual fossil-fuelled practices, there are corporate and military fossil-fuel practices and electoral practices in democracies of voting for governments that refuse to implement carbon taxes.

If little effort is made to educate the population into realizing how much carbon dioxide results from their fossil-fuelled activities, then there will indeed be resistance to paying the full costs of fossil fuels. The prophecy of the iron law will be fulfilled. That will result, not from the impossibility of getting an informed population to accept prices for carbon pollution adequate to mitigate climate change, but rather from lack of effort to explain to the population in understandable terms how their fossil-fuelled practices are causing long-term climate problems. In principle, this generation of the affluent shouldn't object to paying the full cost of fossil fuels through the proxy of a tax on carbon pollution if they understand they have hitherto enjoyed a discount that will have to be paid later in terms of the costs of climate change. But in practice, the population will not like to hear that the relatively inexpensive fossil-fuelled activities, conveniences, and consumer goods they enjoy are causing those problems and accumulating fossil-fuelled debt. They will not like to discover that therefore future generations including their grandchildren, will have to pay additional costs of floods, wildfires, and belatedly high prices for mitigation when the atmospheric carbon sink is full. Nevertheless, shying away from speaking truth to power and to the population in the form of the best available scientific and economic understanding will itself transform a contingent unawareness of the consequences of their fossil-fuelled practices into the fulfillment of an iron law, which maintains dangerous consumption practices, voting practices, etc. The next chapter assesses solutions for this and other sociocultural remedies.

References

Diamond, Jared. 2019. *Upheaval: Turning Points for Nations in Crisis*. New York: Little, Brown.

Dyer, Gwynne. 2008. *Climate Wars: The Fight for Survival as the World Overheats*. Toronto: Random House.

Flannery, Tim. 2015. *Atmosphere of Hope: Searching for Solutions to the Climate Crisis*. New York: Atlantic Monthly Press.

Freudenburg, W. 2006. Environmental Degradation, Disproportionality, and the Double Diversion. *Rural Sociology* 71 (1): 3–32.

Gorman, John. 2019. Nuclear Energy Is an Important Part of Solving the Climate Crisis. *Globe and Mail*, 24 October: B4.

Harvey, Hal, and Robbie Orbis. 2018. *Designing Climate Solutions: A Policy Guide for Low-Carbon Energy*. Washington, DC: Island Press.

Hasegawa, Koichi. 2012. Facing Nuclear Risks: Lessons from the Fukushima Nuclear Disaster. *International Journal of Japanese Sociology* 21 (1): 84–91. https://doi.org/10.1111/j.1475-6781.2012.01164.x.

Hasegawa, Koichi. 2015. *Beyond Fukushima*. Melbourne: Trans Pacific Press.

Hasegawa, Koichi. 2016. *Sociology of Climate Change Policies* (co-edited, in Japanese).

Hawken, Paul. 2017. *Drawdown: The Most Comprehensive Plan Ever Proposed to Reverse Global Warming*. New York: Penguin.

Heal, Geoffrey. 2017. *Endangered Economies: How the Neglect of Nature Threatens Our Prosperity*. New York: Columbia University Press.

Healing, Dan. 2019. Renewable-Power Developers Find Mixing Energy Sources a Winning Strategy. *Globe and Mail*, 28 October: A6.

IEA International Energy Agency. 2018. Global Energy Demand Grew by 2.1% in 2017, and Carbon Emissions Rose for the First Time Since 2014. *International Energy Agency*, 22 March. IEA: Paris. https://www.iea.org/newsroom/news/2018/march/global-energy-demand-grew-by-21-in-2017-and-carbon-emissions-rose-for-the-firs.html. Accessed 23 March 2018.

Jaccard, M. 2005. *Sustainable Fossil Fuels*. Cambridge: Cambridge University Press.

Jaccard, M. 2009. Peak Oil and Market Feedbacks. In *Carbon Shift*, ed. T. Homer Dixon, 96–131. Toronto: Random House.

Kahn, Mizan R., and J. Timmons Roberts. 2013. Towards a Binding Adaptation Regime: Three Levers and Two Instruments. In *Successful Adaptation to*

Climate Change: Linking Science and Policy in a Rapidly Changing World, ed. Susanne C. Moser and Maxwell T. Boykoff, Chapter 8. London: Routledge.

Keith, David, and Edward Parson. 2017. Science Fiction—Or Saviour? *The Globe and Mail*, 9 December: O5.

Keith, David W., Geoffrey Holmes, David St. Angelo, and Kenton Heidel. 2018. A Process for Capturing CO_2 from the Atmosphere. *Joule* 2 (June 7): 1–22. https://doi.org/10.1016/j.joule.2018.05.006.

Kennedy, Christopher. 2015. Key Threshold for Electricity Emissions. *Nature Climate Change* 5 (3): 179–181.

Kennedy, Christopher, et al. 2009. Greenhouse Gas Emissions from Global Cities. *Environmental Science & Technology* 43 (19): 7297–7302.

Leach, Andrew. 2019. How the Commuter vs. Polluter Narrative Could Backfire on Alberta. *Policy Options*, 15 July. https://policyoptions.irpp.org/magaxines/july-2019/how-the-commuter-vs-polluter-narrative-could-backfire-on-alberta/. Accessed 13 November 2019.

Littlemore, Richard. 2019. Catching Air. *Report on Business*, October: 28–34.

McCarthy, Shawn. 2018a. Startup Looks to Slash Carbon-Capture Costs. *Globe and Mail*, 8 June: B6.

McCarthy, Shawn. 2018b. Quebec Startup's Biofuel Powers Transoceanic Flight. *The Globe and Mail*, 29 January: B1, B7.

Nordhaus, William. 2013. *The Climate Casino: Risk, Uncertainty, and Economics for a Warming World*. New Haven: Yale University Press.

Pielke, Roger, Jr. 2010. *The Climate Fix*. New York: Basic Books.

Raymond, Leigh. 2016. *Reclaiming the Atmospheric Commons*. Cambridge, MA: MIT Press.

Reguly, Eric. 2019. Imagine If Suncor Energy Used Canadian Uranium to Clean Up Its Act. *Globe and Mail*, 19 October: B4.

Rhodes, Richard. 2018. *Energy: A Human History*. New York: Simon & Schuster.

Suzuki, David, and Ian Hanington. 2017. *Just Cool It: The Climate Crisis and What We Can Do*. Vancouver and Berkeley: Greystone Books.

Thornes, John. 2010. Atmospheric Services. *Issues in Environmental Science and Technology* 30: 70–104.

Wagner, Gernot, and Martin Weitzman. 2015. *Climate Shock: The Economic Consequences of a Hotter Planet*. Princeton: Princeton University Press.

World Nuclear Association. 2019. Nuclear Energy and Climate Change. *World Nuclear Association*. London: World Nuclear Association.

Zebrowski, Ernest, Jr. 1997. *Perils of a Restless Planet: Scientific Perspectives on Natural Disasters*. Cambridge: Cambridge University Press.

11

Foresight or Discounting Danger?

Observing the drought and wildfires in Australia and California, floods in Houston, sea level rise in Venice, drought in Africa, and intensifying cyclones in many parts of the world, the temptation is strong to despair. It is to think: we have not seen anything yet if the world insists on combusting fossil fuels and emitting greenhouse gases at the present pace. Victims will eventually pay the full cost of harm if polluters refuse to pay it as they combust fossil fuels. However, it does not have to be that way. Payment upon pollution is needed to discourage pollution, to have the finances to innovate carbon-free alternatives, and repay the environmental debt being left to future generations. 'At present we are stealing the future, selling it in the present, and calling it gross domestic product. … We can just as easily have an economy that is based on healing the future instead of stealing it. We can either create assets for the future or take the assets of the future. One is called restoration and the other exploitation. And whenever we exploit the earth, we exploit people and cause untold suffering' (Hawken 2010: Introduction).

Hawken's quotation contains much wisdom, foresight, and hope, but there are weaknesses. First, the 'we' ignores the difference between

R. Murphy, *The Fossil-Fuelled Climate Crisis*,
https://doi.org/10.1007/978-3-030-53325-0_11

principal polluters and minor ones, and between (i) principal decision-makers promoting carbon-polluting practices and discounting future harm[1] compared with (ii) those influenced by fossil-fuel promotion. It is easy to miss these distinctions since the social practices and decisions of everyone contribute some carbon pollution because it is currently impossible to avoid fossil fuels entirely. Second, the transition from a carbon-polluting economy to a restorative economy will not be easy nor cheap. The threat is global, creeping yet urgent, and largely imperceptible by the senses. No technological miracles are in sight, despite faith in them to avoid having to change fossil-fuelled practices or paying their full cost. The longer carbon emissions exceed withdrawals, the more danger intensifies, and the harder and more costly restoration will be.

Fossil-fuelled climate change is a multifaceted global problem: 'The problem of climate change is so massive that it requires a whole range of solutions' (Suzuki and Hanington 2017: 236–237). The numerous remedies correspond to its various aspects. Previous chapters in this section assessed sociocultural and technological remedies that have been proposed. This final chapter assesses further socioeconomic solutions. The goal is not to choose one solution over others, but instead to analyse the possibilities and weaknesses of each in order to prioritize potentials and diminish flaws. Although there are no magic bullet solutions for fossil-fuelled climate change, there have been many valuable suggestions. The evidence informs this assessment, but the analyses and conclusions are my own.

Suggestions to Enhance Foresight

The chapter begins with general suggestions flowing directly from the analysis in this book.

Understand the Science

Speth (2012: 16) argues that 'any hope of changing the system to better serve people and the planet rests on an accurate assessment of the full scope of our problems. … We will do what is needed only when we fully appreciate the situation we face'. Flannery (2015: ix) begins his book entitled *Atmosphere of Hope: Searching for Solutions to the Climate Crisis* by arguing that 'if we are to have real hope, we must first accept reality'. The reality is that, unless the present treadmill of fossil-fuelled social practices is changed to bring emissions in line with carbon withdrawals, societies will flounder on false hopes and be unlikely to prevent a climate catastrophe. Carbon dioxide emissions causing a cumulative, long-lasting greenhouse effect in the atmosphere are imperceptible by the senses. This results in blind spots in perception and comprehension, what Tong (2019) calls 'the reality bubble'. Nevertheless, hidden truths come out of hiding because of scientific work. If the scientific understanding is accepted and acted upon, otherwise imperceptible environmental problems can be solved. The only thing more imperceptible than carbon dioxide and atmospheric carbon is the ozone layer and its depletion, but they were revealed by science resulting in international agreements that largely solved the problem. The same can be true for fossil-fuelled climate change.

Communicate the Science in Ways Non-scientists Can Understand

There exists a paradox concerning the fossil-fuelled climate crisis. Scientific knowledge about it abounds in scientific journals and the media, but ignorance concerning it abounds too. There is a lack of practical understanding by the population of how fossil-fuelled social practices contribute to the climate crisis, and this affects not only their material practices but also their voting practices. Why? Scientists communicate in gigatonnes of carbon dioxide equivalents, 2 °C global temperature increase, global consequences for future generations, etc. Although correct, this appears abstract to the non-scientific layperson. It is not propitious for inciting willingness to bear the costs of mitigation and change long-standing fossil-fuelled social practices. The 2 °C global average temperature increase appears trivial. Nevertheless, it is enough to melt glaciers, permafrost, the ice cover of the Arctic Ocean, and intensifies drought, wildfires, floods, hurricanes and other extreme weather. The scientific expression that these calamities 'are consistent with' fossil-fuelled global warming also appears abstract and needs to be fleshed out.

Furthermore, the global scientific findings need to be disaggregated into an enormous number of mundane social practices on the local level. The Nordhaus (2013: 19) example given at the beginning of this book bears repeating. If he drives 100 miles in his car, 'I consume 5 gallons of gasoline. This will produce about 100 pounds of CO_2, which will come out of the tailpipe and go into the atmosphere. I can't see it or hear it or smell it, and I generally do not even think about it. If I am like most people, I will probably assume that my trip will have no effect on the world's climate, and so I will ignore the consequences'. Ignorance of emissions imperceptible by the senses and their long-term harm resulting from driving, flying, shipping, cruising, etc., foster these carbon polluting practices and support for fossil-fuel companies and political parties that favour fossil fuels. This contrasts with the visibility of wind farms, which incites fear of declining property values and opposition.

It is essential to alert people to the danger of colourless, odourless carbon dioxide. That is done technically for colourless, odourless carbon monoxide by adding mercaptan to natural gas, which produces a potent rotten-egg odour if there is a leak. Something similar could be added to gasoline for vehicles so that the carbon dioxide emitted would stink to alert drivers to the greenhouse gases they are emitting: heavy gas-guzzling vehicles would stink more than light, fuel-efficient ones. A jet-fuel additive could produce a huge black cloud when combusted by a plane. An additive to bunker fuel or other fuels used to propel ships could produce a blood-red stain on the water in its wake. Additives to fossil fuels could make the greenhouse gases that factories, companies, and buildings emit visible to the eye and perceptible to the nose in proportion to their emissions. Cement factories and fossil-fuelled social media servers would thereby result in ugly big stinks.

However, I have no illusions technological innovations will be steered in this direction. The example demonstrates that technological innovation is not neutral and is steered by cultural and economic decisions. The illustrations are given to drive home that an informed citizenry needs a practical understanding of the consequences of fossil-fuelled practices in order to improve them, including voting practices, investing practices, etc. It is crucial to illustrate in everyday terms the otherwise imperceptible greenhouse gases and their consequences revealed by science. Carbon dioxide emissions from driving, flying, taking a cruise on a huge boat, cremation, producing cement for buildings, and electricity for air conditioning and social media servers need to be understood in pounds or kilograms. This would bring home to people the consequences of social practices much more than speaking of a few gigatonnes emitted by the totality of these practices. This is especially true for carbon pollution by big oligopolistic polluters. Translating the science of global warming into everyday units and illustrations would make them more meaningful to laypeople, whose support is needed if the climate crisis is to be mitigated. Disseminating examples like these would give the population a clear, practical understanding of the problem. The goal is to promote restraint, support for a price on carbon pollution, stimulate low-carbon technological innovations, and voting for political parties that will enact legislation to implement mitigation and adaptation. A practical understanding would also allow people to remedy their own contributions to fossil-fuelled climate change, and avoid feeling guilty later when they, their children and grandchildren, belatedly discover the consequences of their social practices.

Similarly, understandable analogies are needed to enhance comprehension of the dynamics involved. A good example is the atmosphere visualized as a bathtub shown in Fig. 11.1. It is taken from the US Environmental Protection Agency (EPA 2017), but similar sketches can be found in Pielke (2010: 9), Harvey and Orbis (2018: 5), and National Geographic. The US Environmental Protection Agency explains it as follows. 'If the amount of water flowing into a bathtub is greater than the amount of water leaving through the drain, the water level will rise. CO_2 emissions are like the flow of water into the world's carbon bathtub.

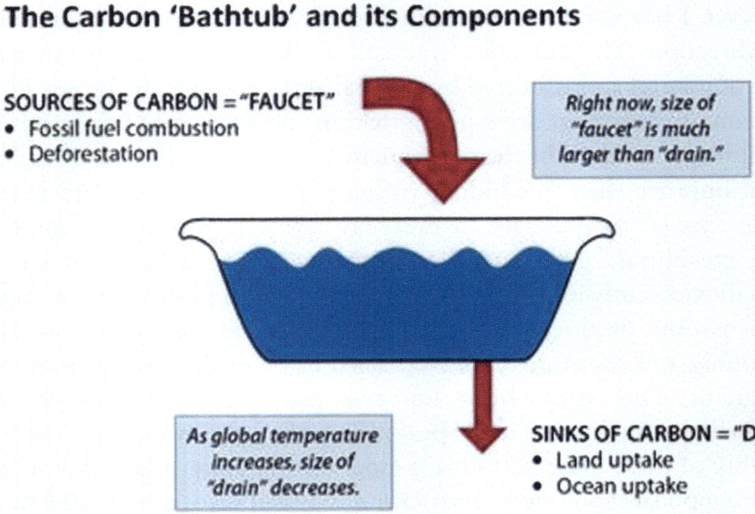

Fig. 11.1 The bathtub analogy for emissions and withdrawals of carbon dioxide from the atmosphere (EPA 2017)

"Sources" of CO_2 emissions such as fossil-fuel burning, cement manufacture, and land use are like the bathtub's faucet. "Sinks" of CO_2 in the ocean and on land (such as plants) that take up CO_2 are like the drain. Today, human activities have turned up the flow from the CO_2 "faucet", which is much larger than the "drain" can cope with, and the level of CO_2 in the atmosphere (like the level of water in a bathtub) is rising' (EPA 2017).

Clear analogies like these, showing the importance of the net change in atmospheric carbon, have been used in specialized books but are not widely disseminated. They are needed to help people understand that global warming worsens if emissions exceed carbon withdrawals and to counter misleading claims of improvement, such as fewer emissions per unit of GDP or per barrel of oil, where the increase of GDP and the number of barrels worsen the fossil-fuelled climate crisis.

Knowledge itself does not guarantee that the climate problem will be solved, since it confronts powerful fossil-fuel oligopolies, long-standing social practices and habits, cultural predispositions, etc. Nevertheless, a practical understanding by the population and decision-makers is a necessary condition for mitigating it, although not a sufficient condition. Awareness of causal social practices is one significant part of transcending present predispositions of consumers and voters, namely solve the climate problem without changing fossil-fuelled practices or making them more expensive.

Make the Fossil-Fuelled Threat Concrete

Framing the fossil-fuelled climate crisis as a threat to the planet is unconvincing and misleading. Everyone knows the planet will not explode because of the combustion of fossil fuels. That framing is a sloppy way of speaking that fails to specify what is at stake. Even if combustion of fossil fuels unleashed forces of nature so destructive they transform Earth into a planet as barren as Mars, our planet would still exist. However, its services to humanity would be destroyed by human socioeconomic practices. This loss of autonomous nature's beneficial services has to be emphasized. What may not survive is the beneficial habitat humans have enjoyed because of nature's services (IPBES 2019). Framing the crisis as degrading the beneficial habitat for ourselves, our grandchildren, and by extension for humanity and as making nature more dangerous is more accurate and more likely to prompt corrective action.

Similarly, 'future generations' seems like an abstraction. People are concerned more concretely about themselves and their loved ones. They would likely be more motivated to change their fossil-fuelled social practices if the hazard were expressed in terms of danger for their significant others, as sociologists call them. It is important to emphasize how climate change will affect grandchildren born today who will have to live in a degraded habitat when they become older, will have to pay for floods and wildfires, and for solutions made more difficult by filling the atmospheric carbon dump. Paying small carbon taxes now could help our grandchildren avoid much greater ones later. If a country, company, or person does not do its share to reduce emissions, then grandchildren will be forced to pay the unpaid externalized costs. Explanations of how the monopolization of the atmospheric carbon sink by the fossil-fuelled practices of the present generation are leading to the exclusion of grandchildren from living standards their grandparents enjoy should not be avoided. People may be annoyed to learn their social practices are causing the problem, but awareness is a necessary condition for mitigating it.

Although the destructiveness of fossil-fuelled climate change will be greatest in the future and affect the most vulnerable who contributed least to it, the biggest carbon polluters are also being affected now. Australia is a wealthy country and one of the principal carbon polluters in absolute and per capita terms because of reliance on its coal industry. In 2019 it suffered severe drought and in early 2020 catastrophic wildfires, as did California earlier. The wealthy Canadian province of Alberta is one of the world's largest carbon emitters because its economy is based on extracting and upgrading oil from its tar sands. Drought in the Boreal forest near the tar sands operations resulted in a catastrophic wildfire in 2016 that destroyed part of the city of Fort McMurray. Drought-fostered wildfires in the area also devastated the communities of La Ronge in 2015 and Slave Lake in 2011. Any one disaster has complex causes, but shying away from explaining connections between fossil-fuel extraction and combustion, drought, and wildfires sustains support for fossil fuels.

There Could Be an App for Carbon Pollution

Technological innovation is amazing. Smartphones know where and how far we are driving, flying and cruising, and how many steps we take when walking. Ride-hailing apps like Uber have options to withdraw payment directly from one's account. Dealing with a problem requires measuring it and making its magnitude visible, particularly to those causing the problem. Until now, however, innovation has failed to make visible the harmful consequences of combusting fossil fuels. Most people do not understand how much their fossil-fuelled social practices contribute to the crisis. Margaret Munro (2019: O3) suggests there should be what she calls 'a planetary conscience in every phone'. It would note whether a fuel-efficient lightweight vehicle or a gas-guzzling SUV or pickup truck is being driven, or the type of plane or cruise ship, the number of passengers, the distance travelled. Then it would calculate the number of pounds (or kilograms) of carbon dioxide equivalent emissions and each passenger's share. Her flight from Vancouver to Montreal emitted an average of 1445 kilograms of CO_2 per passenger according to a carbon tracker. Making this visible with a little green icon that pops up and goes *bing* would inform people of the consequences of their flight, hopefully encouraging them to exercise fossil-fuel restraint when deciding whether to fly and purchase fuel-efficient vehicles. But it could go further. Like Uber's payment system, the icon could deduct payment to offset the emissions produced and place it directly into a certified carbon offset account. Payment could be compulsory like the charge for Uber's ride-hailing, with everyone paying the full cost of their fossil-fuelled travel. This would implement the polluter pays principle and undermine the erroneous belief that carbon-dioxide pollution is free because it can't be seen or smelled.

Munro's idea could be extended to pollution meters in vehicles, indicating pounds of carbon dioxide emitted. Technologically this is not beyond the capacity of manufacturers, but innovation needs to be steered in this pollution-indicating direction. Speedometers incite drivers to restrain speeding when above the speed limit. Similarly, pollution meters could prompt people to buy light fuel-efficient vehicles, use public transit, fly only when necessary, etc. Visibility would extend to whether people are serious about mitigating climate change by exercising fossil-fuel restraint or whether they practice what Thunberg calls 'empty words'. Smartphones themselves could be made smart enough to indicate users' share of emissions generated by the massive fossil-fuelled servers used to make them function. There would be a debate about how to share responsibility for carbon pollution: drivers combusting fossil fuels, passengers, companies enticing purchases of vehicles, flights, cruises, etc. That is where debate belongs, rather than ignoring the consequences of fossil-fuelled practices.

Promote Inclusion and Equality of Opportunity

Using the atmosphere as a carbon pollution dump causes a greenhouse effect, costly wildfires, floods, and extreme weather. A significant threat is that many of nature's services, such as glacier runoff, carbon storage in permafrost, reflection by white cover of the Arctic Ocean, will be closed off to latecomers, whether they be future generations or the poor whose countries develop later. They risk being excluded from opportunities that the affluent in this and previous generations have enjoyed. The threat of fossil-fuelled intergenerational downward mobility is emerging. Prevention of fossil-fuelled climate change requires a more cosmopolitan orientation of keeping opportunities open to others, particularly future generations, and including them in benefits from the advantageous natural environment earlier generations have enjoyed. Societies, such as Nordic social democratic ones, which have had the most success in spreading opportunities to all members of society, have also been the best environmental performers in fostering inclusion and opportunities across space and between generations. This needs to be enhanced and inspire other societies.

Focus on Social Practices Not Just Discourse

Carbon polluters and supporting governments typically do not admit they are at fault. They advance the discourse that they are decreasing the carbon intensity of production (less emissions per unit of GDP). Even if true, that is a false positive indicator of improvement. In detective work, the advice is: follow the money. Similarly, for detecting the source of global warming, trace carbon-emitting social practices up the chain of command and determine whether emissions exceed withdrawals. This is how to transcend greenwashing.

Timeliness

The timeliness of taking action is all-important for a cumulative threat like fossil-fuelled climate change. It is easier, less expensive, and more effective to deal with it early, much like a cancerous tumour, the COVID-19 virus, and the fire at Notre Dame Cathedral. Nature's threatening dynamics have a runaway capacity if they are not nipped in the bud. Fossil-fuel infrastructures like coal-burning electricity generating plants, extraction mines, pipelines, etc., constitute major investments, have long lifetimes, and are costly to shutter, hence they lock in emissions for decades. Nordhaus calculated that the cost of delaying measures to mitigate fossil-fuelled climate change for fifty years would be around \$6.5 trillion. Research by 'virtually all other economic modellers, shows that acting now rather than waiting 50 years has substantial net benefits' (Nordhaus 2013: 300). The key word is 'net': reducing costs over the long term requires a willingness to pay the upfront full costs of fossil fuels and of replacing them.

Combat the Fallacy that Placing a Price on Carbon Pollution Is a Job Killer

Many decision-makers frame the global warming issue as 'economy versus the environment', with claims they are trying to reconcile the two. These claims are sometimes well-intentioned but uninformed whereas others involve deception. Economic growth that causes climate change will harm the future economy (Wagner and Weitzman 2015) and will have long-term costs far exceeding near-term benefits of fossil fuels. The environmental economist Heal (2017) analyses *Endangered Economies: How the Neglect of Nature Threatens Our Prosperity*. He recognizes there were costs to preventing acid rain, but documented that the economic benefits were ten times greater. Similarly, 'switching from fossil to carbon-free power need not raise the cost of power, as often asserted by the fossil fuel industry' (Heal 2017: 189). Opponents of clean energy focus on its costs to the exclusion of benefits. Upfront costs of mitigation bring long-term economic gain (Stern 2009). Hawken (2017: 220) agrees: 'We also look at the *net operating cost or savings* from climate solutions compared to continuing business as usual. The *net operating savings* is \$74 trillion over thirty years'. The Canadian province of British Columbia implemented a carbon tax, which decreased emissions, and it led all provinces in economic growth, partly because the money raised reduced other taxes. Money collected through pricing carbon pollution can stimulate the economy. Sweden introduced the world's highest carbon tax in 1991. Since 1995, its total emissions decreased by 25%, its intensity emissions per unit of GDP decreased by 65%, and its economic growth was 12% higher than the European average (Ragan 2019: A11). Australia had a carbon tax from 2012 to 2014, which reduced its emissions, and its GDP grew by 3% per year (Flannery 2015: 70). Germany innovated feed-in tariffs to develop wind and solar energy and reduce dependence on coal and nuclear energy, which has not prevented it from being an economic powerhouse. Economists agree that taxing bads like pollution is better for the economy and job creation than taxing goods like payrolls, income, profits, etc. (Nordhaus 2013). It promotes innovation, and shifts well-paying jobs to the renewable green energy sector.

It is necessary to underscore the climate-change costs of combusting fossil fuels (wildfires, floods, extreme weather, ocean level rise, etc.). Immediate tiny savings by rejecting carbon taxes will be made at the expense of much greater perpetual long-term costs. Costs are merely displaced to paying for disasters. This is like refusing to pay for disaster mitigation, preparedness, robustness, etc., and then becoming compelled to pay far greater costs later when disaster strikes. Carbon taxes are the most effective, transparent, and least costly way to decrease carbon pollution. The environmental economist Heal (2017: 62) concludes that it 'is ironic and perhaps a comment on our political maturity that this is probably the least popular of all approaches'.

Carbon taxes will make fossil fuels less competitive against low-carbon energy, and will result in fewer jobs in the carbon-polluting sector. Jobs will be transferred to the low-carbon energy sector. The process involves job shifting not job loss. Heal (2017: 202) concludes that 'keeping the natural world intact is not expensive – it in fact yields a generous dividend. It's the destruction of the natural world that will cost us massively in the long run'. Jobs were lost in the asbestos industries when it was found to be dangerous, but alternative products were innovated and safer jobs created. After it was confirmed that smoking cigarettes causes lung cancer, tobacco farmers could no longer sell their crop. They wisely shifted to vegetable and fruit crops, including winemaking, and prospered. There was initial resistance to changing long-standing practices, but the transition to safety was made and prosperity maintained.

Value Services Rendered by Nature's Dynamics

Ecologists underscore the importance of ecosystem services to humanity. Economists refer to them as 'natural capital' (Heal 2017). Many of these services are being degraded by human practices (Thornes 2010). Forests suck carbon dioxide from the atmosphere, but many were clear-cut and others cleared for agriculture. Oceans also absorb carbon dioxide, but as they become warmer and acidic because of emissions, they are less capable of providing this essential service. Global warming melts glaciers, which previously supplied drinking water year-round in mountainous regions. It also melts the Arctic Ocean's ice cover, reducing the valuable Albedo effect of white ice reflecting the sun's rays thereby mitigating warming. It melts permafrost, where methane is safely stored by nature, resulting in emissions from this powerful greenhouse gas, and possible runaway climate change. Fossil fuels themselves were produced by the sun's energy and stored safely underground. The atmosphere is providing the service of a carbon sink, but it is limited. It is important to recognize, value, and protect these services of nature from being degraded by fossil-fuelled climate change.

Aiming for What Is Sociopolitically Acceptable

The chapter turns now to assessing other suggestions. Pielke (2010: 228–229, 231) states 'the only way to a high carbon tax is to start low. … By explicitly connecting carbon pricing with energy innovation, a virtuous circle is enabled that allows those asked to pay the tax to see the benefits and thus builds the support necessary to sustain investments over decades and longer'. He gives the example of a fuel tax on gasoline that enabled the Eisenhower Administration to build the American interstate highway network. Pielke's carbon tax would be dedicated to raising revenue for research to innovate energy sources cheaper than fossil fuels. It would be set upstream at extraction of fossil fuels. If his approach 'were to succeed, then decades hence the world will have a high carbon tax, widespread deployment of low-carbon technologies, and a decarbonized economy' (Pielke 2010: 232).

Although Pielke correctly sees the political difficulty of implementing carbon taxes, starting low ensures that fossil-fuelled climate change will worsen (Latin 2012) because emissions are currently high. Unfortunately, the world would have a carbonized atmosphere, a greenhouse effect, global warming, and degraded oceans because the further accumulation of carbon emissions would intensify the vicious circle of melting Arctic ice reducing its reflective capacity, melting permafrost releasing methane, etc. There would be no disincentive against combusting fossil fuels until research leads to innovations making low-carbon energy less expensive than fossil fuels, which could take a long time. A low, incrementally increasing carbon tax fails the urgency test. Moreover, a low-carbon tax provides little money for the massive research needed, and whether desired innovations would result is uncertain. Pielke argues that the carbon tax should be set at the highest price politically possible. Agreed. However, that would be zero in the current US 'Make America Great Again' administration. It would also yield zero monies for research in countries where the tax would have to be revenue-neutral to be accepted. Furthermore, it is hard to find the climate equivalent of Eisenhower building a small stretch of highway that motivates the population to pay taxes to build more.

When a Canadian province fought a national carbon tax in its supreme court, the latter concluded it is not a tax in a legal sense but instead a regulatory charge on emissions where revenues are not grabbed by the national government but rather returned to individuals and businesses in the province where it is raised. Hence, instead of using the scary word 'tax', one suggestion is to use labels that are more acceptable: 'charge', 'levy', 'fee', or 'fine for pollution'.

Whether government regulations and taxes are acceptable to the population depends on the sociopolitical context. During the Second World War, high taxes were imposed, the 'government enacted – and Americans accepted – price controls. … Gasoline was severely regulated, and a speed limit of 35 miles per hour was imposed nationally to reduce gas and rubber consumption' (Safran Foer 2019: 8). This illustration demonstrates that acceptability of taxes and regulations depends on the gravity and immediacy of the crisis.

Another possibility that could be popular would be for governments to subsidize retrofits of residential, commercial and public buildings to decrease fossil-fuel use for heating, air conditioning, and increase energy efficiency. Resistance to deep retrofits is usually based on the upfront costs, and if these were subsidized, then they would be financially beneficial for owners after a few years. Installing heat pumps where electricity is based on a low-carbon primary source and thermal retrofits to improve insulation could reduce fossil-fuel use. Together with mandated codes for new buildings, this would decrease emissions significantly, since about 13% come from buildings. Retrofits would have the co-benefit of increasing jobs to do the work. However, they would be expensive for governments, which would try to ensure additionality, namely finance the cost only if the retrofit would not be done without a subsidy.

Divestment

The fossil-fuel industry is arguably the wealthiest to have existed. Its investors and shareholders profited from this carbon-polluting industry. There are now attempts to decelerate extraction and combustion by divesting from that industry. The best known promoter of divestment is former Bank of England governor Mark Carney (Carbon Tracker 2015). The prototype was the successful divestment campaigns that contributed to bringing down South Africa's apartheid regime. Divestment is supported by (i) moral reasoning that fossil-fuel combustion degrades the environment future generations will need, and (ii) financial arguments that such investment will lose value as dangers of fossil-fuelled climate change become increasingly recognized (Flannery 2015: 105–112), much like what happened to investment in asbestos companies. If climate-protection legislation is enacted, the decline in value will be accelerated. However, such legislation is uncertain because divestment is opposed by powerful fossil-fuel lobbies. The institutions divesting have typically been environmental organizations and religious groups, which do not have much money invested. Student organizations have attempted to convince university administrators to divest their endowment funds. The best-known case was at Harvard where a student referendum was held, 1300 faculty and alumni signed a petition, and the president's office was occupied twice (Rowe et al. 2016). The administration staunchly opposed divestment from fossil fuels, which bring high returns, so it had limited success. Investors who discount danger snap up the underpriced shares thereby counteracting divestment by the few investors with climate foresight (Willis 2019: B1, B5). Divestment has a role to play in mitigating carbon pollution, but it has limitations.

Lawsuits Against Carbon Pollution

Litigation concerning harm undermined the asbestos industry and was instrumental in keeping asbestos stored safely in the ground. Litigation is having similar success in regulating cigarettes and the tobacco industry. Although at its beginning stages, litigation to compensate harm by the fossil-fuel industry and act as a deterrent to future harm could be part of mitigating the fossil-fuelled climate crisis. In the Deepwater Horizon case, lawsuits 'reduced the company's value by about $30 billion – a good example of the legal liability system operating to internalize some of the external costs that BP generated .... with BP agreeing to pay $18.7 billion in damages to the federal government and its agencies, and another $5.6 billion in damages to the various states affected' (Heal 2017: 56). Environmental lawyers (Wood 2013) have formulated a general legal argument based on the public trust doctrine. They contend all citizens have a constitutional right to a livable environment, and a democratic government has a duty to protect it. Specifically, that includes water, forests, wildlife, air and the atmosphere. They argue that it refers to preventing future harm and repairing past damage threatening future generations. This is the legal basis of ongoing lawsuits. However, litigation is expensive, slow, and has unpredictable results. Courts awarded $5 billion in punitive damages against Exxon for its 1989 Valdez oil spill in Alaska, the largest oil spill ever, but the company has been appealing ever since and the money has not been paid.

Reducing Consumption Practices

A technically simple solution to the fossil-fuelled climate crisis would be to use less fossil fuels even if they are not replaced by low-carbon energy. More generally, reducing overall consumption would mitigate most environmental problems. It would diminish the threat of a draconian exclusion from consumption for future generations (read grandchildren) forced upon them by global warming caused by the present generation's excessive consumption of carbon polluting fossil fuels. Speth (2009: 115–116) argues that the 'powerful forces driving the clash of economy and environment thus will continue, and that makes it necessary to address those forces – growth, consumerism, corporate behavior among them. So it makes very good sense to question economic growth and the growth imperative. Right now and for the foreseeable future, there's a trade-off: economy versus environment'. Carbon emissions are closely related to economic growth: when it is strong, so are carbon emissions. Global economic growth was relatively strong in 2017, and the International Energy Agency documented that 1.6 million more barrels of oil were combusted daily in 2017 than in 2016, which increased emissions and gave incentives for even more extraction.

Thus, one suggestion is to practice a bold cultural change of what is called 'degrowth' (Klein 2014). Economic growth would no longer be the goal; instead, new social practices would enable populations to live well where consumption of material resources, especially fossil fuels, would be sharply reduced. In plain English, this implies conservation, economic contraction instead of growth, downsizing production, and eliminating perceived entitlements to which the population has become accustomed. Advocates of degrowth and post-growth argue it would lead to greater well-being by pursuing happiness in terms of more time spent on family and community relations, nature, art, etc. It 'envisions major changes in values, lifestyles, and human behavior. It involves a deep change in social values – away from ever-increasing material consumption and toward close community and personal relationships, social solidarity, and a strong connection to nature' (Speth 2009: 44). Enjoy nature in a nearby park rather than travelling to a distant nature reserve, visit local art galleries rather than the Louvre, see local attractions instead of the world. Decreasing consumption of fossil fuels implies decreasing consumer goods, often transported long distances in fossil-fuelled cargo ships, trains, transport trucks, etc., and reducing travel practices in automobiles, planes, and cruise ships, which are increasing elements of consumption in the modern world. To be effective, degrowth needs to be done collectively rather than be only an individual choice by a few ecological saints.

Aviation is one area where reduction of fossil-fuelled social practices is the only remedy for its contribution to the climate crisis for the foreseeable future. Presently, there is no technological alternative to jet fuel to power planes for rapid, comfortable flights. Commercial aviation is currently experiencing rapid growth globally, resulting in more carbon emitted into the atmosphere. Degrowth would change this. Flying for pleasure would be curtailed. Conferences would be transformed: jet-fuelled participation would be replaced by tele-conferences; ideas would travel, but not people. The young climate activist Greta Thunberg's transatlantic trips highlighted that an important cause of climate change consists of combusting jet fuel. Her extreme example could motivate less flying. Will people be inspired by Thunberg's actions, or will they dismiss her as a cute kid saying nice words and doing courageous actions?

Carolan argues in favour of a circular, sharing economy where people would be 'unburdened' by ownership. Although there would be more time free from work, the desire to 'get away' would be reduced, with associated reductions in driving, flying, cruising, etc., and hence in emissions. Very little would stay the same in this social reorganization of 'consuming collaboratively; working less; taxing bads/rewarding goods; democracy; creating walkable communities; making repairable goods; enforcing antitrust laws; a maximum wage (and on and on)' (Carolan 2014: 208). He adds that 'we have to realize that freedom ultimately requires restraint' (Carolan 2014: 183) by consuming and wasting less, but this does not rest entirely on sacrifice (Carolan 2014: 169).

Limiting consumption and economic growth have appeal in theory, but have had little success in practice, and remain utopian. In the half-century since *Limits to Growth* (Meadows et al. 1972) and *Overshoot* (Catton 1980) were published, global consumption and economic growth have accelerated. Furthermore, increasing consumption and economic growth have been central to raising populations out of extreme poverty, for example in China. It is problematic to convince populations to reduce consumption of fossil fuels because they perceive it as sacrificing things they need or enjoy. It goes against the trend to greater consumption whether it be the poor acquiring air conditioning, the affluent combusting jet fuel to enjoy intercontinental tourism, and both using energy-glutton, carbon-emissions-intensive computer servers for data storage, social media mobile devices, etc. Reducing consumption is strongly resisted by powerful, prosperous corporations and countries, by the middle class everywhere, and by poor countries and the poor in developed countries who see in economic growth a solution to their economic problems. Pielke (2010: 219) contends that 'for the foreseeable future, efforts to reduce emissions through a willful contraction of economic activity are simply out of the cards. Countries around the world – rich and poor, North and South – have expressed a commitment to sustaining economic growth, and these commitments are not going to change anytime soon, no matter how much activists, idealists, or dreamers complain to the contrary'. There is little chance that degrowth will be implemented at the scale and timeliness needed to deal with the urgent fossil-fuelled climate crisis. Typically degrowth is involuntary, occurs only by means of recessions (USA 2008), wars (Iraq, Syria), civil unrest (South Sudan, Venezuela, Nigeria), pandemics (USA, Brazil, Italy, Spain 2020), and causes much suffering.

Nevertheless, restraint concerning fossil-fuelled social practices and fossil-fuel consumption is needed to mitigate global warming, especially with regard to aviation. Fly when necessary, but limit flying for pleasure. If restraint is not practiced, then faith that a technological carbon-free alternative to jet fuel can be found is relied upon, but it is nowhere in sight.

Compassion and Moral Suasion Impelling Actions

Another solution would be to develop compassion, empathy, and a cosmopolitan orientation (Roberts and Parks 2007; Beck 2015). This would involve an appeal to the better angels of humanity and to altruism. 'At present, there is already a strong movement for realization of a third generation of human rights, which includes the right to a safe environment' (Kahn and Roberts 2013: 137). Human rights, empathy, and ethics to deal with the fossil-fuelled climate crisis are certainly worth promoting and maximizing. A concept of 'ethical energy' could be defined as energy that has the least harmful impact on future generations and vulnerable poor countries. Unethical energy would be defined as that type which results in proportionately more greenhouse-gas emissions and hence greater adverse consequences for others. By this definition, wind, solar, and other types of low-carbon energy would be the most ethical, whereas coal, heavy oil from Venezuela, and oil from Alberta's tar sands would be the most unethical, as shown in Table 2.1 of Chapter 2. However, ethics confronts interests and ingrained habitus, and has had limited success in environmental matters. In the USA, where Kahn and Roberts are located, a strong movement in 2016 elected Donald Trump, who increased carbon polluters' privileges and de-emphasized compassion, empathy, cosmopolitan orientations, and a safe environment. Those virtues can be hoped for, but mitigating fossil-fuelled climate change should not have to depend on them.

Protecting and Growing Forests

The greenhouse effect can be mitigated by drawing down carbon from the atmosphere. The growth of tropical forests is particularly effective at withdrawing atmospheric carbon and sequestering it in trees, roots, and other vegetation. Much of this will be done by nature's processes if those forests are protected. Brazil dramatically cut deforestation rates in the Amazon, and was restoring the forest, at least until Bolsonaro was elected president in 2019. Hawken's (2017: 114–116) team estimates that protection and restoration of tropical forests over the next thirty years can draw 61 gigatonnes of carbon dioxide out of the atmosphere. Temperate forests are not as good at withdrawing carbon, but are still good. Over the same period, if protected and restored, they can withdraw 22.6 gigatonnes (Hawken 2017: 128–129). Afforestation is the term used for creating new forests in areas where there were none for fifty years because of pasture lands, mining, etc. Planting forests there can result in a new sink drawing down 18 gigatonnes of carbon over the next thirty years (Hawken 2017: 132–134).

Offsetting

Despite deficiencies of offsetting indicated in Chapter 4, it has an important role to play for compensating emissions from social practices for which there is no alternative to fossil fuels in the current state of technology, such as aviation. For example, Swiss researchers calculated that a return flight between New York and London covering 11,100 km results in 1.8 tonnes of greenhouse gases per passenger in economy class, which is 3800 pounds or 20 times the weight of an average 190-pound passenger, and the offsetting cost would be US$54 if spent in developing countries (Myclimate 2020). This would not be an exorbitant increase in the ticket price of a return flight between New York and London. The equivalent estimate for more spacious business class shared by fewer passengers was 3.4 tonnes of greenhouse gases per passenger, which is 6800 pounds and the offsetting cost would be US$103. Even more spacious first class would produce 5.3 tonnes of greenhouse gases per passenger or 10,600 pounds and the offsetting cost would be US$160. Tickets are presently cheap because the cost of damage to the environment is not included in the price and is left to be paid belatedly by others in the form of costly global warming and extreme weather. Since there were over 4 billion passengers flying in the world in 2019 and the number is growing, aviation's contribution to fossil-fuelled global warming and its related costs are enormous. If offsetting were included in the price of tickets, it would provide a huge increase in financing for mitigating fossil-fuelled climate change, for example not only afforestation but also financing solar energy in poor southern countries to replace coal thereby mitigating global warming and benefiting everyone everywhere. Note that if spent in wealthy countries with high labour costs, the same mitigation benefits of offsetting would add twice as much to the ticket price.

Offsets should be certified to ensure they result in additional emissions reductions that would not have occurred without them. Most importantly, offsets should be mandatory and increased to include the full cost of damaging emissions rather than being voluntary and tiny, as done presently for aviation. Offsetting that finances additional solar and wind energy and energy efficiency retrofits are better than planting trees because a wildfire or logging can put all the carbon back into the atmosphere (Suzuki and Hanington 2017: 234–237). Monies raised could be used to finance (i) the replacement of fossil-fuel-powered electricity generation by low-carbon energy and (ii) the transition to electrical vehicles, networks of charging stations, and especially (iii) research into improving the storage of emissions-free energy. This would increase the price of plane tickets, result in a decrease in air traffic, and mitigate aviation's contribution to fossil-fuelled climate change, but it would face considerable opposition. Hence, effective offsetting has been shelved and token voluntary offsetting persists. This needs to be changed.

State offsetting is already being done by Norway, which is using part of its oil wealth to finance afforestation in developing countries, thereby compensating some emissions from its North Sea oil. In 2008, Norway agreed to pay Brazil $1 billion to reduce its rate of deforestation of the Amazon rainforests 75% by 2015; Brazil achieved the goal and received Norway's payment. Since states that base their economies on oil extraction are unlikely to keep the oil in the ground, offsetting is a needed alternative, with social democratic Norway being an exemplary role model. However, there is resistance in most oil extracting and consuming countries to sending money abroad because it reduces profits, royalties, and jobs.

Educating Girls

Global warming must not be mitigated by keeping people in extreme poverty. Economic growth of poor countries is necessary. But imagine the global warming impact if poor countries, with their high populations and birth rates, reached the same per capita level of fossil-fuel consumption as wealthy ones, for example, if Madagascar's annual emissions of 0.1 tonnes per person increased to the American rate of 18 tonnes (Hawken 2017: 81). Hence Hawken (2017: 78–82) argues that educating girls and family planning constitute important means of restraining emissions and mitigating global warming.

Governance

Far from government and market action being mutually exclusive, a government cap on emissions or carbon tax is needed to harness the market to solve this urgent global problem. Apparent market-based remedies are actually hybrid market-state solutions. Government regulations forcing polluters to reduce emissions, or prohibiting the worse carbon-polluting practices and types of fossil fuels, is also necessary (Jaccard 2018). Currently, coal-fired power provides 40% of the world's electricity. 'A dramatic change of course is required if the world is to avoid a climate-change disaster. And, of course, dramatic shifts require decisive government action – to limit coal use and encourage new technologies by putting an effective price on carbon. Otherwise cheap, dirty coal, with its vast already-built and paid-off infrastructure, will prevail' (Flannery 2015: 186). The closing of coal-fired electricity generation by the Ontario government reduced greenhouse-gas emissions and had co-benefits of decreasing urban smog, air pollution, and fatalities among asthma sufferers. Harvey and Orbis (2018: 12) documented that 'a strong building code that continuously strengthens over time and has a strong monitoring and enforcement mechanism, as in California, can dramatically reduce energy use and emissions. And we know fuel economy standards for vehicles, when designed well, can dramatically improve fuel efficiency'. Germany increased its use of renewable energy for electricity through government-mandated feed-in tariffs by reverse auctions where the lowest bid wins the supply contract. To reduce emissions, governments 'removing subsidies for fossil fuels is the first step – though still widely ignored. Next, policymakers must incorporate the cost of externalities, such as adding a carefully derived social cost of carbon or setting a carbon cap' (Harvey and Orbis 2018: 18). Government action is required to promote densification of cities to reduce commuting emissions, construct dedicated bicycle lanes to foster this carbon-free transportation, build rapid mass transit and high-speed electric rail to reduce short-haul plane travel and jet-fuel emissions, improve fuel-efficiency standards, etc. (Flannery 2015: 91–92). Harvey and Orbis (2018: 300) assessed the efficacy of government policies to reduce greenhouse-gas emissions. In a country that cannot afford electrified subways, they documented how bus rapid transit on dedicated lanes and high-density development near stations decreased automobile use and emissions in Bogota, Columbia. Some remedies require changes in cultural aesthetics. Governments could paint the infrastructure and its roofs white so that the sun's rays would be reflected back into space, which many southern European countries do to some extent (Flannery 2015: 137). Governments promoting vegetarianism and best practices management of livestock would also reduce emissions (Flannery 2015: 152).

Conservative governments, called different names in different countries, are typically the most reluctant to mitigate fossil-fuelled climate change. Nevertheless, the urgency of the fossil-fuelled climate crisis requires that all forms of governance act to mitigate it, rather than having left-wing governments acting then having it undone by right-wing ones. And conservative governments have mitigated climate change in some cases. Angela Merkel's conservative Christian Democrats led the way with wind and solar energy by innovating feed-in tariffs. UK Conservative governments pushed forward wind energy. In France, Macron's government is not left-wing nonetheless attempted to implement carbon taxes. In the USA, international agreements were implemented during Republican regimes to phase out CFCs to diminish ozone layer depletion, leaded gasoline was banned under Reagan, and George H. W.

Bush initiated cap-and-trade to alleviate acid rain. In 1988, the Canadian Conservative Party held the first international meeting on climate change.

Most remarkable is the 1989 speech before the United Nations General Assembly by the icon of conservatism, Margaret Thatcher, best known for promoting individual rights, free markets, deregulation, etc. 'The problem of global climate change is one that affects us all, and action will only be effective if it is taken at the international level. It is no good squabbling over who is responsible or who should pay. … Put in its bluntest form: The main threat to our environment is more and more people and their activities. The land they cultivate more intensively. The forests they cut down and burn. The mountainsides they lay bare. The fossil fuels they burn. The rivers and seas they pollute. The result is that change in future is likely to be more fundamental and more widespread than anything we have known hitherto. Change to the sea around us, change to the atmosphere above, leading in turn to change in the world's climate, which could alter the way we live in the most fundamental way. … There will be no profit or satisfaction for anyone if pollution continues to destroy our planet' (quoted in Reguly 2020: B4). She called for a framework convention on climate change, which she branded 'a good conduct guide for all nations'. Her knowledge of science, having studied chemistry at Oxford, likely explains her early awareness of the danger of global warming, since its source is the chemical reaction of carbon with oxygen. Unfortunately, conservatives ignored Thatcher's climate warnings. The American Republican Party long denied anthropogenic climate change, attacked the science, and refused to change fossil-fuelled social practices (Jacques et al. 2008; Dunlap and Jacques 2013; Dunlap et al. 2016). Conservatives everywhere should heed Thatcher's 'good conduct guide' to mitigate climate change, otherwise little progress will be made given the swings in democratic electoral politics.

Every country has to lower its carbon pollution through legally binding, enforced international agreements between governments, which would amount to Paris Agreements plus sanctions for recalcitrants. An international rules-based governance order mandating carbon pollution restraint is necessary. The key concept is disproportionately (Freudenburg 2006): countries that are disproportionately high per capita emitters have the most responsibility to lower their emissions; otherwise wealthy, high-emitting, low population countries like Canada will blame poor, populous countries like India as an excuse for keeping emissions high. High per capita emissions, wealthy societies like Luxembourg (CDIAC 2018) need to lead in reducing their emissions, even if they account for a small percentage of overall emissions, because they are disproportionately causing harm to future generations.

Transcending Capitalism as We Know It

Speth (2009: 116) argues that 'the planet cannot sustain capitalism as we know it'. His solution is to reinvent the political economy constructing 'America the possible' (Speth 2012: xiii) by transforming the capitalist market and politics, decreasing consumption and inequality, simple living, respecting nature, and pursuing happiness in non-material pleasures. 'The future described herein rests on massive change that will be realized only if the American people insist on it. So the book is more about what can be than what will be' (Speth 2012: preface). It is true that capitalism, as we know it, has engendered the fossil-fuelled climate crisis, must be transformed, and political economy reinvented (Foster, Clark, and York 2011). But what will that involve?

A leap into the unknown of massive change can be done quickly in theory, but its implementation in the messy real-world of power, interests, and practices is likely too slow, too partial, and too problematic for the urgent climate crisis. It would probably have less chance of success than a merely good solution. Harvey and Orbis (2018: 63) argue that a 'highly abating and perfectly designed policy is not worth pursuing if there is no chance it can be implemented'. Speth's proposal resembles that of Bernie Sanders, which even primary voters in the American Democratic Party did not accept. That a massive change of capitalism 'will be' in practice in the urgent time frame required for dealing with the climate crisis is a long shot.

Speth is not suggesting replacing 'capitalism as we know it' with socialism 'as we know it'. But it is a misleading comparison to contrast the worst excesses of capitalism with the best ideals of theoretical socialism which has never existed. What does his proposed transformation of the market consist of? Ambiguity is an escape hatch in Speth's theory. His proposal might consist of a utopia never before seen and likely will not be seen. Or the transformation of 'capitalism as we know it' could mean implementing 'capitalism as we don't know it in America', namely social democratic capitalism inspired by the Nordic model. It could consist of improving and scaling up real-world political systems that have been the best performers in mitigating environmental problems, monopolization, exclusion, and inequality of opportunity, thereby reducing social closure. That would be social democracy.

A Social Democratic Proposal

Fairbrother notes that neo-Marxist approaches see no significant reduction in externalities unless the capitalist market is abolished, and that ecological modernization (Mol et al. 2009; Sonnenfeld 2009) is based on vague, questionable indicators of improvement (for critiques see York and Rosa 2003; York 2012; York and McGee 2016). Fairbrother (2016: 380) argues that 'an externalities perspective focuses on the political question of why the state may or may not take regulatory action against externalities in a given case'. This converges with the focus of closure theory on specific political action to mitigate exclusion, monopolization, and environmental problems. Such political action is typically, though not uniquely nor always, accomplished through social democratic governance.

Evidence demonstrates that social democracy is the governance with the best environmental performance record, including for reducing greenhouse-gas emissions (Germanwatch 2012, 2019; Yale University 2012, 2018; CDIAC 2018). Minimizing monopolization of wealth and mitigating exclusion, which are more characteristic of social democracy than of other forms of governance, has been extended to future generations and poor countries by reducing impacts on the environment they will need. This involves a more cosmopolitan orientation, as Beck (2015) recommended, and greater empathy for others distant in time or space. Social democracy has the best record of creating institutions, values, and social practices that foster equality of opportunity, inclusion, environmental justice, and intergenerational justice.

For example, states with large oil reserves typically elect conservative governments opposed to carbon taxes and restraints on carbon pollution. Oil extraction has become the principal basis of prosperity for Norway, but it is done in social democratic ways that obtain the most value for the least emissions. Norway formed a state-owned company to exploit this resource so that it would benefit all Norwegians. It allowed private companies to exploit North Sea oil only by paying huge royalties (75%), and created a sovereign wealth fund of savings to benefit future generations. It introduced carbon taxes to decrease emissions and promote technological innovations, built an electric transit network, and financed reforestation of Brazil's and Indonesia's tropical rainforests to offset Norway's carbon emissions. This is not as good for mitigating fossil-fuelled global warming as keeping its oil in the ground under the sea. Nonetheless, it is the best any major oil extracting country has done. There is a need to learn from Nordic social democracies, notwithstanding that their collectivist values are not easily transferable to individualistically oriented states, such as the USA.

Social democracy is the form of governance most closely related to inclusionary solutions deduced from the social closure theoretical framework. That framework takes radical analyses into account, such as those of Speth (2009, 2012), but redirects them towards real world, social democratic resolution of fossil-fuelled climate change holding inequalities and processes of monopolization in check, especially fossil-fuel oligopolies. For the urgent crisis of fossil-fuelled climate change, it stresses the need for prompt implementation of solutions.

Energy Futures Under the Threat of Climate Change

The fossil-fuelled climate crisis has been caused socioeconomically, will have socioeconomic consequences, and will need socioeconomic remedies if it is to be mitigated.

Three possible outcomes of the crisis are foreseeable.

a. A dangerous fossil-fuelled global warming energy future
One possibility is that societies, prodded on by fossil-fuel industries, refuse to change their fossil-fuelled social practices. Emissions continue to exceed carbon withdrawals and more and more carbon is transferred from ground to sky. Impact science's warnings of future harm are discounted. High-carbon emitting countries legitimate their pollution by claiming it is pointless to stop because other countries will continue theirs. Near-term economic benefits are enjoyed, with faith future generations will implement just-in-time technological solutions even though the present generation has not. Societies believe they can adapt to anything nature throws at them, be resilient, and ride runaway global warming, which is like assuming there is no need to close the faucet because adaptation can succeed by mopping up the overflow from the bathtub. This constitutes a failure of foresight, a race to the bottom, and the incubation of disaster. The biggest danger is that fossil-fuelled global warming unleashes second-order global warming by nature's dynamics and uncontrollable climate change. Even if societies belatedly stop emissions, the evil genie is out and vicious feedback forces of nature are let loose. Insufficient mitigation is a possible energy future, with future generations paying the environmental costs of previous generations' refusal to pay the full cost of their fossil-fuelled social practices. Foresight requires that this possible energy future not be dismissed in order to take action to avoid it.

b. A safe, low-carbon energy future based on reduced consumption
There is awareness that if this generation does not reduce consumption voluntarily, runaway climate change will impose it on future generations. Moral considerations replace individualistic orientations with

cosmopolitan ones of caring for future generations and poor countries. Wealth is shared with poor societies. Consumption willingly decreases, especially fossil-fuel consumption and corresponding social practices. People drive less, fly less, cruise less, cremate less, use less cement, less air conditioning, less social media servers, etc. The emphasis is on enjoying non-fossil-fuelled pleasures and nature, sights, sports, art galleries, and the like near home. Urban sprawl and commuting are decreased by choosing to live near work in dense cities, like Seoul and Amsterdam, and raising children in apartment blocks rather than in detached houses with a yard. Reduced consumption means economic growth stops and degrowth begins, but sharing work avoids unemployment, and people need less money because travel and consumption decrease. Fossil fuels remain safely underground. Opposition from fossil-fuel industries and states is overcome. The aspirations of Beck (2015), Speth (2009, 2012) and Klein (2014) become reality. This bold massive change mitigates global warming. Unless catastrophic consequences become visibly imminent as in wartime, this energy future is unlikely to occur because it will be perceived as incurring major sacrifices of social practices people enjoy. By the time consequences become grave, it will be too late to undo the accumulation of carbon pollution in the sky, as Giddens states.

c. A safe, low-carbon energy future with growth

The full cost of fossil fuels is paid upon purchase because governments impose a price on carbon pollution through taxes, cap-and-trade, or regulations. This incentivizes using solar, wind, tidal, hydro, and geothermal energy, and provides financing for technological innovations improving energy storage, electric vehicles, and low-carbon energy. Emissions are decreased to the level of carbon withdrawals. Most fossil fuels remain safely stored underground. Internationally enforced agreements ensure compliance by all nations. Although costly to pay the hitherto unpaid externalized costs of fossil-fuel combustion, this option consists of only sacrificing cheap fossil fuels. Carbon-free energy, which may become cheap in the future, replaces them. There would be little sacrifice of consumption and social practices, with social media servers, air conditioning, and transportation powered by carbon-free energy. Economic growth continues with global warming mitigated. Although technically challenging and facing opposition from consumers who have a sense of entitlement to cheap fossil fuels, as well as from fossil-fuel industries and states, it is the most promising possibility to replace the dangerous fossil-fuelled energy future with a safe, low-carbon energy future.

Which of these energy futures will predominate is presently unforeseeable because of uncertainties concerning whether decision-makers and populations will respond (i) with foresight to the well-documented fossil-fuelled global warming or (ii) by discounting the danger.

Foresight or Discounting Danger?

Researchers documented that prior to disasters there is typically a period where signs of danger are discounted, which they call a 'failure of foresight' resulting in the 'incubation' of 'man-made disasters' (Turner and Pidgeon 1978). Scientific evidence indicates that humanity is incubating the slow onset, cumulative global catastrophe of fossil-fuelled climate change. It is largely imperceptible by the senses and risks closing off resources and opportunities to groups distant in space and time from those causing the problem. The objective of this book has been to analyse this dangerous incubation and assess solutions in order to enhance foresight and promote socioeconomic practices in harmony with nature's dynamics.

Analyzing the depth and gravity of the fossil-fuelled climate crisis and translating the findings of science into words and analogies understandable by non-scientists will hopefully encourage low-carbon social practices, including voting practices. Hope, as opposed to blind faith in a last-minute technological cure-all, rests on evidence-based assessments. None of the solutions and remedies appraised in this chapter and book is perfect, and all have varying gradations of value. A whole range of solutions was assessed to indicate the value, limitations, and possibilities of each so that they could be used appropriately and in combination to meet this multifaceted challenge. Mitigation has to occur quickly because the longer climate change worsens, the more draconian the required solution. To accomplish this and promote carbon-free energy, the polluter pays principle has to be implemented in terms of users of fossil fuels paying their full cost by including the cost of harm they cause in their price. This will make low-carbon energy more competitive and finance its development to the long-run benefit of everyone.

I resist the temptation to end the book with a pre-determined happy conclusion or an apocalypse. Human agents choosing their social practices individually and collectively in a structure of social and biophysical constraints will determine which result prevails. This is not as emotionally satisfying as enunciating perfect solutions in the abstract, but it is the only candid conclusion. The priority given to foresight versus discounting danger will determine the outcome of the fossil-fuelled climate crisis and its creeping threat to the sustainability of the beneficial habitat for humanity.

Note

1. Note that discounting here is used in the broad sociological sense rather than in the technical way economists use the concept. See Heal (2017: 145–149) for a good discussion of its use in environmental economics.

References

Beck, U. 2015. Emancipatory Catastrophism. *Current Sociology* 63 (1): 75–88.

Carbon Tracker. 2015. *Mark Carney warns Investors Face Huge Climate Change Losses*. 30 September. https://carbontracker.org/mark-carney-warns-investors-face-huge-climate-change-losses/. Accessed 15 April 2020.

Carolan, Michael. 2014. *Cheaponomics: The High Cost of Low Prices*. Abingdon, UK: Earthscan.

Catton Jr., William R. 1980. *Overshoot: The Ecological Basis of Revolutionary Change*. Urbana and Chicago: University of Illinois Press.

CDIAC. 2018. Carbon Dioxide Information Analysis Center. U.S. Department of Energy. Berkley. cdiac.Ess-dive.lbl.gov.

Dunlap, Riley E., and Peter J. Jacques. 2013. Climate Change Denial Books and Conservative Think Tanks: Exploring the Connection. *American Behavioral Scientist* 57: 699–731.

Dunlap, R., A. McCright, and J. Yarosh. 2016. The Political Divide on Climate Change: Partisan Polarization Widens in the U.S. *Environment Science and Policy for Sustainable Development* 58 (5): 4–23.

EPA. 2017. Causes of Climate Change. *United States Environmental Protection Agency*. https://19january2017snapshot.epa.gov/climate-change-science/causes-climate-change_.html. Accessed 17 April 2020.

Fairbrother, Malcolm. 2016. Externalities: Why Environmental Sociology Should Bring Them. In *Environmental Sociology* 2 (4): 375–384.

Flannery, Tim. 2015. *Atmosphere of Hope: Searching for Solutions to the Climate Crisis*. New York: Atlantic Monthly Press.

Foster, John Bellamy, Brett Clark, and Richard York. 2011. *The Ecological Rift: Capitalism's War on the Earth*. New York: Monthly Review Press.

Freudenburg, W. 2006. Environmental Degradation, Disproportionality, and the Double Diversion. *Rural Sociology* 71 (1): 3–32.

Germanwatch. 2012. *CCPI The Climate Change Performance Index Results 2012*. http://germanwatch.org/klima/ccpi.pdf. Accessed 5 November 2012.

Germanwatch. 2019. *CCPI: The Climate Change Performance Index Results 2019*. https://germanwatch.org/sites/germanwatch.org/files/CCPI-2019-Results-190614-WEB-A4.pdf. Accessed 22 September 2019.

Harvey, Hal, and Robbie Orbis. 2018. *Designing Climate Solutions: A Policy Guide for Low-Carbon Energy*. Washington, DC: Island Press.

Hawken, Paul. 2010. *The Ecology of Commerce: A Declaration of Sustainability*, rev. ed. New York: Harper Business.

Hawken, Paul. 2017. *Drawdown: The Most Comprehensive Plan Ever Proposed to Reverse Global Warming*. New York: Penguin.

Heal, Geoffrey. 2017. *Endangered Economies: How the Neglect of Nature Threatens Our Prosperity*. New York: Columbia University Press.

IPBES. 2019. Intergovernmental Science-Policy Platform on Biodiversity and Ecosystem Services. *Global Assessment Report on Biodiversity and Ecosystem Services*. Bonn: IPBES. https://ipbes.net/sites/default/files/2020-02/ipbes_global_assessment_report_summary_for_policymakers_en.pdf. Accessed 10 April 2020.

Jaccard, Mark. 2018. Divisive Carbon Taxes Are Much Ado About Nothing. *The Globe and Mail*, 15 December: O4.

Jacques, Peter, Riley E. Dunlap, and Mark Freeman. 2008. The Organization of Denial: Conservative Think Tanks and Environmental Scepticism. *Environmental Politics* 17: 349–385.

Kahn, Mizan R., and J. Timmons Roberts. 2013. Towards a Binding Adaptation Regime: Three Levers and Two Instruments. In *Successful Adaptation to Climate Change: Linking Science and Policy in a Rapidly Changing World*, ed. Susanne C. Moser and Maxwell T. Boykoff, Chapter 8. London: Routledge.

Klein, Naomi. 2014. *This Changes Everything: Capitalism vs the Climate*. New York: Simon & Schuster.

Latin, H. 2012. *Climate Change Policy Failures*. Singapore: World Scientific Publishing.

Meadows, Donella, Dennis Meadows, Jorgen Randers, and William Behrens. 1972. *The Limits to Growth*. New York: Universe Books.

Mol, A.P.J., D. Sonnenfeld, and G. Spaargaren. 2009. *The Ecological Modernisation Reader: Environmental Reform in Theory and Practice*. London: Routledge.

Munro, Margaret. 2019. Green Screen: A Planetary Conscience in Every Phone? There Should Be an App for That. *Globe and Mail*, 26 October: O3.

Myclimate. 2020. *My Climate: Shape Our Future*. Zurich: Foundation Myclimate. https://co2.myclimate.org/en/portfolios?calculation_id=3492359&localized_currency=USD. Accessed 30 April 2020.

Nordhaus, William. 2013. *The Climate Casino: Risk, Uncertainty, and Economics for a Warming World*. New Haven: Yale University Press.

Pielke Jr., Roger. 2010. *The Climate Fix*. New York: Basic Books.

Ragan, Chris. 2019. Borrowing Policies Is Canada's Best Policy. *Globe and Mail*, 12 August: A11.

Reguly, Eric. 2020. Trump's Davos Talk Shows He's No Thatcher. *The Globe and Mail*, 22 January: B4.

Roberts, J. Timmons, and Bradley Parks. 2007. *A Climate of Injustice: Global Inequality, North-South Politics, and Climate Policy*. Cambridge, MA: MIT Press.

Rowe, James, Jessica Dempsey, and Peter Gibbs. 2016. The Power of Fossil Fuel Divestment (and Its Secret). In *A World to Win: Contemporary Social Movements and Counter-Hegemony*, ed. William R. Carroll and Kanchan Sarker, 233–249. Winnipeg: ARP Books.

Safran Foer, Jonathan. 2019. *We Are the Weather: Saving the Planet Begins at Breakfast*. New York: Farrar, Strauss and Giroux.

Sonnenfeld, D. 2009. Contradictions of Ecological Modernisation. In *The Ecological Modernisation Reader*, ed. A. Mol, D. Sonnenfeld, and G. Spaargaren, 372–390. Abingdon: Routledge.
Speth, James Gustave. 2009. *The Bridge at the Edge of the World: Capitalism, the Environment, and Crossing from Crisis to Sustainability*. New Haven: Yale University Press.
Speth, James Gustave. 2012. *America the Possible: Manifest for a New Economy*. New Haven: Yale University Press.
Stern, Nicholas. 2009. *A Blueprint for a Safer Planet*. London: Random House.
Suzuki, David, and Ian Hanington. 2017. *Just Cool It: The Climate Crisis and What We Can Do*. Vancouver and Berkeley: Greystone Books.
Thornes, John. 2010. Atmospheric Services. *Issues in Environmental Science and Technology* 30: 70–104.
Tong, Ziya. 2019. *The Reality Bubble:Blind Spots, Hidden Truths, and the Dangerous Illusions that Shape Our World*. Toronto: Penguin.
Turner, B., and N. Pidgeon. 1978. *Man-Made Disasters*. London: Wykeham.
Wagner, Gernot, and Martin Weitzman. 2015. *Climate Shock: The Economic Consequences of a Hotter Planet*. Princeton: Princeton University Press.
Willis, Andrew. 2019. The Future Is Plastics: Murray Edwards, Li Ka-Shing Add to Oil Patch Holdings as Others Flee. *Globe and Mail*, 7 July: B1, B5.
Wood, Mary Christina. 2013. *Nature's Trust: Environmental Law for a New Ecological Age*. New York: Cambridge University Press.
Yale University. 2012. *EPI Environmental Performance Index 2012*. http://epi.yale.edu. Accessed 8 November 2012.
Yale University. 2018. *EPI Environmental Performance Index 2018*. https://epi.envirocenter.yale.edu/epi-country-report/CAN. Accessed 22 September 2019.
York, R. 2012. Do Alternative Energy Sources Displace Fossil Fuels? *Nature Climate Change* 2 (6): 441–443.
York, R., and J.A. McGee. 2016. Understanding the Jevons Paradox. *Environmental Sociology* 2 (1): 77–87. https://doi.org/10.1080/23251042.2015.1106060.
York, R., and E. Rosa. 2003. Key Challenges to Ecological Modernization Theory. *Organization & Environment* 16 (3): 273–288.

Index

R. Murphy, *The Fossil-Fuelled Climate Crisis*,
https://doi.org/10.1007/978-3-030-53325-0

J

S

T

The manufacturer's authorised representative in the EU is Springer Nature Customer Service Centre GmbH, Europaplatz 3, 69115 Heidelberg, Germany. If you have any concerns regarding our products, please contact ProductSafety@springernature.com

Printed and bound by CPI Group (UK) Ltd, Croydon, CR0 4YY

30/09/2026

02225689-0005